Autodesk Fusion 360
Black Book

By
Gaurav Verma
Samar
(CADCAMCAE Works)

ISBN # 978-1-988722-17-7

NOTICE TO THE READER

DEDICATION

To teachers, who make it possible to disseminate knowledge
to enlighten the young and curious minds
of our future generations

To students, who are the future of the world

THANKS

To my friends and colleagues

To my family for their love and support

Training and Consultant Services

At CADCAMCAE Works, we provides effective and affordable one to one online training on various software packages in Computer Aided Design(CAD), Computer Aided Manufacturing(CAM), Computer Aided Engineering (CAE), Computer programming languages(C/C++, Java, .NET, Android, Javascript, HTML and so on). The training is delivered through remote access to your system and voice chat via Internet at any time, any place, and at any pace to individuals, groups, students of colleges/universities, and CAD/CAM/CAE training centers. The main features of this program are:

Training as per your need

Highly experienced Engineers and Technician conduct the classes on the software applications used in the industries. The methodology adopted to teach the software is totally practical based, so that the learner can adapt to the design and development industries in almost no time. The efforts are to make the training process cost effective and time saving while you have the comfort of your time and place, thereby relieving you from the hassles of traveling to training centers or rearranging your time table.

Software Packages on which we provide basic and advanced training are:

CAD/CAM/CAE: CATIA, Creo Parametric, Creo Direct, SolidWorks, Autodesk Inventor, Solid Edge, UG NX, AutoCAD, AutoCAD LT, EdgeCAM, MasterCAM, SolidCAM, DelCAM, BOBCAM, UG NX Manufacturing, UG Mold Wizard, UG Progressive Die, UG Die Design, SolidWorks Mold, Creo Manufacturing, Creo Expert Machinist, NX Nastran, Hypermesh, SolidWorks Simulation, Autodesk Simulation Mechanical, Creo Simulate, Gambit, ANSYS and many others.

Computer Programming Languages: C++, VB.NET, HTML, Android, Javascript and so on.

Game Designing: Unity.

Civil Engineering: AutoCAD MEP, Revit Structure, Revit Architecture, AutoCAD Map 3D and so on.

We also provide consultant services for Design and development on the above mentioned software packages

For more information you can mail us at:
cadcamcaeworks@gmail.com

Table of Contents

Chapter 2 : Sketching

Chapter 3 : 3D Sketch and SolidModeling

Chapter 4 : Advanced 3D Modeling

Chapter 5 : Practical and Practice

Chapter 6 : Solid Editing

Chapter 7 : Assembly Design

Chapter 11 : Drawing

Chapter 12 : Sculpting

Chapter 13 : Sculpting-2

Chapter 14 : Mesh Design

Chapter 15 : CAM

Chapter 16 : Generating Milling Toolpaths - 1

Chapter 21 : Simulation Studies in Fusion 360

Preface

Autodesk Fusion 360 is a product of Autodesk Inc. Fusion 360 is the first of its kind software which combine 3D CAD, CAM, and CAE tool in single package. It connects your entire product development process in a single cloud-based platform that works on both Mac and PC. In CAD environment, you can create the model with parametric designing and dimensioning. The CAD environment is equally applicable for assembly design. The CAE environment facilitates to analysis the model under real-world load conditions. Once the model is as per your requirement then generate the NC program using the CAM environment.

The **Autodesk Fusion 360 Black Book** is the first edition of our series on Autodesk Fusion 360 Ultimate. The book is written on Autodesk Fusion 360 V 2.0.3174 Ultimate. With lots of features and thorough review, we present a book to help professionals as well as beginners in creating some of the most complex solid models. The book follows a step by step methodology. In this book, we have tried to give real-world examples with real challenges in designing. We have tried to reduce the gap between educational use of Autodesk Fusion 360 and industrial use of Autodesk Fusion 360. In this edition of book, we have included topics on Sketching, 3D Part Designing, Assembly Design, Rendering & Animation, Sculpting, Mesh Design, CAM, Simulation, 3D printing, 3D PDFs, and many other topics. The book covers almost all the information required by a learner to master the Autodesk Fusion 360. The book starts with sketching and ends at advanced topics like CAM, Simulation, and Mesh Design. Some of the salient features of this book are :

In-Depth explanation of concepts

Every new topic of this book starts with the explanation of the basic concepts. In this way, the user becomes capable of relating the things with real world.

Topics Covered

Every chapter starts with a list of topics being covered in that chapter. In this way, the user can easy find the topic of his/her interest easily.

Instruction through illustration

The instructions to perform any action are provided by maximum number of illustrations so that the user can perform the actions discussed in the book easily and effectively. There are about **1500** small and large illustrations that make the learning process effective.

Tutorial point of view

At the end of concept's explanation, the tutorial make the understanding of users firm and long lasting. Almost each chapter of the book has tutorials that are real world projects. Moreover most of the tools in this book are discussed in the form of tutorials.

Project

Free projects and exercises are provided to students for practicing.

For Faculty

If you are a faculty member, then you can ask for video tutorials on any of the topic, exercise, tutorial, or concept.

Formatting Conventions Used in the Text

All the key terms like name of button, tool, drop-down etc. are kept bold.

Free Resources

Link to the resources used in this book are provided to the users via email. To get the resources, mail us at ***cadcamcaeworks@gmail.com*** with your contact information. With your contact record with us, you will be provided latest updates and informations regarding various technologies. The format to write us mail for resources is as follows:

Subject of E-mail as ***Application for resources of _____book.***
Also, given your information like
Name:
Course pursuing/Profession:
E-mail ID:

Note: We respect your privacy and value it. If you do not want to give your personal informations then you can ask for resources without giving your information.

About Authors

The author of this book, Gaurav Verma, has authored and assisted in more than 16 titles in CAD/CAM/CAE which are already available in market. He has authored **AutoCAD Electrical Black Books** which are available in both **English** and **Russian** language. He has also authored books on various modules of Creo Parametric and SolidWorks. He has provided consultant services to many industries in US, Greece, Canada, and UK. He has assisted in preparing many Government aided skill development programs. He has been speaker for Autodesk University, Russia 2014. He has assisted in preparing AutoCAD Electrical course for Autodesk Design Academy. He has worked on Sheetmetal, Forging, Machining, and Casting in Design and Development department.

If you have any query/doubt in any CAD/CAM/CAE package, then you can contact the authors by writing at cadcamcaeworks@gmail.com

For Any query or suggestion

If you have any query or suggestion, please let us know by mailing us on *cadcamcaeworks@gmail.com*. Your valuable constructive suggestions will be incorporated in our books.

Chapter 1

Starting with Autodesk Fusion 360

Topics Covered

The major topics covered in this chapter are:

- *Overview of Autodesk Fusion 360*
- *Installing Autodesk Fusion 360 (Educational)*
- *Starting Autodesk Fusion 360*
- *File Menu*
- *User Account drop-down*
- *Data Panel*
- *Simulation Browser*

OVERVIEW OF AUTODESK FUSION 360

Autodesk Fusion 360 is a new Autodesk product which is designed to be a powerful 3D Modeling software package with an integrated, parametric, feature based CAM module built into the software. Autodesk Fusion 360 is the first 3D CAD, CAM and CAE tool of its kind. It connects your entire product development process in a single cloud-based platform that works on both Mac and PC; refer to Figure-1.

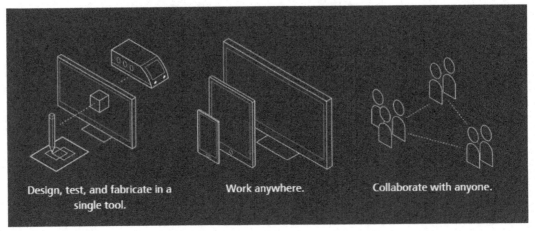

Figure-1. Overview

It combines industrial and mechanical design, collaboration, simulation, and machining in a single software. The tools in Fusion 360 enable rapid and easy exploration of design ideas with an integrated concept to production toolset. This software needs a good network connection to work in collaboration with other team members.

This is a good software for those individual and companies who wants to develop themselves. This software is much affordable than any other software offered by Autodesk. To use this software one has to pay the monthly subscription of Fusion 360. You can work offline in this software and later save the file on Autodesk Server. User can access this software from anywhere with an internet connection. The user is able to open his saved file and also able to share files with anyone from anywhere as long as he has the software and good internet connection. Also, the pricing of this software is cost effective so anyone can uses it for manufacturing of tools and parts. Autodesk Fusion 360 has been built to work in multi body manner: both parts and assemblies built in a single file. The procedure to install the software is given next.

INSTALLING AUTODESK FUSION 360 (EDUCATIONAL)

• Connect your pc with the internet connection and then login to **http://www.autodesk.com/products/fusion-360/students-teachers-educators** as shown in Figure-2.

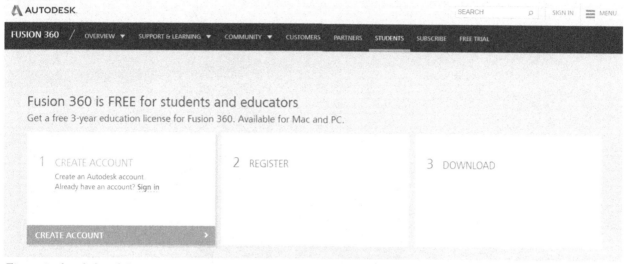

Figure-2. Autodesk website

• Click on the **Create Account** button and on next page, select your Country and Education role. In **Country** drop-down, select your Country and in **Education role** drop-down, you need to select **Student**(There is free subscription for students with license term of 3 years)as shown in Figure-3.

Figure-3. Select country and education role

• After filling the details, click on **NEXT** button. The next page of this website opens. In this page, you need to fill your personal details and then click on **Create Account** as shown in Figure-4.

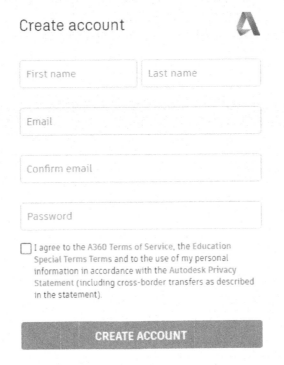

Figure-4. Creating account

- After creating account you need to verify your E-mail address by login into your E-mail account. After verifying E-mail address, you need to give your **Education details**.
- After completion of these processes, you need to sign-in into your Autodesk Account and search the **Autodesk Fusion 360 software**.
- Download the **Autodesk Fusion 360 software**.
- Open the downloaded setup file and follow the instructions as per the setup instruction.
- This software will be installed in couple of minutes.

STARTING AUTODESK FUSION 360

- To start **Autocad Fusion 360** from **Start** menu, click on the **Start** button in the **Taskbar** at the bottom left corner, click on **All Programs** folder and then on **Autodesk folder,** select the Autocad Fusion 360 icon; refer to Figure-5.

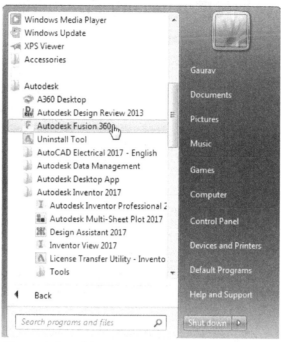

Figure-5. Start Menu

- While installing the software, if you have selected the check box to create a desktop icon, then you can double-click on that icon to run the software.
- If you have not selected the check box to create the desktop icon but want to create the icon on desktop, then right-click on the **Autodesk Fusion 360** icon in the **Start** menu and select the **Send To-> Desktop(Create icon)** option from the Shortcut menu displayed.

After you perform the above steps, the Autodesk Fusion 360 software window will be displayed; refer to Figure-6.

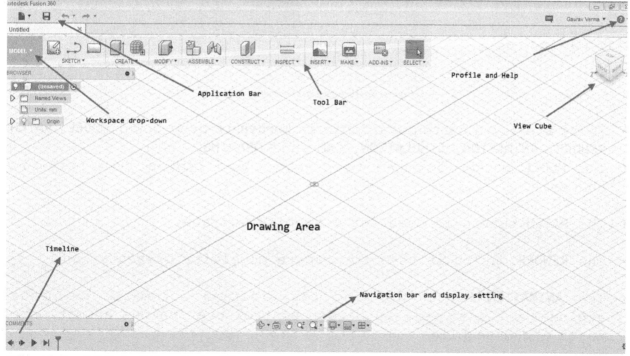

Figure-6. Autodesk Fusion 360 application Window

STARTING A NEW DOCUMENT

- Click on the **File** drop-down and select the **New design** tool; as shown in Figure-7.

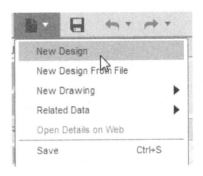

Figure-7. File Menu

- After performing the above step, a new document will open.
- Select the desired workspace from the **Workspace** drop-down as per your need; refer to Figure-8.

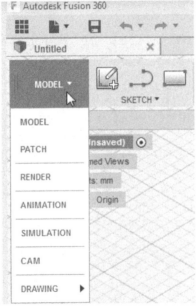

Figure-8. Workspace drop-down

- There are seven button available in **Workspace** drop-down; **MODEL, PATCH, RENDER, ANIMATION, SIMULATION, CAM,** and **DRAWING.**
- The **MODEL** button is used to create solid bodies from primitive shapes.

- The **PATCH** button is used for designing surface.

- The **RENDER** button is used to prepare model for presentation.

- The **ANIMATION** button is used for creating automatic or manual exploded views as well as direct control over unique animation of parts and assemblies.

- The **SIMULATION** button is used for Engineering Analysis.

- The **CAM** button is used for producing CNC codes.

- The **DRAWING** button is used for generating drawing and sketch.

You will learn more about these workspace later in this book.

FILE MENU

The options in the **File** menu are used to manage files and related parameters. The various tools of **File** menu are discussed next.

New Design From File

This is used to import the file into Fusion 360 software. The steps to do so are given next.

- Click on the **New Design From File** tool from the **File** menu; refer to Figure-9. The **Open** dialog box will be displayed.

*Figure-9. New Design
From File tool*

- Select the desired format from the **File Type** drop-down; refer to Figure-10 and double-click on the file to be imported.

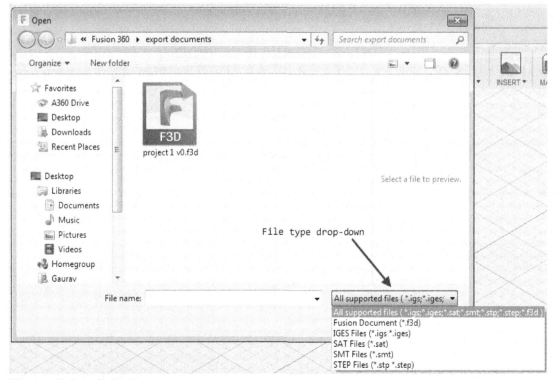

Figure-10. Open dialog box

This command supports file formats:

- Autodesk Fusion 360 Archive Files (*.f3d)
- IGES (*ige, *iges, *igs)
- SAT/SMT Files (*.sat, *.smt)
- STEP Files (*.step, *.stp)

New Drawing

The **New Drawing** tool is used for initiating a new drawing from animation or design; refer to Figure-11. The method to use this tool will be discussed later in the book.

Figure-11. New Drawing

Related data

Files and object that are related to current file are displayed in **Related Data** cascading menu; refer to Figure-12. You will learn about this option later in this book.

Figure-12. Related Data

Open details On Web

The **Open Details On Web** tool is used to display the details of current file in web browser. To use this option, you must have saved your file on Autodesk cloud.

Save

The **Save** tool is used to save the current file. You can also press **CTRL+S** key to save file.

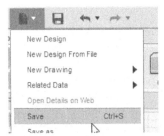

Figure-13. Save tool

- Click on the **Save** tool from **File** menu or select the **Save** button from application bar as shown in Figure-13. The **Save As** dialog box will be displayed asking you to specify the **Name** and **Location** of your file; refer to Figure-14 in case if you are offline. If you are online then the **Save As** dialog box will be displayed; refer to Figure-15.

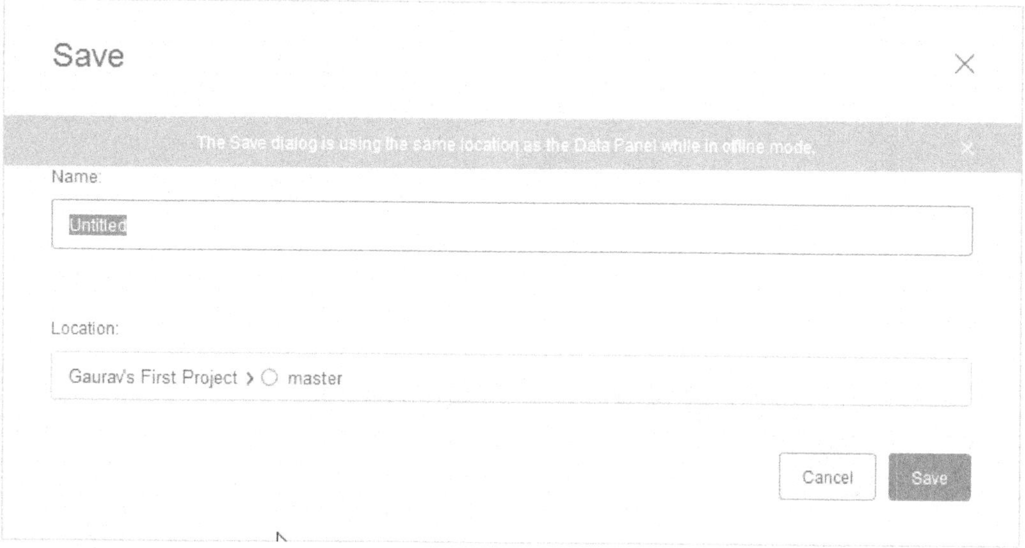

Figure-14. Save As Dialoge box offline

- Click on the down button next to Location field and select the desired location to save the file. Note that you can create a new project or new folder by using respective button in the dialog box; refer to Figure-15.

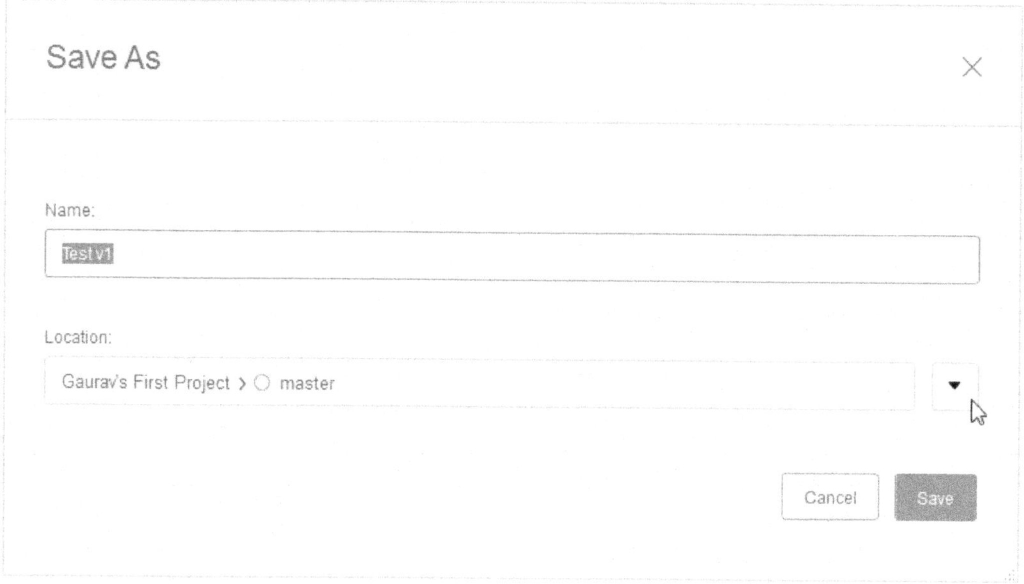

Figure-15. Save As online

The **Save As** dialog box is displayed when you save the file for first time using this tool. After that if you press **CTRL+S** or click on the **Save** button then file will be saved without displaying any dialog box.

Save As

Using the **Save As** tool, you can save the file in desired format. The procedure to use this tool is discussed next.

- Click on **Save As** tool from **File** menu; refer to Figure-16.

Figure-16. Save As

- On selecting this tool, the **Save As** dialog box will be displayed which has been discussed earlier.

Recover Documents

The **Recover Documents** tool is used to recover your last unsaved file; refer to Figure-17. Note that this tool will be active only when you have some unsaved work and software closes unexpectedly. The procedure to use this tool is given next.

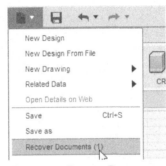

Figure-17. Recover Document tool

- Click on the **Recover Documents** tool from the **File** menu. The **File Recovery** dialog box will be displayed where you need to select your unsaved file; refer to Figure-18.

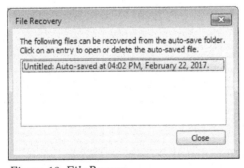

Figure-18. File Recover

- Double-click on the file that you want to be recovered. The file will be recovered and will open in **Autodesk Fusion 360**.

Export

The **Export** tool is used to export the file in various different formats.

By using this tool, you can share or send the file to anyone you want.
The procedure to use this tool is given next.

- Click on the **Export** tool from the **File** menu; refer to Figure-19. The
  Export dialog box will be displayed; refer to Figure-20.

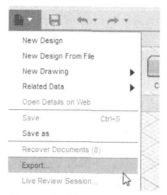

Figure-19. Export tool

- In the **Type** drop-down, there are five options to specify export
  format; **F3D, IGES, SAT, STM,** and **STEP.** Select the desired format from
  the drop-down.

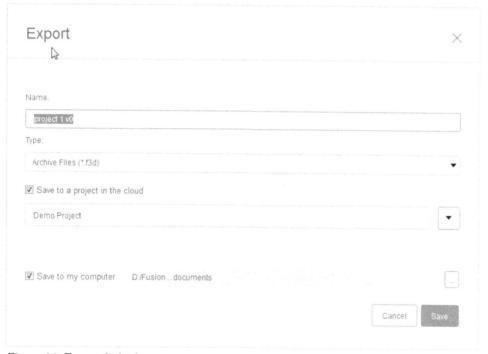

Figure-20. Export dialog box

- Clear the **Save to a project in cloud** check box if you do not want
  to save the file on cloud otherwise make it selected. If you are
  saving this file for the first time on cloud after selecting this
  check box then the **Save to a project in the cloud** dialog box will be
  displayed; refer to Figure-21. Select the desired project from the
  Project drop-down in the dialog box and click on the **Select Project.**
  The project will get selected.

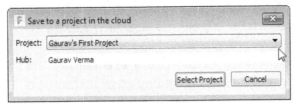

Figure-21. Save to a project in the cloud

- You can also save this file to your computer by selecting the **Save To My Computer** check box. On selecting this option, you need to specify the location of file where you want to save the file in field next to the check box.

Live Review Session

Till the time of writing this book, the **Live Review Session** tool is under preview mode and not an active part of software. To use this tool, you must be in online mode with file saved on cloud. The procedure to use this tool is given next.

- Click on the **Live Review Session** tool from the **File** menu; refer to Figure-23. The **Fusion 360** dialog box will be displayed asking you to enable preview of the feature if you have not opted to display preview tools earlier while installing. Click on the **Enable** button from the dialog box. The **LIVE REVIEW SESSION** dialog box will be displayed in the left of application window; refer to Figure-22.

Figure-22. Live Review Session dialog box

- Copy the link in **Share this link to invite others** edit box and distribute it to participants via E-mail or other methods. (Your participants need to open the link in their web browser to participate. Note that their web browser should support OpenGL to display model.)

- Type desired comments in the input box at the bottom and press **ENTER** key. The message will be displayed to all the participants in **Chat history** area of the box.
- After completing session, click on the **Stop** button and then click on the **End Session** button to close live review session.

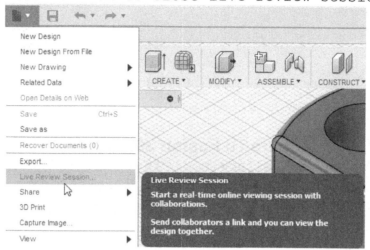

Figure-23. Live Review Session

Share

The **Share** tool is used to share the file or project. Click on the **Share** tool from **File** menu. The three options will be displayed in **Share** cascading menu; refer to Figure-24.

Figure-24. Share cascading menu

Share Public Link

The **Share Public Link** tool is used to share the project link with anyone. The procedure to use this tool is discussed next.

- Click on **Share Public Link** tool of **Share** cascading menu from **File** Menu.
- On selecting this tool, a **Share Public Link** dialog box will be displayed, refer to Figure-25.
- Copy the link from **Copy** box and share the link via E-mail or other method with anyone. Select the **Allow item to be downloaded** check box to download the file.

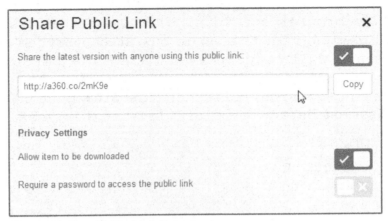

Figure-25. Share Public Link dialog box

Share To Fusion 360 Gallery

The **Share To Fusion 360 Gallery** tool is used to share the file online to the gallery of Fusion 360 where the file is accessible to everyone registered with the gallery. The procedure to use this tool is given next.

- Click on **Share To Fusion 360 Gallery** tool from **Share** cascading menu. The **SHARE TO FUSION 360 GALLERY** dialog box will be displayed; refer to Figure-26.

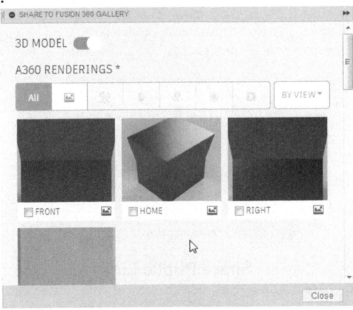

Figure-26. Share To Fusion 360 Gallery dialog box

- Select the project from the gallery and click on **Publish** button to upload.

Publish to GrabCad

Autodesk Fusion 360 allows you to publish your file directly to **GrabCad**. Using this tool there is no need to manually Export or Upload the project.

- Click on **Publish to GrabCad** tool from **Share** cascading menu. The **PUBLISH**

TO GRABCAD dialog box will be displayed; refer to Figure-27.

- In this dialog, box you need to login to GrabCad platform with your GrabCad account details to upload or share the project.
- If you don't have an account on GrabCad then, you can create the account by clicking on **Sign Up** button.

Figure-27. Publish To GrabCad dialog box

3D Print

The **3D Print** tool is used to prepare and send current model for 3D printing. The procedure to use this tool is given next.

- Click on the **3D Print** tool from the **File** menu or click on the **3D Print** tool from the **Make** panel in the **Toolbar**; refer to Figure-28. The **3D-PRINT** dialog box will be displayed; refer to Figure-29.

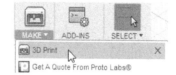

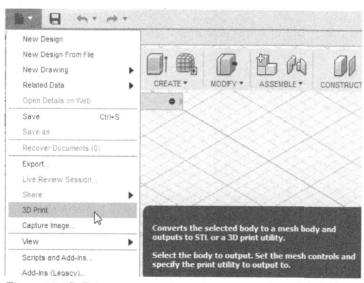

Figure-28. 3D Print

- This tool convert the selected body to Mesh Body and sends the output to **3D-Print utility** and **STL**.
- By default, the Selection button is active in this dialog box and you are prompted to select model to be 3D printed. Click on the model to select for 3D Print.
- Now, select the **Preview Mesh** check box to see the Number of triangles forming in the selected body.

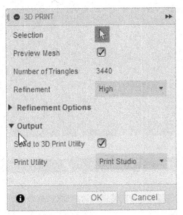

Figure-29. 3D-Print Dialog box

- Select the resolution of printing in **Refinement** drop-down menu. There are four resolution in this drop-down. If you select the **High,Low** and **Medium** command then the **Surface deviation, Normal deviation,Maximum Edge Length** and **Aspect Ratio** will be adjusted automatically. If you want to adjust these parameters manually then you need to select the **Custom** option in **Refinement** drop-down; refer to Figure-30.

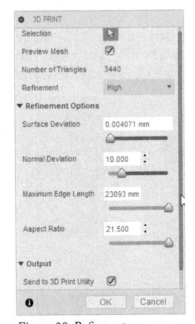

Figure-30. Refinment

- Select the **Send to 3D Print Utility** check box in **Output** section to further edit the model in 3D printing utility. Specify the utility in **Print Utility** drop-down; refer to Figure-31.

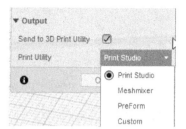

Figure-31. Output 3D-Print

- Clear the **Send To 3D Print Utility** check box, if you want to save the project file in **STL** format so that you can open this file in any 3D printing software.
- For printing the model in 3D, select any software according to your need. Now, download and install the desired software in your PC.
- If you select the **Print Studio** Utility software from Autodesk Fusion 360, click on the **OK** button. The **Print Studio** application window will be displayed; refer to Figure-32.

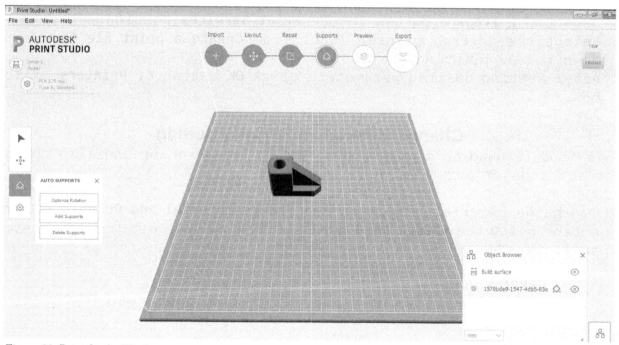

Figure-32. Print Studio Window

The tools of this application are discussed next.

Print studio

On selecting the **OK** button from **Print Studio** dialog box, the file is exported from Autodesk Fusion 360 and imported to Autodesk Print Studio.

Change the Printer

- To change the **Printer**, select the button at upper left corner; refer to Figure-27. The **Printers** dialog box will be displayed. In this dialog box, you can select the printer to connect via **WI-FI** network or you can **plug** your printer into your PC.

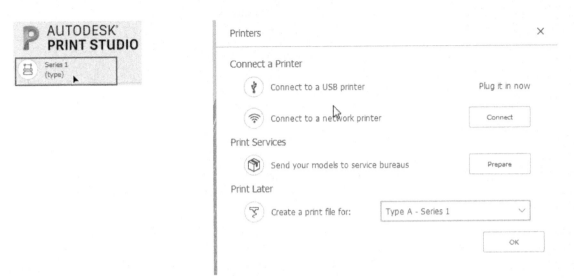

Figure-33. Selecting a 3D printer

- If you want to use the **Print Services** tools to modify parts, then select the **Prepare** button in the **Print Services** Section.
- Select the desired printer model from **Create a print file for** drop-down to for printing.
- After setting desired parameters click **OK** button of **Printers** dialog box.

Change Material and Printer Setting

This tool is used to change the material for printing and also change the settings of printing.

- To change material, click on the **Change Material and Printer Setting** button below the **Change the Printer** button in the left toolbar; refer to Figure-34.

Figure-34. Changing Material and Printer Setting

- On selecting this button, the **Settings** dialog box will be displayed;

refer to Figure-34. Select the **Profile** according to your needs in **Profiles** drop-down. In **Basic Setting** section, specify the desired parameters.
- If you want to specify advanced parameters like, support dimensions, Infill pattern etc. then, click on **Advanced Setting** button and specify the desired parameters.
- After setting the desired parameters, click on **OK** button from **Settings** dialog box.

Toolbar

The tools in the **Toolbar** of **Print Studio** application are used to finalize the project for 3D printer; refer to Figure-35.

Figure-35. Print Studio Toolbar

- **Import** tool is used to import the file to Autodesk Print Studio. We have already imported the file so the button is displayed selected in the toolbar in the Figure-35.
- **Layout** tool is used to position the model on printing tray. On selecting this tool, the toolbar for layout will be displayed on the left of the application window; refer to Figure-36.

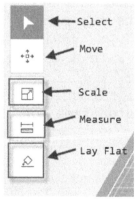

Figure-36. Layout

- Select **Scale** button, if you want to scale up or scale down the model. Select **Measure** button, if you want to measure your model before printing. If you want to lay your model face on surface, then select **Lay Flat** button.
- On setting the parameters of **layout** tool, next step is **Repair** the model. On selecting the **Repair** tool from **Toolbar** section, a repair tool box will be displayed at the left of screen. This tool detects the model issues like holes and other geometry defects. You can

resolve these issues manually or automatically. When issues are resolved, this tool display a notification i.e. **No Problems**; refer to Figure-37.

Figure-37. Repair in Print Studio

- Next tool in this section is **Supports** tool. This tool is used for giving the support to project in 3D-print. To apply **supports**, click on the Support tool from the toolbar. The **AUTO SUPPORTS** toolbox will be displayed in the left of screen; refer to Figure-38.
- In this Tool you can Optimize the Rotation using the **Optimize Rotation** tool. If you want to add or remove the support then select the respective button i.e. **Add supports** and **Delete Supports.**
- Now, select the **Preview** tool from **Toolbar,** a **Preview Slices** Box will be displayed at the left of the screen; refer to Figure-39. This tool generate the **Toolpaths** for model. This tool preview the model into slices. Use this tool to review the model before sending the information to 3D-Printer. This tool give you a chance to resolve any issue before printing.
- In **Preview Slices** Box, you can check the time taken by printer and volume of model in **Estimated time** and **Estimated volume** section. By moving the **Slice Preview** button, you can see the formation of model layer by layer.

Figure-38. Support Print Studio

Figure-39. Preview tool

After adjusting all the parameters, select the **Export** tool from **Toolbar** to print the model in printer.

Capture Image

The **Capture Image** tool is used to capture the image of current model in **Autodesk Fusion 360** screen. The procedure to use this tool is discussed next.

- Click **Capture Image** tool from **File** menu; refer to Figure-40. The **Image Options** dialog box will be displayed where you need to specify the size of image in **Image resolution** section.

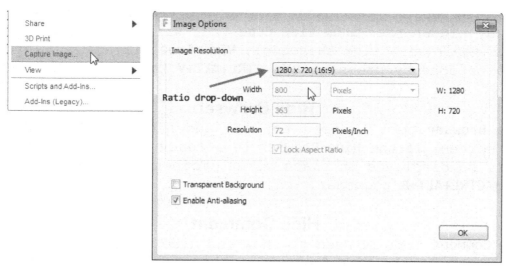

Figure-40. Capture Image tool

- On selecting the size of image from **Ratio** drop-down, all other parameters like Width, Height and resolution will be adjusted automatically.
- If you want to select these parameters manually then, select **Custom** in **Current Document Window Size** drop-down. Select **Transparent Background** check box for setting the background of the image to be transparent.
- After adjusting all the parameters, select the **OK** button of **Image Options** dialog box. The **Save As** dialog box will be displayed which has been explained earlier.

View

The **View** tool is used to show and hide various elements of Fusion 360 application window. The procedure to use these tools are given next.

- Click on **View** tool from **File** Menu. The **View** cascading menu will be displayed; refer to Figure-41.

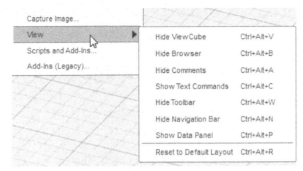

Figure-41. View tool

Hide View Cube

The **Hide View Cube** tool is used to hide the **View Cube**. If the **View Cube** is already hidden then **Show View Cube** tool will be displayed in place of it. On again selecting this tool, **View Cube** will be displayed. This can also be done by pressing the **CTRL+ALT+V** keys.

Hide Browser

The **Hide Browser** tool is used to view and hide the **Browser** from the current screen. If the **Hide Browser** is already hidden then **Show Browser** tool will be displayed in place of it. This tool can also be used by pressing **CTRL+ALT+B** together.

Hide Comment

The **Hide Comment** tool is used to show and hide the **Comments** from Fusion 360 software. If the **Hide Comment** is already hidden then **Show Comment** tool will be displayed in place of it. This tool can also be used by pressing **CTRL+ALT+A** together.

Show Text Commands

The **Show Text Commands** tool is used to show and hide the **Text Commands Bar** from Fusion 360 software. If the **Show Text Commands** is already hidden then **Hide Text Commands** tool will be displayed in place of it. This tool can also be used by pressing **CTRL+ALT+C** together.

Hide Toolbar

The **Hide Toolbar** tool is used to show and hide the **Toolbar** from Fusion 360 software. If the **Hide Toolbar** is already hidden then **Show Toolbar** tool will be displayed in place of it. This tool can also be used by pressing **CTRL+ALT+W** together.

Hide Navigation Bar

The **Hide Navigation Bar** tool is used to show and hide the **Navigation bar** from Fusion 360 software. If the **Hide Navigation Bar** is already hidden then **Show Navigation Bar** tool will be displayed in place of it. This tool can also be used by pressing **CTRL+ALT+N** together.

Show Data Panel

The **Show Data Panel** tool is used to show and hide the **Data Panel** from Fusion 360 software. If the **Show Data Panel** is already hidden then **Hide Data Panel** tool will be displayed in place of it. This tool can also be used by pressing **CTRL+ALT+P** together.

Reset To Default Layout

The **Reset To Default Layout** tool is used to reset all the user defined setting. This tool can also be used by pressing **CTRL+ALT+R** together.

Script and Add-Ins

The **Script and Add-Ins** tool is used to manage **Script** and **Add-Ins**. The procedure to use this tool is discussed next.

- Click on **Script and Add-Ins** tool from **File** menu. The **Script and Add-Ins** dialog box will displayed; refer to Figure-42.

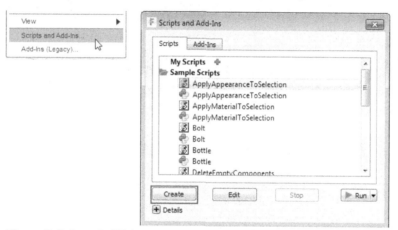

Figure-42. Script and Add-ins

In this dialog box, there are two sections i.e. **Scripts** and **Add-ins**.
- **Script** is used to automate the design process.
- **Add-Ins** is used to run the third party application in Autodesk Fusion 360.

Creating Script and Add-Ins

- To create the script and add-Ins for the current project, click on **Create** button from **Script and Add-ins** dialog box. The **Create New Script or Add-Ins** dialog box will be displayed.
- Specify the parameters in this dialog box and click on **Create** button.
- A **Hello Script** dialog box will be displayed if you are using the default script. Select **OK** button from **Hello Script** dialog box to complete the process.
- If you want to Debug the new script, then again open the **Script and Add-Ins** dialog box and select the **Debug** button in **Run** drop-down refer to Figure-43.

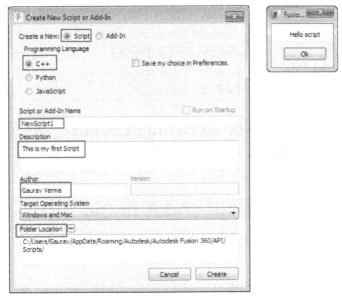

Figure-43. Creating New Script

The process of creating the new **Add-Ins** is similar to creating the **Script**.

Add-ins (Legacy)

Similarly, you can use the **Add-ins(Legacy)** tool to add apps created in earlier versions.

UNDO AND REDO BUTTON

The **Undo** tool is used to undo an action. The procedure to use this tool is discussed next.

* Click on **Undo** tool in **Application bar** or you can press **CTRL+Z** keys to do the same action; refer to Figure-44.
* You can't undo some actions, like clicking commands on the File tab or saving a file. If you can't undo an action, the Undo command changes to Can't Undo.

Figure-44. Undo and Redo

* Redo tool is used to redo an action. To use this tool, select the **Redo** button in **Application bar** or you can press **CTRL+Y** keys to do the same action.
* The Redo button only appears after you've undone an action.

USER ACCOUNT DROP-DOWN

There are various tool in **User Account** drop-down menu which are discussed next.

Preferences

The **Preferences** tool is used to set the **preferences** for the file or project. The procedure to use this tool is given next.

- Click on **Preferences** tool under **User Account** drop-down from **Help and Profile** section; refer to Figure-45. The **Preferences** dialog box will be displayed where you need to specify the various parameters for the project; refer to Figure-46.

Figure-45. Preferences

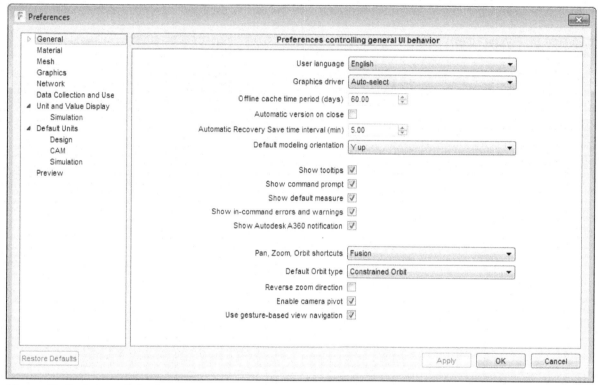

Figure-46. Preferences dialog box

- In **General** node, you need to specify the preferences for **Material, Mesh, graphics, Network,** and **Data collection and Use.**
- In **Unit and Value Display** node, you need to specify the preference of **Units and Values** and **display for Simulation.**
- In **Default Units** node, you need to specify the preference of **Default Unit, Design, CAM** and **Simulation.**

Work Offline

The **Work Offline** tool is used when you want to work in Offline mode. With the help of this tool, you are able to work anywhere and anytime without the need of an internet connection. The procedure to use this tool is discussed next.

- Click on **Work Offline** tool from **User Account** drop-down; refer to Figure-47.

Figure-47. Work Offline

- If you want to return to the online mode, then un-check the **Work Offline** tool. The documents changed in Offline mode will automatically updated and uploaded to Data Panel.

Similarly, you can set the other options in this drop-down.

HELP DROP-DOWN

When you face any problem working with Autodesk Fusion 360, this tool will help you. In **Help** drop-down, there are various tool; refer to Figure-48 which are discussed next.

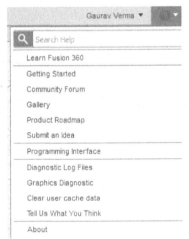

Figure-48. Help

Search Help Box

When you are facing any problem regarding your Sketch or software, then you need to type your problem in **Search Box** and find a suitable solution for your problem; refer to Figure-49.

Figure-49. Search box

Learn Fusion 360

The **Learn Fusion 360** tool can be used to learn about the software tools and process being used. On clicking this button, website of Autodesk Fusion 360 will be opened, where you can find the information; refer to Figure-50.

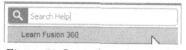

Figure-50. Learn fusion 360

Getting Started

When you are new to Autodesk Fusion 360 then this tool will help you to understand this software terminology and basic settings.

- Click on **Getting Started** tool from **Help** drop-down. The **GETTING STARTED** dialog box will be displayed.
- The navigation functions of mouse can be set as Autodesk Inventor, SolidWorks or Alias by using this tool and also you can set the units for your project; refer to Figure-51.

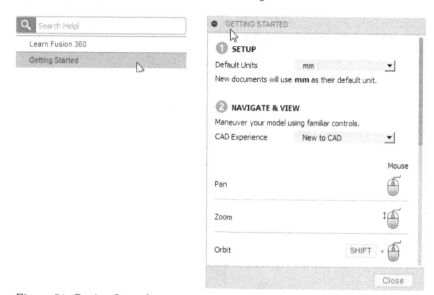

Figure-51. Getting Started

- If you are new to CAD, then you can select the **New to Cad** in **Cad Experience** drop-down menu.

Similarly, you can use **Community Forum, gallery, Product Roadmap, Programming Interface** and **submit an idea** tool to get help from Autodesk Fusion 360 or provide feedback.

Diagnostic log Files

When you are having a problem in Fusion 360 and want **Technical support** team to look at software problem data, then you can create the **Log Files** and send it to the support team of Autodesk.

- To create log files, select **Diagnostic Log Files** tool in **Help** drop-down. A **Diagnostic Log Files** dialog box will be displayed; refer to Figure-52.

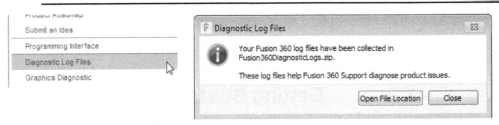

Figure-52. Daignostic Log Files

- Click on **Open File Location** button and select the log file which you want to send to technical support team for the solution of problem.

Similarly, you can use **Graphics Diagnostic, Clear user cache data,** and **About** tool from **Help** drop-down in Autodesk Fusion 360.

DATA PANEL

Data Panel is used to manage the fusion 360 project. The project you saved in cloud will be shown here. You can easily access the saved project files from anywhere and anytime with the help of internet access.

- To show and hide the **Data Panel**, click the button at upper left corner of the Fusion 360 window; refer to Figure-53.

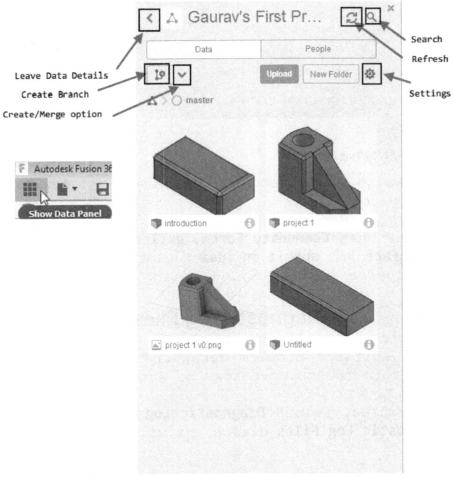

Figure-53. Data Panel

- The **User's First Project** Box will be displayed. In this, there are two sections i.e. **Data** and **People**.

- In **Data section**, the project file which are saved earlier by you are displayed. You are able to open any project by double-clicking on the file.
- In **People section**, the information of the user will be displayed refer to Figure-54.

Figure-54. People section in Data panel

Create and Merge Branch

The tools in this drop-down are used to create various categories of the current project like, we have master as our main project and we are creating mechanical tools as a branch. At any point in entire project, you are able to create the Branch. This help you to manage your projects in a different category; refer to Figure-53. The steps to use this tool are given next.

- Click on **Create Branch** button from **Create and Merge Branch** drop-down; refer to Figure-55.

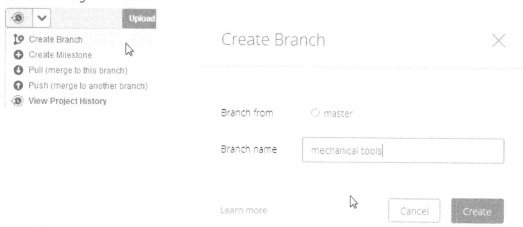

Figure-55. Create Branch

- On selecting the **Create Branch** button, the **Create Branch** dialog box will be displayed. Where you need to specify the location for branch in **Branch from** section and name under **Branch name** section. Note that once you have created a branch then you will not be able to rename or delete any created branch. File in the branch cannot be moved

outside the branch or project. You can only move files from branch to master or master to master.
- After specifying the parameters, click on the **Create** button of **Create Branch** dialog box to create the Branch.

Merge Push or **Pull** commands are only available, when there are more than one branch in your project. At least one file should be available in the folder to merge branch. Empty folder will not merge.

Upload

The **Upload** tool is used to upload the project to cloud. The procedure to use this tool is discussed next.

- Click on **Upload** button from **Data Panel.** The **Upload** dialog box will be displayed; refer to Figure-56 where you need to select the file for uploading.
- If you want to change the location of uploading, then you can also select the desired location under **Location** Section.
- After specifying the desired parameters, click on **Upload** button from **Upload** dialog box. File will be uploaded soon.

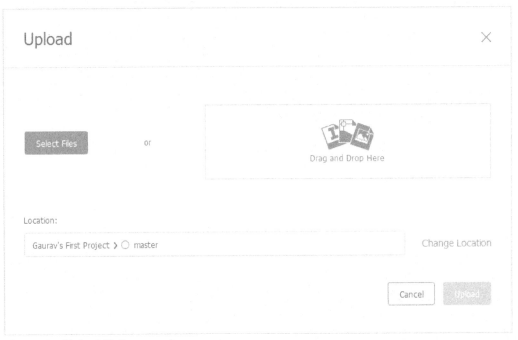

Figure-56. Upload file Data panel

- **New Folder** button is used to create a new folder. With the help of this tool you can locate similar files to the specific folder.

Leave Data Details

To open the data, you need to select the arrow button which is displayed at the upper left corner of the Data Panel box. On selecting this button, an another box will be displayed; refer to Figure-50.

- If you want to return to the previous screen, then click on the **Back** button which is highlighted in the figure.
- In **Projects** section, your projects are mentioned. If you want to select the **My Recent Data, Demo Project** and **User's first project** then double-click on any of the link.
- In **Samples** section, there are various samples and tutorial of design and project. To select any one of these samples double-click on the link.

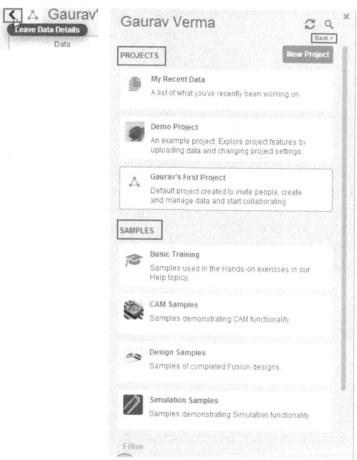

Figure-57. Leave Data Details

BROWSER

The **Browser** presents an organized view of your design steps in a tree like structure on the left of the Fusion 360 screen. When you select a design or model in the browser, it is also highlighted in the graphics window; refer to Figure-51.

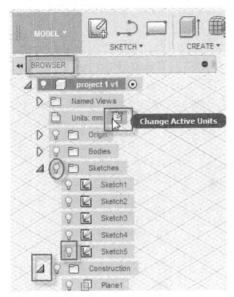

Figure-58. Browser

- To access context menu commands, right click on the **Browser** node.
- Click on the **light bulb** button to toggle the visibility of the design. If any of the color of header's background is **blue** in browser then it indicates the **Active** node.
- To change the Units of the design or sketch, select the **Edit** button of **Change and Active Units**.

NAVIGATION BAR

The **Navigation Bar** is positioned at the bottom of the window of Fusion 360. It provide access of navigation commands for design. To start a navigation command, click on any tool from the **Navigation Bar**; refer to Figure-59.

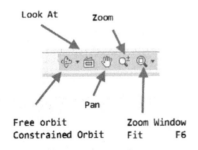

Figure-59. Navigation Bar

There are various tool in **Navigation Bar**, which are discussed next.

Orbit- The **Orbit** tools are used for rotating the current view of project. There are two types of orbit tools i.e. **Free Orbit** and **Constrained Orbit**. **Free Orbit** is used to rotate the design view **freely** and **constrained Orbit** is used to rotate the view in **constrained** motion.

Look At- This tool is used to view the faces of a model from a selected plane.

Pan- It is used to move the design parallel to the screen.

Zoom- It is used to increase or decrease the magnification of the current view.

Zoom Window and Fit- **Zoom window** tools are used for **magnification** of the selected area. **Fit tool** is used to position the entire design on the screen.

DISPLAY BAR

The tools of **Display Bar** are used to visualize the design from different viewports; refer to Figure-60.

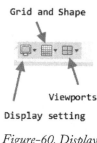

Figure-60. Display Setting

Display Setting- The **Display Setting** tool is used to enable or disable various commands like **Desired Visual Style, Visibility of objects,** and **Camera settings.**

Grid And shapes- The **Grid and Shapes** tool is used to show or hide various tool like **Layout Grid, Layout Grid Lock, Snap to Grid** etc.

Viewports- The **Viewports** tool is used to display four **viewports** in same time on the fusion 360 screen. This tool also allows to work in one view and enable the change is another view.

Viewcube

The **ViewCube** tool is used to rotate the view of camera. Drag the viewCube to perform a free orbit. There are six faces in viewcube i.e. **Top, Bottom, Front, Back, Right,** and **Left.** To access the standard orthographic and isometric views, click on the faces and corners of the cube; refer to Figure-54.

Figure-61. View-cube

Mouse

- Use the mouse shortcuts to zoom in/out, pan the view, and orbit the view refer Figure-62.
- Scroll on the middle mouse button to zoom in or zoom out.
- Click and hold the middle mouse button to pan the view.
- Use the Shift + middle mouse button to orbit the view.

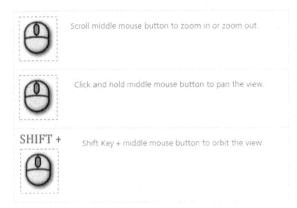

Figure-62. Mouse shourtcut keys

Timeline

The **Timeline** tool is very useful to save time in editing. It records the design feature in chronological order; refer to Figure-56.

Figure-63. Timeline

If you want to edit the model, then double-click on tool from **Timeline** and make the necessary changes. The final design will automatically generated with the changes; refer to Figure-64.

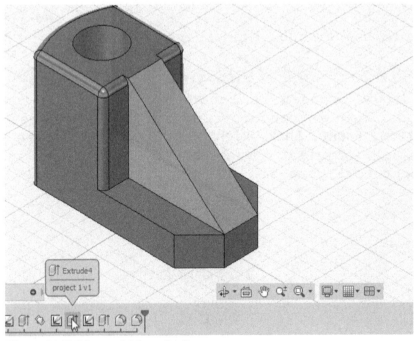

Figure-64. Editing model from timeline

SELF ASSESSMENT

Q1. ---------- is used to manage the fusion 360 project.

Q2. Which tool is used to share the files to GrabCad.

Q3. The ----- button is used for producing CNC codes.

Q4. Which of the following workspace is not available in Autodesk Fusion 360 application.

a. Patch Workspace
b. Model Workspace
c. Assemble Workspace
d. Drawing Workspace

Q5. Discuss the use of **Save** tool with example.

Q6. Discuss the process of exporting a project in computer with example.

Q7. Discuss the use of **3D Print** tool with example.

Q8. Which tool is used to generate **Toolpaths** for model?

Q9. How to reset the default layout?

Q10. Explain the process of creating **Script** with example.

FOR STUDENT NOTES

Chapter 2

Sketching

The major topics covered in this chapter are:

- *Starting Sketch*
- *Sketch Creation Tools*
- *Sketch Editing Tools*
- *Sketch Palette*
- *Constraints*
- *Sketching Practical and Practice*

INTRODUCTION

In engineering, sketches are based on real dimensions of real world objects. In this chapter, we will be working with geometric entities like; Line, Circle, Arc, Polygon, Ellipse and so on. Note that the sketching environment is the building base of 3D Models so you should be proficient in sketching.

In this chapter, we will be working in **Model Workspace**. We will learn about various tools of toolbar used in sketching and 3D Sketching.

STARTING SKETCH

- To start a new sketch, click on any tool of **Sketch** drop-down from **Toolbar** or you can also select the **Create Sketch** button from the **Sketch** drop-down; refer to Figure-1. The three main plane or planer face will be displayed on canvas screen; refer to Figure-2.

Figure-1. Create Sketch

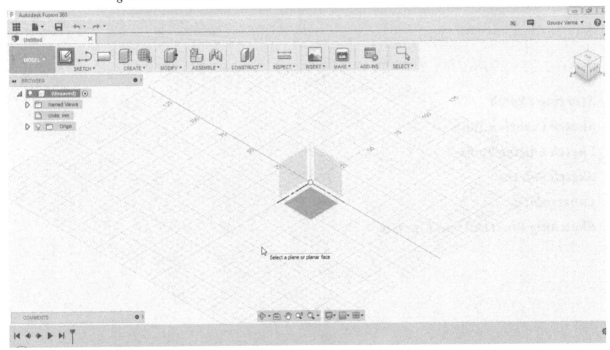

Figure-2. Select Plane for Sketch

- Click on the desired plane from the canvas screen. The selected plane will become parallel to the screen. Now, we are ready to draw sketch on the selected plane.

First, we will start with the sketch creation tools and later, we will discuss the other tools.

SKETCH CREATION TOOLS

In **Sketch** drop-down there are various sketching tools which are discussed next.

Line

The **Line** tool is used to create a line. The procedure to use this tool is discussed next.

* Click on **Line** tool of **Sketch** drop-down from **Toolbar**; refer to Figure-3. You can also press **L** key to select the **Line** tool.

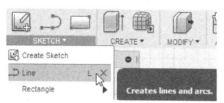

Figure-3. line

* Select a point on screen with the help of cursor.
* Enter the specify value of length in the length edit box. You can press **TAB** key to toggle between **Angle** edit box and **Length** edit box.
* If you want to create an angular line then, press **TAB** key to toggle to **Angle** edit box and enter the desired value of angle; refer to Figure-4.

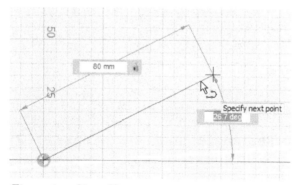

Figure-4. making of line

* If you are making a freestyle line then, click on the screen to select the first and last point for the line.
* To create a line for construction purpose, you need to right-click on the existing line and click on **Normal Construction** button from **Marking menu**; refer to Figure-5.

Marking menu

It is the radical display of most frequently used tool. This menu also provide a quick access to the tools of the toolbar. This menu is the fastest way to input the tool. To access this menu, right-click anywhere on the screen within the sketch canvas. To select any tool of this menu, move the cursor towards the tool and the tool will be highlighted. Click on it to activate the tool.

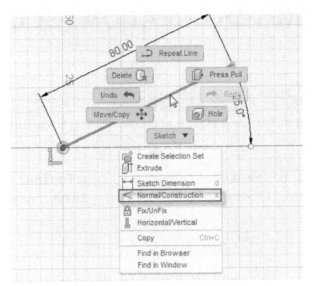

Figure-5. marking menu

For example, select the **Sketch Dimension** tool to apply the dimension.

Rectangle

The **Rectangle** tool is used to create rectangle. There are three tools in **Rectangle** cascading menu i.e. **2-Point Rectangle, 3-Point Rectangle,** and **Center Rectangle**; refer to Figure-6.

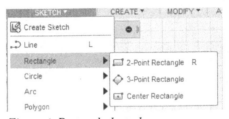

Figure-6. Rectangle drop-down

You can also select **Rectangle** tool by pressing **R** key. The procedure to use these tools are discussed next.

2-Point Rectangle

- Click on the **2-Point Rectangle** tool of the **Rectangle** cascading menu from **Sketch** drop-down. You are asked to specify the first point.
- Click on the sketch canvas to specify first point of rectangle. You will be asked to specify the other corner point; refer to Figure-7.

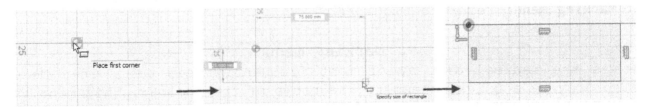

Figure-7. 2-Point Rectangle tool

- Specify the length and width of rectangle in the dynamic input boxes. To switch between the length dynamic input box and width dynamic input box, press **TAB** key.

- After specifying the parameters, press **ENTER** to create the rectangle.
- If you want to create a freestyle rectangle then, click on the screen to select the first corner point and last corner point of the rectangle on the sketch canvas.
- To add dimension right-click on the line of rectangle and click on **Sketch Dimension** tool from the **Marking menu.** You will learn more about dimensioning later in this chapter.

3-Point Rectangle

- Click **3-Point Rectangle** tool of **Rectangle** cascading menu from **Sketch** drop-down.
- Now, click to specify the first point of the rectangle on the sketch canvas; refer to Figure-8.

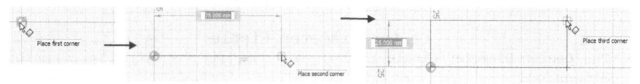

Figure-8. Creation of 3-point rectangle

- Click to specify the second point on the canvas or specify the dimension in dimension box.
- Click to specify the third point of the rectangle to create the complete rectangle.

Center rectangle

- Click on the **Center Rectangle** tool of the **Rectangle** cascading menu from **Sketch** drop-down.
- Click on the canvas to specify the center point of the rectangle; refer to Figure-9.
- Enter the dimension for rectangle in the particular edit box and press **ENTER** to create the rectangle.

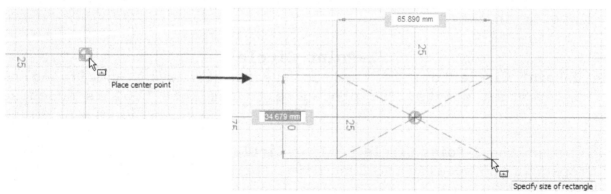

Figure-9. Creation of center rectangle

- If you want to create the free style rectangle then, click on the screen to select the first point and second point to create rectangle or specify the desired dimensions in the dynamic input boxes.

Circle

There are five tools in **Circle** cascading menu; **Center Diameter Circle,**

2-Point Circle, **3-Point Circle**, **2-Tangent circle**, and **3-Tangent circle**; refer to Figure-10.

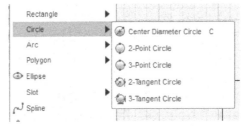

Figure-10. Circle tool

The shortcut key to use the circle tool is pressing **C** key from keyboard. The tools to create circle are discussed next.

Center Diameter Circle

- The **Center Diameter Circle** tool create the circle with the diameter dimension.
- Click on the **Center Diameter Circle** tool from **Circle** cascading menu.
- Click to specify the center point for the circle; refer to Figure-11.

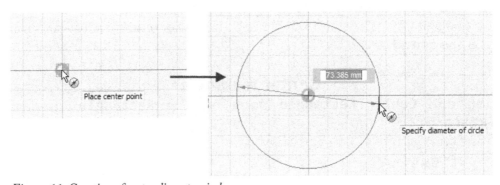

Figure-11. Creation of center diameter circle

- Click to specify the diameter point or enter desired value of diameter in the dynamic input box.

2-Point Circle

The **2-Point Circle** tool is used to create circle by specifying two circumferential points. The procedure to use this tool is discussed next.

- Click on the **2-Point Circle** tool of **Circle** cascading menu from **Sketch** drop-down and specify the first point for circle; refer to Figure-12.

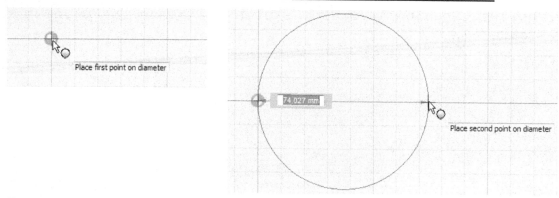

Figure-12. Creation of 2-point circle

- Specify the second point on the canvas screen or enter the dimension in the floating window and press **ENTER** key.

3-Point Circle

The **3-Point Circle** tool is used to create a circle using 3 points as references. The procedure to use this tool is discussed next.

- Click on **3-Point Circle** tool of **Circle** cascading menu from **Sketch** drop-down. Check again you are already in sketching and have selected plane.
- Click on the screen to specify the first point of circle; refer to Figure-13.

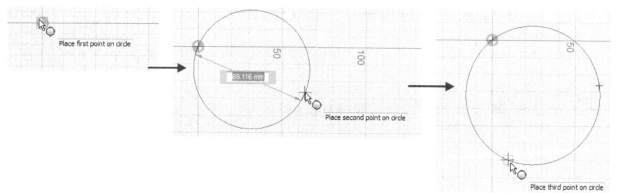

Figure-13. Creation of 3-point circle

- Click on the screen to specify the second and third point of the circle.
- Right-click on the screen and click **OK** button from **Marking menu** to complete the process. You can also press **ESC** key to exit the circle tool.
- If you want to add dimensions then, right-click on the circle and select the **Sketch Dimension** tool from **Marking menu**. You can also select **Sketch Dimension** tool from **Create** drop-down.

2-Tangent Circle

The **2-Tangent Circle** tool is used to create a circle using two tangent lines. The procedure to use this tool is discussed next.

- Select **2-Tangent Circle** tool of **Circle** cascading menu from **Sketch** drop-down.

- Select two tangent lines from sketch to create the circle; refer to Figure-14.

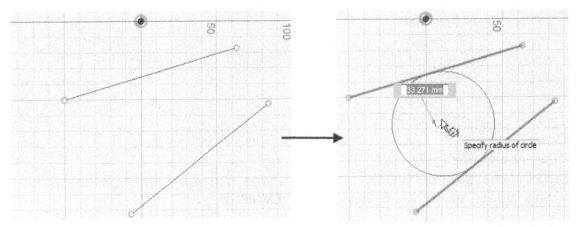

Figure-14. Creation of 2-tangent circle

- A preview of circle will displayed on screen. Enter the desired radius of the circle in the floating window and press **ENTER** to create a circle.

3-Tangent Circle

The **3-Tangent Circle** tool is used to create a circle using three tangent lines. The procedure to use this tool is discussed next.

- Click on **3-Tangent Circle** tool of **Circle** cascading menu from **Sketch** drop-down.
- Select three lines to be tangent to the circle or you can use window selection; refer to Figure-15.
- A preview of circle will be displayed. Right-click on the screen and click **OK** button from **Marking menu** to complete the process. You can also press **ESC** key to exit the circle tool.

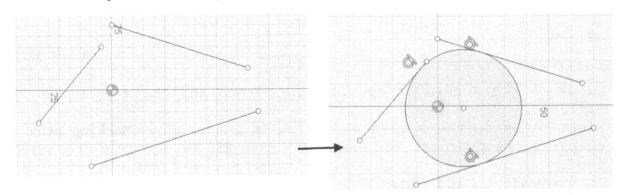

Figure-15. Creation of 3-tangent circle

Arc

The **Arc** tool is used to create arcs. There are three tools in the **Arc** cascading menu; **3-Point Arc, Center Point Arc,** and **Tangent Arc;** refer to Figure-16.

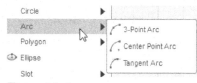
Figure-16. Arc

The tools of **Arc** cascading menu are discussed next.

3-Point Arc

The **3-Point Arc** tool is used to create a arc using three points. To use this tool, you need to specify two end points and one circumferential point of the arc. The procedure to use this tool is discussed next.

- Click on **3-Point Arc** tool of **Arc** cascading menu from **Sketch** drop-down. Select the desired plane or face from canvas screen if not selected yet.
- Click on the screen to specify the start point of the arc; refer to Figure-17.

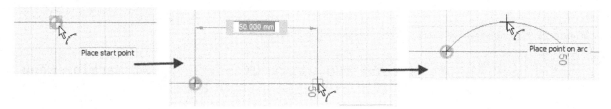

Figure-17. Creation of 3-point arc

- Click on the screen to specify the end point of the arc.
- Click a point on the arc to set the radius.
- Now, right-click on the arc and select **OK** button to complete the arc.

Center Point Arc

- Click **Center Point Arc** tool of **Arc** cascading menu from **Sketch** drop-down.
- Click on the screen canvas to specify the center point of arc.
- Click to specify the start point of the arc.
- Enter the radius in floating window or click on the screen to specify the end point of the arc; refer to Figure-18.

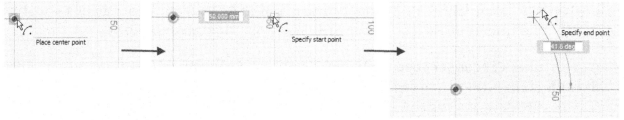

Figure-18. Creation of center point arc

- A preview of arc will be displayed. Right-click on the screen and click **OK** button from **Marking menu** to complete the process. You can also press **ESC** key to exit the circle tool.

Tangent Arc

- Click on the **Tangent Arc** tool of **Arc** cascading menu from **Sketch** drop-down.
- Click on an entity (line, arc or curve) from screen to specify the start point of the arc; refer to Figure-19.
- Click on the screen or entity to specify the end point of the arc.

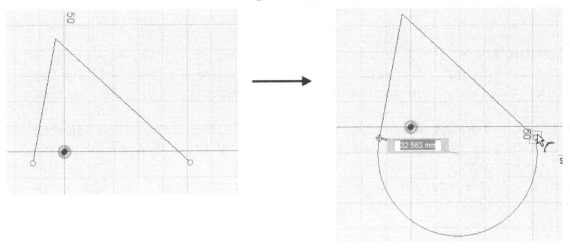

Figure-19. Creation of tangent arc

- Right-click on the screen and click **OK** button from **Marking menu** to complete the process. You can also press **ESC** key to exit the arc tool.

Polygon

The **Polygon** tool is used to create polygon of desired number of sides. There are three tools to create the polygon in **Polygon** cascading menu i.e. **Circumscribed Polygon, Inscribed Polygon,** and **Edge Polygon;** refer to Figure-20. The tools to create the polygon are discussed next.

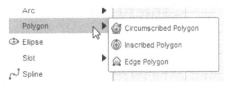

Figure-20. Polygon

Circumscribed Polygon

Use the **Circumscribed Polygon** tool to create a polygon formed outside the circle. Note that polygon corner points lie on the Circumscribed circle. The procedure to use this tool is discussed next.

- Click the **Circumscribed Polygon** tool of **Polygon** cascading menu from **Sketch** drop-down; refer to Figure-20.
- Click on the screen to specify the center point of the polygon; refer to Figure-17.

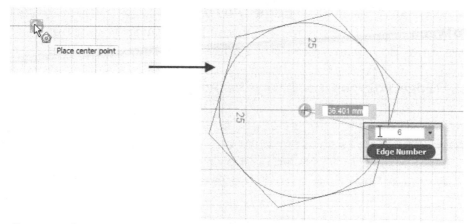

Figure-21. Creation of circumscribed polygon

- Enter the distance between center point of polygon and mid point of polygon edge in floating window.
- Select the number of edges of the polygon from **Edge Number** edit box. You can use the **TAB** key to switch between the floating window and **Edge Number** box.
- After specifying the parameters, right-click on the screen and click **OK** button to exit the polygon tool.

Inscribed Polygon

Use the **Inscribed Polygon** tool to create a polygon formed inside the circle. The procedure to use this tool is discussed next.

- Click on **Inscribed Polygon** of the **Polygon** cascading menu from **Sketch** drop-down.
- Click on the screen to specify the center point of the circle; refer to Figure-22.
- Enter the distance between center point of polygon and vertex of polygon edge in floating window.

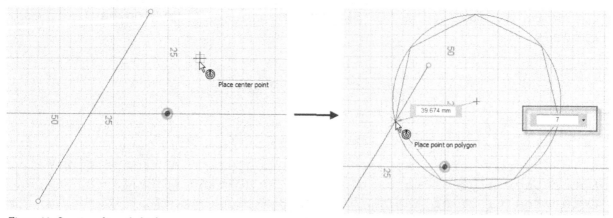

Figure-22. Creation of inscribed polygon

- Specify the value of number of edges in **Edge Number** edit Box.
- After specifying the parameters, right-click on the screen and click **OK** button from **Marking menu** to exit the polygon tool.

Edge Polygon

The **Edge Polygon** tool create a polygon by defining a single edge and the position of the polygon. The procedure to use this tool is discussed next.

• Click the **Edge Polygon** tool of the **Polygon** cascading menu from **Sketch** drop-down.

• Click on the screen to select the start point and end point of edge. You can also specify the value of angle and distance of the edge in their respective edit box; refer to Figure-23.

• Enter the value of number of edges of polygon in **Edge Number** edit box.

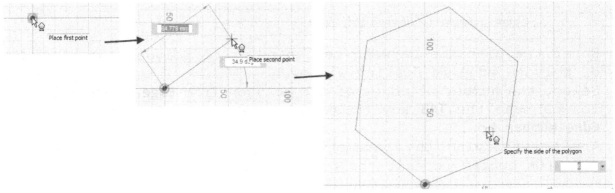

Figure-23. Creation of edge polygon

• Move the mouse and click on the screen to specify the orientation of the polygon.

• Right-click on the screen and click **OK** button from **Marking menu** to exit this tool.

Ellipse

The **Ellipse** tool is used to create ellipse. The procedure to create ellipse is discussed next.

• Click on **Ellipse** tool of **Sketch** drop-down from **Toolbar**; refer to Figure-24.

Figure-24. Ellipse

• Click on the screen to specify the center point for ellipse.

• Enter the value of major diameter in floating window. You can also enter the value of angle in angle floating window.

• Click on the screen and hover the mouse to enter the value of minor diameter in the floating window. Enter the value and press **ENTER** to create a ellipse; refer to Figure-25.

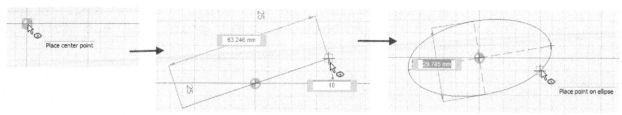

Figure-25. Making of ellipse

- Right-click on the screen and click **OK** button to exit this tool.

Slot

The **Slot** tool is used to create slot. There are three tools in **Slot** cascading menu i.e. **Center to Center Slot, Overall Slot,** and **Center Point Slot**; refer to Figure-26.

Figure-26. Slot cascading menu

The tools of **Slot** cascading menu are described next.

Center To Center Slot

The **Center To Center Slot** tool is used to create the linear slot defined by the distance of slot arc centers. The procedure to use this tool is discussed next.

- Click **Center To center slot** of **Slot** cascading menu from **Sketch** drop-down.
- Click on the screen to specify the start point of slot.
- Enter the radius in the floating window or click on the screen to specify the start point.
- Move the cursor to enter the diameter of slot in floating window or you can also click on the screen to specify the end point; refer to Figure-27.

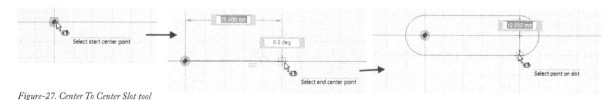

Figure-27. Center To Center Slot tool

Overall Slot

The **Overall Slot** tool creates a linear slot which is defined by orientation, length and width. The procedure to use this tool is discussed next.

- Click **Overall slot** of **Slot** cascading menu from **Sketch** drop-down.
- Click on the screen to specify the start point of the slot; refer to Figure-28.

- Click on the screen to specify the end point of the slot or you can also enter the distance between start point and end point in the floating window.

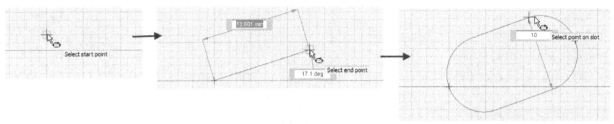

Figure-28. Making of overall slot

- Move the mouse to enter the diameter in the floating window or click on the screen.
- Right-click on the screen and click **OK** button to exit this tool.

Center Point Slot

The **Center Point Slot** tool creates a linear slot which is defined by center point, location of arc center and by Slot width. The procedure to use this tool is discussed next.

- Click on **Center Point slot** of **Slot** cascading menu from **Sketch** drop-down.
- Click on the screen to specify the center for slot; refer to Figure-29.
- Click to specify the half length of the center to center distance of the slot.
- Specify the value of width for slot in the floating window and press **ENTER** key.

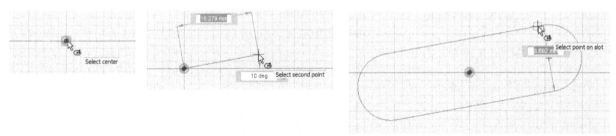

Figure-29. Making of center point slot

- Right-click on the screen and click **OK** button to exit this tool.

Spline

The **Spline** tool is used to create the spline curve through the selected points. It is a freeform curve creation tool. The procedure to use this tool is discussed next.

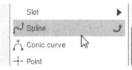

Figure-30. Spline tool

- Click **Spline** tool of **Sketch** drop-down from **Toolbar**; refer to Figure-30.

- Click on the screen to specify the start point for the spline.
- Similarly specify the other points for the spline path; refer to Figure-31.

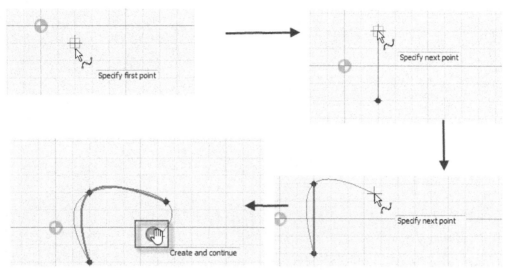

Figure-31. Making of spline

- On creating the path, click on the **Tick** button to end the spline path .
- Right-click on the spline and click **OK** button to create another spline or click on **Cancel** button to end the current command from the **Marking menu**.

Conic Curve

The **Conic Curve** tool create a curve driven by end points and rho value. The shape of the curve will be elliptical, parabolic, or hyperbolic depend on the value of Rho. The procedure to use this tool is discussed next.

- Click on **Spline Curve** tool from the **Sketch** drop-down; refer to Figure-32.

Figure-32. Conic Curve tool

- Click on the screen to specify the start point of the conic curve.
- Click on the screen to specify the end point of the conic curve.
- Click on the screen to specify the vertex point.
- Enter the **Rho** value in rho floating window for the conic curve and press **ENTER** key; refer to Figure-33. The value of rho lies between **0** to **0.99**.

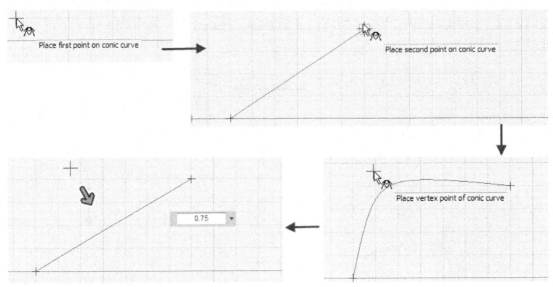

Figure-33. Making of conic curve

- Press **ESC** key to exit this tool.

Point

The **Point** tool is used to create sketch point in a sketch. This tool is very important entity and finds its major usage when creating surfaces. The point give the flexibility to parametrically change the surface design.

- Click on **Point** tool of the **Sketch** drop-down from **Toolbar**; refer to Figure-34.

Figure-34. Point tool

- Click on the sketch to specify the location of the point.
- Press **ESC** key to exit this tool or click **OK** button from **Marking menu**.

Text

The **Text** tool is used to create text which can be used for embossing/engraving on a solid model. The **Text** tool is also used to create notes and other information for the model.

- Click on **Text** tool of the **Sketch** drop-down from **Toolbar**; refer to Figure-35. A **TEXT** dialog box will be displayed; refer to Figure-36.

Figure-35. Text tool

Figure-36. TEXT dialog box

- Click in the **Text** edit box and enter the desired text.
- Click in **Height** edit box and enter the desired value of height for text.
- Click in **Angle** edit box and enter the desired value of angle for text.
- Select the specific style for text in **Text Style** section.
- Click on the **Font** drop-down and select the desired font for text.
- Click **OK** button of the **TEXT** dialog box to complete the process.

Fillet

The **Fillet** tool is used to create fillet at the corners created by intersection of two entities. Fillet is sometimes also referred to as round. The procedure to use this tool is discussed next.

- Click on **Fillet** tool of **Sketch** drop-down from **Toolbar**; refer as Figure-37.

Figure-37. Fillet tool

- Click on the sketch to select the first and second entity for fillet.
- Enter the value of radius in floating window, refer to Figure-38.
- Press **ENTER** key to complete the process.

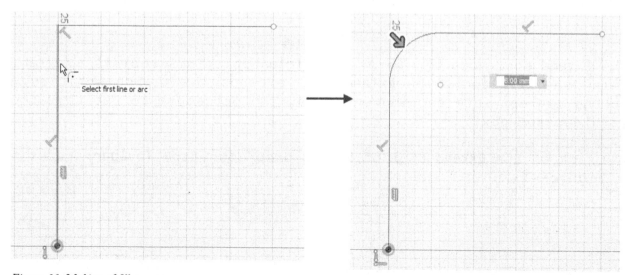
Figure-38. Making of fillet

SKETCH EDITING TOOL

The standard tools to edit sketch entities are categorized in this section. The tools in this section are discussed next.

Trim

The **Trim** tool is used to remove unwanted parts of a sketch entity. While removing the segments, the tool considers intersection point as the reference for trimming. The procedure to use this tool is discussed next.

- Click on **Trim** tool of **Sketch** drop-down from **Toolbar**; refer to Figure-39.

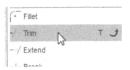

Figure-39. Trim tool

- Click on the entity to trim. You can also trim multiple entities by clicking on them.
- The entity will be trimmed. Figure-40 shows the procedure of trimming.

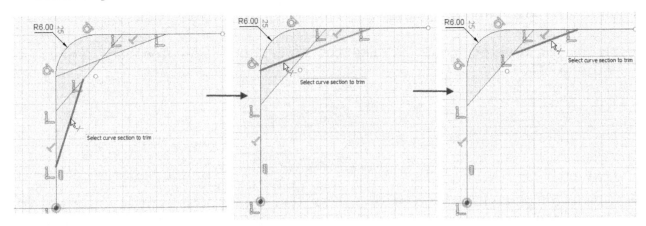

Figure-40. Procedure of fillet

- Press **ESC** key to exit this tool.

Extend

The **Extend** tool does the reverse of **Trim** tool. This tool is available below the **Trim** tool. The **Extend** tool extends the sketch entities up to the nearest intersecting entities. The procedure to use this tool is discussed next.

- Click on the **Extend** tool of **Sketch** drop-down from **Toolbar**; refer to Figure-41.

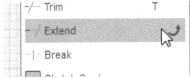

Figure-41. Extend tool

- Hover the cursor on the entity that you want to extend. Preview of the extension will display.
- Click on the entity if, the preview shown is as per your requirement.
- Figure-42 shows the process of extending entities.

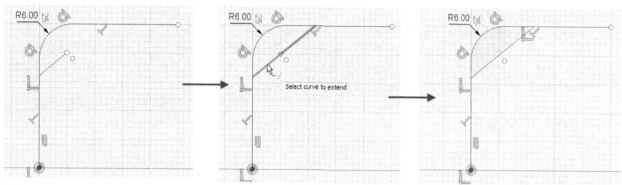

Figure-42. Procedure of Extend tool

- Press **ESC** key to exit this tool.

Break

The **Break** tool is used to break the entity into different entities. The procedure to use this tool is discussed next.

- Click on **Break** tool of **Sketch** drop-down from **Toolbar.**
- Hover the cursor on the entity. The preview will be displayed of break.
- If preview shown is as per your requirement then, click on the entity to break; refer to Figure-43.

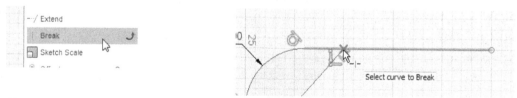

Figure-43. Break

- Press **ESC** key to exit this tool.

Sketch Scale

The **Sketch Scale** tool is used to enlarges or reduces the selected entities, sketches, or components based on a specified scale factor. The procedure to use this tool is discussed next.

- Click on **Sketch Scale** tool of **Sketch** drop-down from **Toolbar.** A **SKETCH SCALE** dialog box will be displayed; refer to Figure-44.

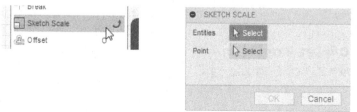

Figure-44. Sketch Scale tool

- **Entities** selection is selected by default. Click on the entities from sketch to scale.
- Click on **Select** button of **Point** section and select the reference point for scale. The updated **SKETCH SCALE** dialog box will be displayed; refer to Figure-45.
- Click in the **Scale Factor** edit box or floating window from screen and enter the desired value for scale. The preview will be displayed. You can also set the scaling value by moving the drag handle from screen.
- If preview is as per your requirement then, click on the **OK** button of **SKETCH SCALE** dialog box to complete the process.

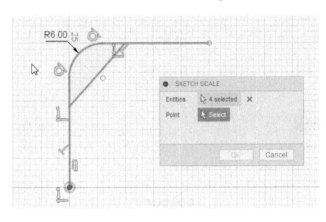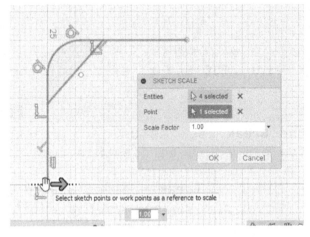

Figure-45. Procedure of Sketch Scale

Offset

The **Offset** tool is used to create copies of the selected entities at a specified distance. If you are the user of **AutoCAD** then this tool is the most common tool bring used while creating layouts. The procedure to use this tool is discussed next.

- Click on **Offset** tool of **Sketch** drop-down from **Toolbar**; refer to Figure-46. A **OFFSET** dialog box will be displayed; refer to Figure-47.

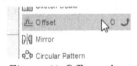

Figure-46. Offset tool

- **Sketch Curves** selection is selected by default. Select the entities for offset. The updated **OFFSET** dialog box will be displayed.
- Check the **Chain Selection** check box, if you want to select the whole entity. Otherwise clear the check box if you want to select only parts of sketch for offset.
- Click in the **Offset Position** edit box and enter the desired distance for offset; refer to Figure-47. You can also set the offset distance by moving the drag-handle from screen.

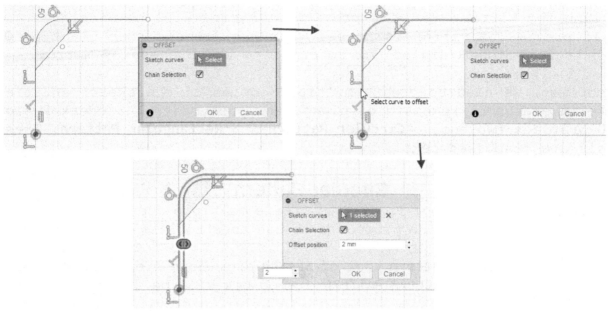

Figure-47. Procedure for Offset

• Click **OK** button of **OFFSET** dialog box to finish the process.

Mirror

The **Mirror** tool is used to create mirror copy of the selected entities with respect to a reference called Mirror line. The procedure to create the mirror entities is given next.

• Click on **Mirror** tool of **Sketch** drop-down from **Toolbar**; refer to Figure-48. A **MIRROR** dialog box will be displayed; refer toFigure-49.

Figure-48. Mirror tool

• **Objects** selection is selected by default. Click on the entities or sketch to create the mirror copy; refer to Figure-49.

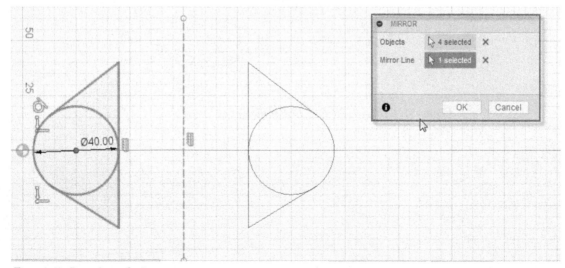

Figure-49. Procedure of mirror

- Click on **Select** button of **Mirror line** section and select the reference line for Mirror copy. The preview of mirror will be displayed.
- If preview of mirror is as per your requirement then click on **OK** button of **MIRROR** dialog box to complete the process of Mirror.

Sometimes, we need to create multiple copies of the sketch entities like in sketch of keyboard or piano. For such cases, Autodesk Fusion 360 provides two tools, **Circular Pattern** and **Rectangular Pattern**. These tools are discussed next.

Circular Pattern

The **Circular pattern** tool is used to create multiple copies of an entity in circular fashion. The procedure to use this tool is discussed next.

- Click on **Circular Pattern** tool of **Sketch** drop-down from **Toolbar**; refer to Figure-50. A **CIRCULAR PATTERN** dialog box will be displayed; refer to Figure-51.

Figure-50. Circular Pattern tool

- **Objects** selection is selected by default. Select the entity or sketch to apply the circular pattern.
- Click on **Select** button of **Center Point** section and select the reference point for circular pattern; refer to Figure-51. The updated **CIRCULAR PATTERN** dialog box will be displayed; refer to Figure-52.

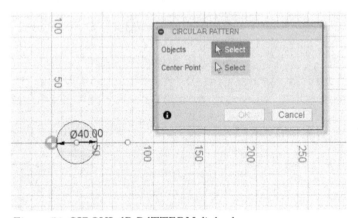

Figure-51. CIRCULAR PATTERN dialog box

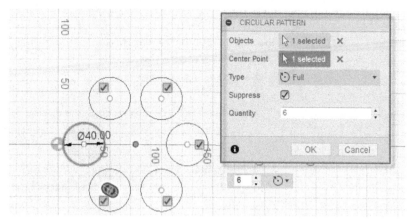

Figure-52. Making of circular patten

- Select **Full** option from **Type** drop-down, if you want to create a pattern over full 360 round. Select the **Angle** option from **Type** drop-down, if you want to define the span of pattern in angle value. Select the **Symmetric** option from **Type** drop-down, if you want if you want to create a circular pattern symmetric to the center line ; refer to Figure-53.

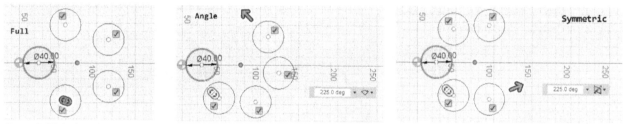

Figure-53. Tools of type drop-down

- Check the **Suppress** check box, if you want to suppress the particular pattern of the entity otherwise clear the check box.
- Click in the **Quantity** edit box and enter the desired value for quantity of circular pattern. You can also set the quantity by moving drag handle.
- After specifying the parameters, click **OK** button form **CIRCULAR PATTERN** dialog box to complete the process.

Rectangular Pattern

The **Rectangular Pattern** tool is used to create multiple copies of an entity in horizontal and vertical direction. You can create pattern in two linear directions at a time. The procedure to create rectangular pattern is given next.

- Select the **Rectangular Pattern** tool of **Sketch** drop-down from **Toolbar**; refer to Figure-54. A **RECTANGULAR PATTERN** dialog box will be displayed; refer to Figure-55.

Figure-54. Rectangular Pattern tool

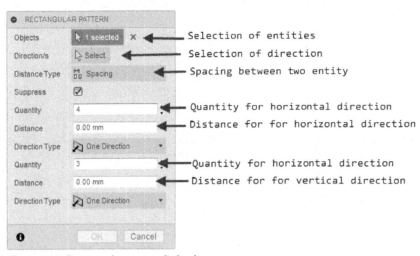

Figure-55. Rectangular patten dialog box

- **Objects** section is selected by default. Select the sketch or entity for rectangular pattern.
- Click on **Select** button of **Direction/s** section and select the direction for rectangular pattern by clicking on the **Arrow** key from screen. You can also select both directions i.e. vertical and horizontal; refer to Figure-56.

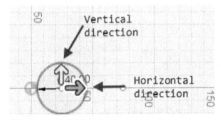

Figure-56. Direction rectangular pattern

- Select **Extent** option from **Distance Type** drop-down, if you want to specify the distance between two entity. Select **Spacing** from **Distance Type** drop-down, to set the total distance for the pattern. In our case we are selecting **Extent** option.
- Check the **Suppress** check box, if you want to suppress the particular pattern of the entity otherwise clear the check box.
- Click in the Quantity edit box and enter the desired value for quantity of copies.
- Click on the **Distance** edit box and enter the desired distance to create copies; refer to Figure-57.

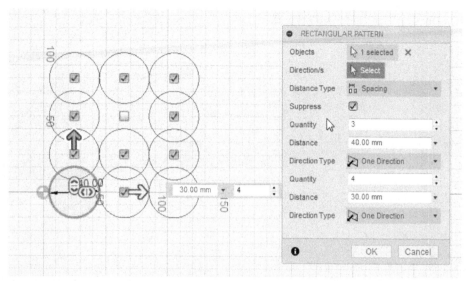

Figure-57. Making of Rectangular Pattern

- Select **One Direction** option from **Direction Type** drop-down, if you want to create copies in only one direction. Select **Symmetric** option form **Direction Type** drop-down, if you want to create copies in both sides.
- Similarly, specify the parameters for other direction.
- After specifying the parameters, click **OK** button from **RECTANGULAR PATTERN** dialog box to complete the process.

Project/Include

This tool is used to project the face/dimension of model/project. This tool will be discussed later in this book.

Sketch Dimension

The **Sketch Dimension** tool is used to dimension the entities automatically. This tool can create different type of dimension like horizontal,vertical or inclined dimensions. The procedure to use this tool is discussed next.

- Click on the **Sketch Dimension** tool of **Sketch** drop-down from **Toolbar**; refer to Figure-58.

Figure-58. Sketch dimension

- Select the entity you want to dimension and then click at the desired distance to place dimension; refer to Figure-59.

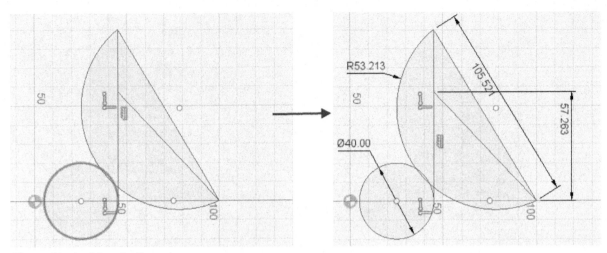

Figure-59. Applying the dimension

SKETCH PALETTE

When sketch is active the **SKETCH PALETTE** dialog box is displayed at bottom right corner of Fusion 360 screen. This dialog box contain commonly used Sketch options and Sketch constraints; refer to Figure-60.

Figure-60. SKETCH PALETTE dialog box

The various tools of Sketch Palette are discussed next.

Options

In **Options** section of **SKETCH PALETTE** dialog box, there are various tools which are discussed next.

- **Look At**- This tool Rotates the camera to look directly at the sketch plane.

- **Sketch Grid**- This tool controls the display of sketch grid. To view grid, select the **Grid** check box otherwise clear the check box to hide.

- **Snap**- This tool controls the grid Snap. To activate Snap, select **Snap** check box and clear the **Snap** check box to disable.

- **Slice**- This tool cuts the object using **Sketch Plane**. This tool does not able to change the geometry because it is only a display option. To enable, select the **Slice** check box and clear the **Slice** check box to disable.

- **Show Profile**- This tool is able to show the closed profile in the sketch. To enable this tool, select the **Show Profile** check box and to disable it clear the check box.

- **Show Constraints**- This tool is used to view and hide the constraints in the sketch. To view, select the **Constraints** check box and clear the check box to hide constraints.

- **3D Sketch**- This tool Controls the ability to pull sketch objects out of the sketch plane to create a 3D sketch. To enable this tool, select the **3D Sketch** check box and to disable it clear the check box.

Constraints

These constraints are used to constrain the shapes/position of sketch entities with respect to other entities.

- To apply the geometric constraints, click on **Show Constraints** check box from **SKETCH PALETTE** dialog box; refer to Figure-61.

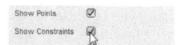

Figure-61. Show Constraints check box

- The list of constraints that can be applied in **Autodesk Fusion 360** is given next; refer to Figure-62.

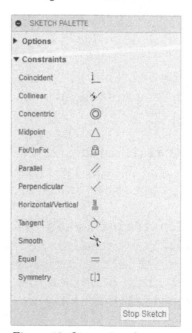

Figure-62. Constraints

└ The **Coincident Constraint** makes a selected point to be coincident with a selected line, arc, circle, or ellipse. To apply this constraint, select the **Coincident** button from **Constraints** section in **SKETCH PALETTE** dialog box and select the entities for coincident.

The **Collinear Constraint** makes the selected lines to lie on the same infinite line; refer to Figure-63 . To use this constraint, select the **Collinear constraint** button and select the entities.

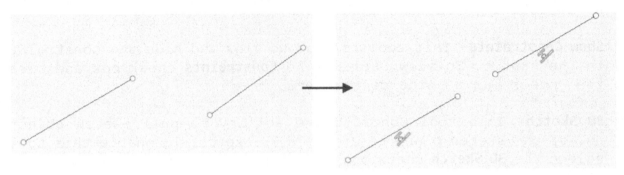

Figure-63. applying collinear constraint

The **Concentric Constraint** makes a selected arc or circle to share the same center point with other arc, circle, point, vertex or circular edge. To apply this constraint, click the **Concentric** button and select the entity to apply the concentric constraint.

The **Midpoint Constraint** makes a selected point to move to the midpoint of a selected line. To apply this constraint, click on the **Midpoint** button. Select the point and the line for which the midpoint constraint has to be applied.

The **Fix/Unfix Constraint** makes the selected entity to be fixed at a specified position. If you apply this constraint to a line or an arc, its location will be fixed but you can change its size by dragging the endpoints. To apply this constraint, click on the **Fix/Unfix** button and select the entities.

The **Parallel Constraint** makes the selected lines to become parallel to each other. To apply this constraint, click on the **Parallel** button and select the desired entity.

The **Perpendicular Constraint** makes the selected lines to become perpendicular to each other. To apply this constraint, click on the **Perpendicular** button and select the entities.

The **Horizontal/Vertical Constraint** makes one or more selected lines to become horizontal/Vertical. You can also select an external entity such as an edge, plane, axis, or sketch curve on an external sketch that will act as a line to apply this constraint. To apply this constraint, click on the **Horizontal/Vertical** button and select the entities.

The **Tangent Constraint** makes a selected arc, circle, spline, or ellipse to become tangent to other arc,circle, spline, ellipse, line or edge. To apply this constraint, click on the **Tangent** button and select the required entity.

The **Smooth Constraint** is used to creates a curvature continuous (G2) condition between a spline and another curve. To use this tool, click on **Smooth** button and select the desired entity.

= The **Equal Constraint** makes the selected lines to have equal length and the selected arc, circles, or arc and circles, or arc and circle to have equal radii. To apply this constraint click on the **Equal** button and select required entity to apply the equal constraint .

[] The **Symmetry Constraint** makes two selected lines, arcs, points, and ellipses to remain equidistant from a center line. This constraint also makes the entities to have the same orientation. To apply this constraint, click on the **Symmetric** button and select the required entities for symmetry constraint.

PRACTICAL 1

Create the Sketch as shown in Figure-64. Also dimension the sketch as per the figure.

Steps to be Performed:

Below is the step by step procedure of creating the sketch shown in the Figure-64.

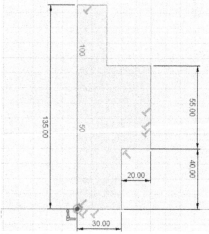

Figure-64. Prectical 1

Starting Sketching Environment

- Start **Autodesk Fusion 360** if not started already.
- Select the **New Design** tool from **File** menu to start a new design; refer to Figure-65.

Figure-65. New Design tool

- Select the **Create Sketch** tool from **Sketch** drop-down; refer to Figure-66.

Figure-66. Create Sketch tool

- On selecting this tool, you need to select the **plane** for creating sketch; refer to Figure-67.

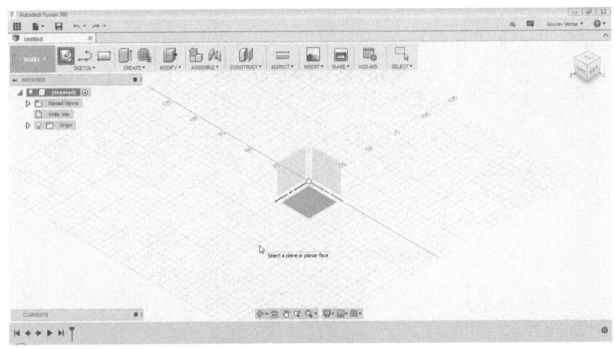

Figure-67. Select Plane for Sketch

- Select the "**Top**"(XY)plane for Sketch.

Creating Lines

- Click on the **Line** button from **Sketch** drop-down or press **L** key from keyboard to select line. The Line tool will become active and you are asked to select the start point.
- Click on the coordinate system and move the cursor towards right; refer to Figure-68.

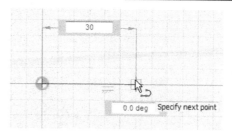

Figure-68. Starting creation of line

- Enter **30** in the dimension box displayed near line and press **TAB** key to lock the dimension and then click on the screen to create a line.
- Move the cursor vertically upwards and enter the value **40** in the dimension box and press **TAB** key to lock the dimension and then click on the screen to complete the line.
- Move the cursor towards right and enter **20** in the dimension box. Lock the dimension as discussed earlier and click on screen; refer to Figure-69.

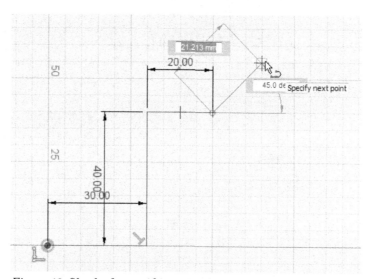

Figure-69. Sketch after specifying 20 value

- Move the cursor upward and specify the value as **55** in the dimension box.
- Move the cursor towards left and specify the value as **30** in the dimension box.
- Move the cursor upwards and specify the value as **40** in the dimension box.
- Move the cursor towards left and specify **20** in the dimension box.
- Move the cursor downward and click on the coordinate system to close the sketch.

The sketch after performing the above steps is displayed as shown in Figure-70.

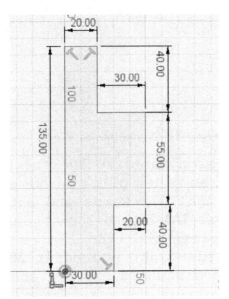

Figure-70. Final Practical 1

PRACTICAL 2

In this practical, we will create the sketch as shown in Figure-71.

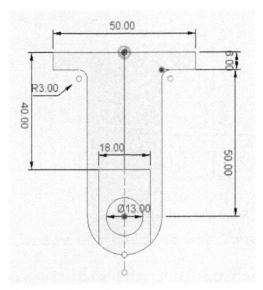

Figure-71. Practical 2

Steps to be performed:

Below is the step by step procedure of creating the sketch shown in the Figure-71.

Starting Sketching Environment

- Start **Autodesk Fusion 360** if not started already.
- Select the **New Design** tool from **File** menu to start a new design.
- Select the **Create Sketch** tool from **Sketch** drop-down.
- On selecting this tool, you need to select the **plane** for creating sketch.
- Select the **Top Plane** to create the sketch.

Creating Lines

- Click on the **line** button from **sketch** drop-down or press **L** key. You are asked to specify the start point of the line.
- Click on the coordinate system and enter **25** in the dimension box and press **TAB** key to lock the dimension. Click on the screen to complete the line.
- Move the cursor down perpendicular to the previous line and enter **6** in the dimension box. Lock the dimension as discussed earlier and click on screen
- Move the cursor to the left and enter **12** in the dimension box.
- Move the cursor downwards and enter **50** in the dimension box. Till this point, our sketch should display like.

Figure-72. Sketch
after creating lines

Creating Arcs

- Select the **Tangent Arc** tool from **Arc** cascading menu in **Sketch** drop-down.
- Click on the screen to create the start point.
- Click on the end point of the vertical line recently created and move the cursor downwards.
- Specify the dimension value **13** in the dimension box and press enter to complete the arc; refer to Figure-73.

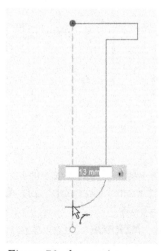

Figure-73. Arc creation

Creating Fillet

- Click on the **Fillet** tool from **sketch** drop-down 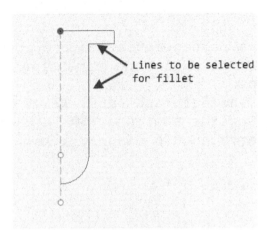.
- Select the lines for fillet; refer to Figure-74.
- Enter the value as **3** in the radius box.

Lines to be selected
for fillet

Figure-74. Line selection for fillet

Creating Mirror Copy

- If **Centerline** is not created yet, then press **L** key and create a line from coordinate system to end point of the Arc.
- Press **ESC** key form keyboard to exit the tool.
- Now, right-click on recently created line and select **Normal Construction** from **Marking menu**.
- Select the **Mirror** tool from **Sketch** drop-down and select all the entities we have sketched except center line in **Object** section; refer to Figure-75.

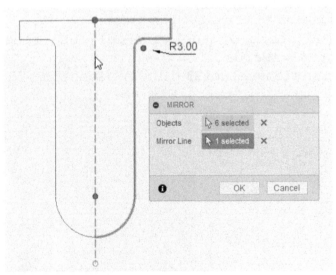

R3.00

MIRROR		
Objects	6 selected	✕
Mirror Line	1 selected	✕
ℹ	OK	Cancel

Figure-75. Entities selected for creating mirror

- Select the center line from sketch in **Centerline** section. Preview of mirror will be displayed.
- Click **OK** button from the **MIRROR** dialog box.

Creating Circle and Lines to complete.

- Click on the **Center Diameter Circle** tool from **Circle** cascading menu.
- Click at the center point of the bottom arc and drag the cursor; refer to Figure-76.

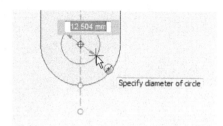

Figure-76. Circle creation

- Enter the dimension as **13** in the input box.
- Press **ENTER** to create the circle.
- Click on the **Line** tool from **Sketch** drop-down and click on the center line to specify the start point of the line; refer to Figure-77.

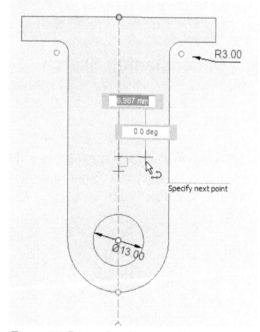

Figure-77. Point selected on centerline

- Move the cursor horizontally towards right and specify the value as **9**.
- Move the cursor vertically downwards and click when the cursor is on arc.
- Press **ESC** to exit the tool.
- Mirror both the lines as we did earlier. The sketch should display as shown in Figure-78.

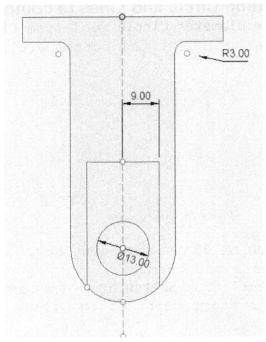

Figure-78. Sketch after all sketching operation

Dimensioning Sketch

If dimension is not applied yet then, Click on the **Sketch Dimension** tool from **Sketch** drop-down and select the arc. Dimension will get attached to cursor.

*   Place the dimension at proper spacing. Press **ENTER** at proper place. Enter the radius in floating window and press **ENTER.**
*   Click on the two lines as shown in Figure-79.

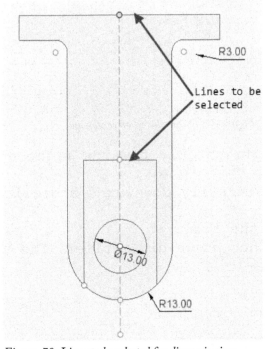

Figure-79. Lines to be selected for dimensioning

• Click to place the dimension at its proper place. In the dimension
  input box, enter the value as 40.

Similarly, dimension all the entity in the sketch until it is fully
defines. The final sketch after dimensioning will be displayed as shown
in Figure-80.

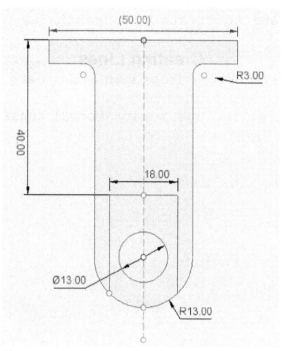

Figure-80. Final sketch

PRACTICAL 3

Create the sketch as shown in Figure-81. Also dimension the sketch
as per the figure.

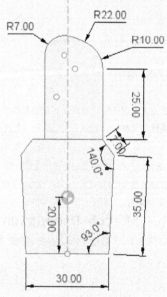

Figure-81. Practical 3

Below is the step by step procedure of creating the sketch shown in
Figure-81.

Starting Sketching Environment

- Start **Autodesk Fusion 360** if not started already.
- Select the **New Design** tool from **File** menu to start a new design.
- Select the **Create Sketch** tool from Sketch drop-down.
- On selecting this tool, you need to select the **plane** for creating sketch.
- Select the **Top Plane** to create the sketch.

Creating Lines

- Select **Line** tool from **Sketch** drop-down and create a line on coordinate axis.
- Right-click on this line and select **Normal construction** to make this line as a center line; refer to Figure-82.

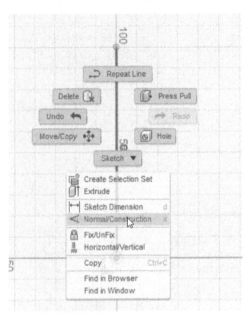

Figure-82. Creating centerline

- Press **L** key to select line and click below the origin on centerline to create a line.
- Move the cursor to the right and enter the value as **15** in the dimension box. Press **TAB** key to lock the dimension and click on the screen.
- Move the cursor upward and create a line with some outward angle. Enter value as **35** in dimension box and lock the dimension with **TAB** key and click on the screen.
- Press **D** key to select the **Sketch Dimension** tool and select the line as shown in Figure-84. Enter the value as **93** degree in the dimension box.

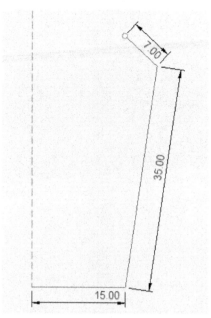

Figure-83. Creating lines with angle

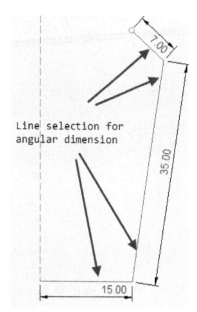

Figure-84. Selection of line for angular dimension

- Now, select **line** tool and move the cursor to the left and enter the value as **7** in the dimension box; as shown in Figure-83.
- Similarly select the upper lines for angular dimension and enter the value as **140** degree in the dimension box.

Creating Mirror copy

- Select the **Mirror** tool from **Sketch** drop-down and select the entities we have sketched except center line for mirror in object section.
- Select the Center line of sketch in **Centerline** section. Preview of mirror will be displayed; Figure-85.

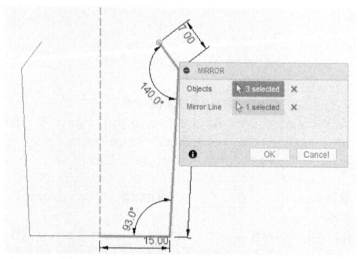

Figure-85. Entities selected for mirror

- On selecting the required entities click **OK** button from **MIRROR** dialog box to complete the mirror copy.
- Now select **Line** tool and create a line joining the open ends of sketch; refer to Figure-86.

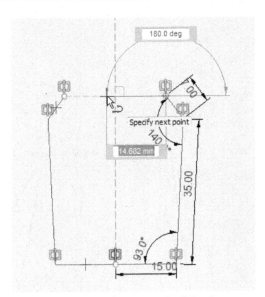

Figure-86. Joining the open ends of sketch

- Create a horizontal line from one end of the sketch. Enter the value as **25** in the edit box; as shown in Figure-87.

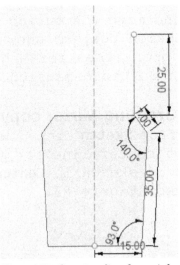

Figure-87. Creating line from right side of sketch

Creating Arcs

- Click **Tangent Arc** tool from **Arc** cascading menu in **Sketch** drop-down.
- Click on the end point of vertical line recently created and move the cursor upwards. Create three consecutive arc joining each other; as shown in Figure-88.
- Select **Sketch Dimension** tool from **Sketch** drop-down or press **D** key to select this tool.
- Apply this tool to recently created three Arc with dimension **10,22** and **7** respectively; as shown in Figure-89.

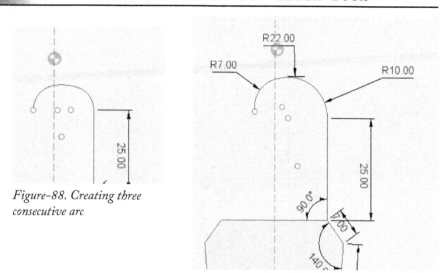

Figure-88. Creating three
consecutive arc

Figure-89. After sketch dimension on arc

- Select the **Line** tool and join a line from the end of recently created arc and sketch and press **ESC** tool to exit the command.

Dimensioning Sketch

- Click on the **Sketch Dimension** tool or press **D** key to select the tool.
- Click on the origin and the point to apply the dimension. Enter value as **20;** as shown in Figure-90.

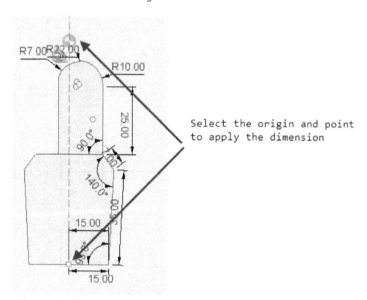

Select the origin and point
to apply the dimension

Figure-90. Select entites for dimension

- Place the dimension at proper spacing from sketch.
- Enter the value of dimension in Dimension box.

Similarly, dimension all the entities in the sketch until it is fully defined. The final sketch after dimensioning will be displayed; as shown in Figure-91.

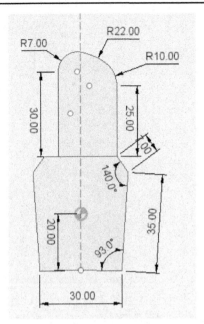

Figure-91. Final sketch of practical 3

Following are some sketches for practice.

PRACTICE 1

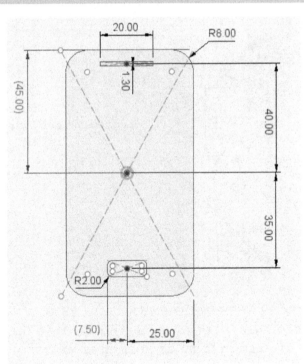

Figure-92. Practice 1

PRACTICE 2

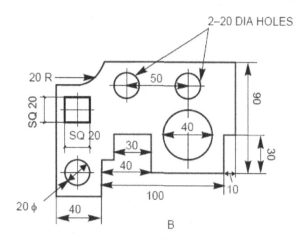

Figure-93. Practice 2

PRACTICE 3

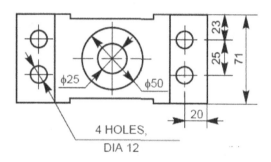

Figure-94. Practice 3

PRACTICE 4

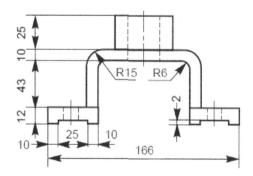

Figure-95. Practice 4

To get more exercise, mail us at cadcamcaeworks@gmail.com

FOR STUDENT NOTES

FOR STUDENT NOTES

Chapter 3

3D Sketch and SolidModeling

Topics Covered

The major topics covered in this chapter are:

- *Solid Modeling*
- *Operation drop-down*
- *Loft Tool*
- *Rib Tool*
- *Multiple Holes*
- *Torus Tool*
- *Circular Pattern*

SOLID MODELING

Till this point, we have worked on 2D drawings and have created on Top or Front plane. Now, we will came out of the 2D window and will explore the 3D world.

Solid modeling is the most advanced method of geometric modeling in three dimensions. Solid modeling is the representation of the solid parts of the object on your computer. In this chapter we will learn about various tools and commands to create solid models which are discussed next.

CREATE DROP-DOWN

In **Create** drop-down, there are various 3D modeling tool which are discussed next.

New Component

Click on **New Component** tool from **Create** drop-down; refer to Figure-1. This tool will be discussed later in Assembly section.

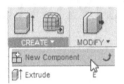

Figure-1. New component

Extrude

Extrude tool is used to create a solid volume by adding height to the selected sketch. In other words, this tool adds material (by using the boundaries of sketch) in the direction perpendicular to the plane of sketch. The procedure to use this tool is discussed next.

- You need to create a sketch by using sketch creation tools which were discussed in the last chapter.
- After creating a Sketch, exit the sketch by selecting the **Stop Sketch** tool from **Toolbar** or **Right-click** outside the sketch and select **Stop Sketch** from the **Marking menu**; refer to Figure-2.

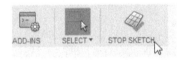
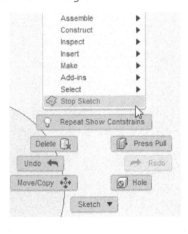

Figure-2. Stop Sketch

- Click on **Extrude** tool of **Create** drop-down from **Toolbar**; refer to Figure-3. The **EXTRUDE** dialog box will be displayed; refer to Figure-4.

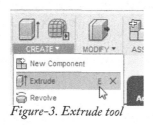

Figure-3. Extrude tool

Figure-4. EXTRUDE dialog box

- Profile selection is selected by default. Select the body or sketch in the **Profile** section by clicking on the sketch.

Start drop-down

- There are three tools In **Start** drop-down; refer to Figure-5.

Figure-5. Start drop-down

- Click on **Profile Plane** tool, if you want to extrude the model from sketch.
- Click on **Offset Plane** tool, if you want to extrude the model from some offset distance to sketch. The updated **EXTRUDE** dialog box will be displayed; refer to Figure-6.

Figure-6. Offset box

- Click in **Offset** edit box and enter **5** as the distance for extrusion between start of extrusion and sketch plane.
- Click on **From Object** option from **Start** section, if you want to extrude the profile from the surface/face of any object.

- Click on **Select** button of **profile** section and specify the sketch profile for extrusion.
- Click on **Select** button of **Object** section and select the Object from where you are going to start the extrude; refer to Figure-7.

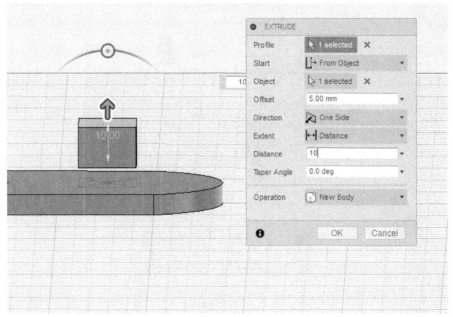

Figure-7. From object tool

- Click in **Distance** edit box and specify the height of extrusion. If you want to an angular extrusion then, specify in the **Taper Angle** edit box.

Direction drop-down

- There are three tools in **Direction** drop-down of **EXTRUDE** dialog box; refer to Figure-8.

Figure-8. Direction drop-down

- Click on **One Side** option from **Direction** drop-down, if you want to specify the height of extrusion in one direction. To specify, click in **Distance** edit box and enter the desired distance. You can also set the height by moving **Drag handle** from canvas screen.
- Click on **Two Sides** option from **Direction** drop-down, if you want to specify the height of extrusion specifically in **both direction** or both side of sketch; refer to Figure-9.

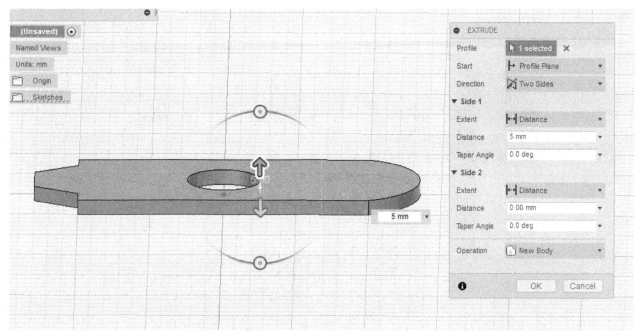

Figure-9. Two sides tool

- Click on **Symmetric** option from **Direction** drop-down, if you want to specify the height of extrusion symmetrically. In this, height of extrusion of one side to the sketch plane is equal to the height of extrusion of other side. Select the **Half length** and **Whole length** option as required; refer to Figure-10.

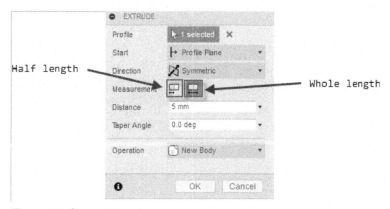

Figure-10. Symmetric tool

Extent drop-down

- Click on **Distance** in **Extent** drop-down from **EXTRUDE** dialog box, if you want to set the height of extrusion manually. Click in the **Height** edit box and specify the desired height.
- Click on **To Object** tool from **Distance** drop-down, if you want to specify the surface/face in **Object** section at which extrusion will end; refer to Figure-11. Click on **Select** button of **Object** section and select the object.

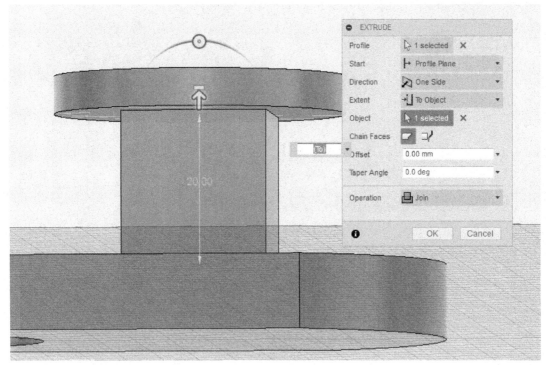

Figure-11. To Object

- You can also set the offset distance and angle in **Offset** and **Taper angle** box respectively.
- Click on **All** tool, if you want to cut through the model or part from extrusion. The extrusion will goes through from the body of model; refer to Figure-12.

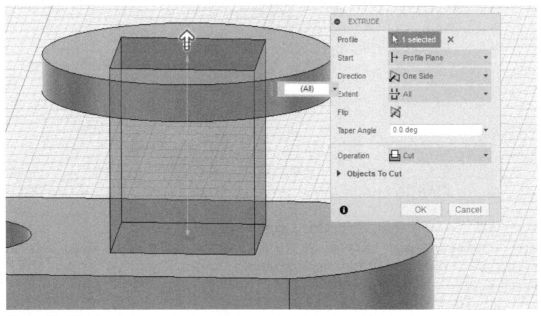

Figure-12. All tool

- On selecting **All** tool, you need to specify all the required parameters in the **EXTRUDE** dialog box. The preview of extrusion will be displayed; refer to Figure-13.

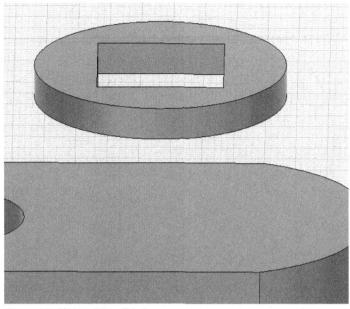

Figure-13. After applying all tool

Operation drop-down

- There are five tools in **Operation** drop-down of **EXTRUDE** dialog box, which are discussed next; refer to Figure-14.

Figure-14. operation dropdown

- Click on **Join** tool of **Operation** drop-down, if you want to combine two bodies; refer to Figure-15.

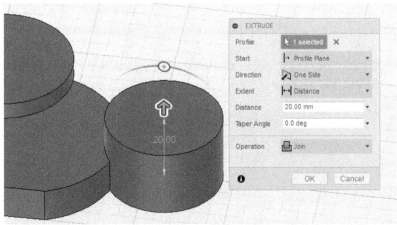

Figure-15. Join tool

- Click on **Cut** tool, if you want to remove material from base body; refer to Figure-16.

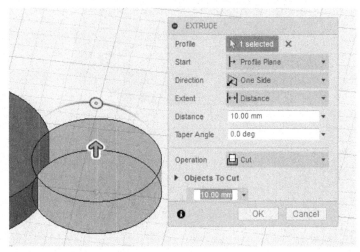

Figure-16. Cut tool

- Click on **Intersect** tool from **Operation** drop-down, if you want to create the region commonly bounded by solid bodies, ; refer to Figure-17.

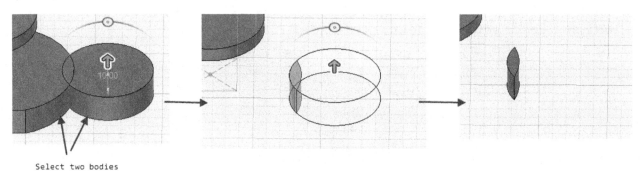

Select two bodies

Figure-17. Intersect tool

- Click on **New body** tool from **Operation** drop-down, If you want to create a new body.

Revolve

Revolve tool is used to create a solid volume by revolving a sketch about selected axis. In other words, If you revolve a sketch about an axis then the volume that is covered by revolved sketch boundary is called revolve feature. The procedure to use this tool is discussed next.

- Click on **Revolve** tool from **Create** drop-down of **Toolbar**; refer to Figure-18. A **REVOLVE** dialog box will be displayed; refer to Figure-19.

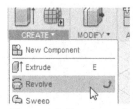

Figure-18. Revolve tool

Figure-19. Revolve dialog box

- If not created yet, you need to create a sketch by using sketch creation tools which were discussed in the last chapter.
- On creating a Sketch, exit the sketch by selecting the **Stop Sketch** tool from **Toolbar** or **Right-click** outside the sketch and select **Stop Sketch** from the **Marking menu**.
- Now, Click on **Select** button for **Profile** selection from **REVOLVE** dialog box and then select the sketch for profile from the screen canvas.
- Click on **Select** button for **Axis** selection from **REVOLVE** dialog box and then select the sketch edge, line or center line from the screen canvas; refer to Figure-19.

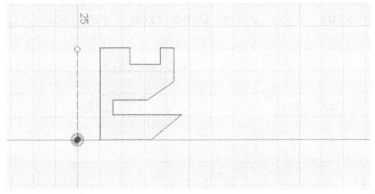

Figure-20. Region selected for revolve

- After selecting the parameters, the preview of model will be displayed; refer to Figure-21.

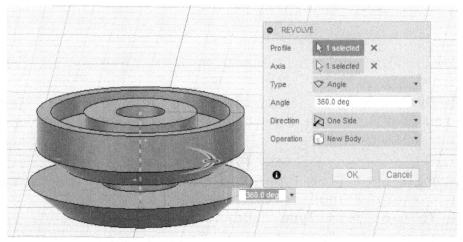

Figure-21. Preview of revolve feature

- If you want to create an angular model then, Click in the **Angle** edit box and enter the desired value in floating window. You can also set the value of angle by moving the drag arrow; refer to Figure-21.

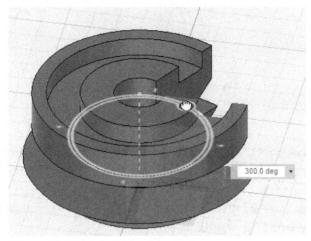

Figure-22. Moving the drag handle

Direction drop-down

The direction option can also be changed to adjust how the model generate.
- Click on **One Side** tool, if you want to revolve the model in one side.
- Click on **Two Side** tool from **Direction** drop-down, if you want to revolve the model in both sides individually; refer to Figure-23.

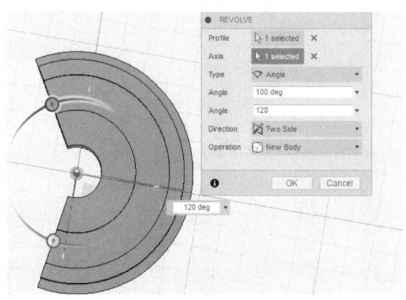

Figure-23. Two side tool

- Click on **Symmetric** tool from **Direction** drop-down, if you want to revolve the model in both direction with same angle or symmetrically.

Type

- Click on **Angle** tool from **Type** drop-down, if you want to revolve the model in a specific angel.
- Click on **To** tool from **Type** drop-down, if you want to revolve the model up to a plane, face or model from origin plane of sketch. On selecting this tool, you need to select a plane; refer to Figure-24.

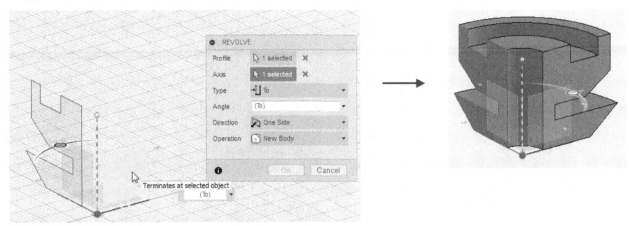

Figure-24. Preview of To tool

- Click on **Full** tool from **Type** drop-down, if you want to rotate the sketch in 360 degree angle.
- Now, click **OK** button from **REVOLVE** dialog box to complete the process.

Operation drop-down

There are various tools in **Operation** drop-down which are discussed next.

New Component

The **New Component** tool is used to create a child component in active component. The procedure to use this tool is discussed next.

- Click on **New Component** tool of **Operation** drop-down from **REVOLVE** dialog box.
- After creating the active component, you need to again select the revolve tool.
- **Right-click** on the screen and select **Repeat Revolve** button from **Marking menu**; refer to Figure-25. A **REVOLVE** dialog box will be displayed.

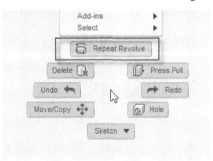

Figure-25. Repeat revolve

- Now, click on the **light bulb** of the sketch in **Simulation Browser** for the preview of the sketch; refer to Figure-26.

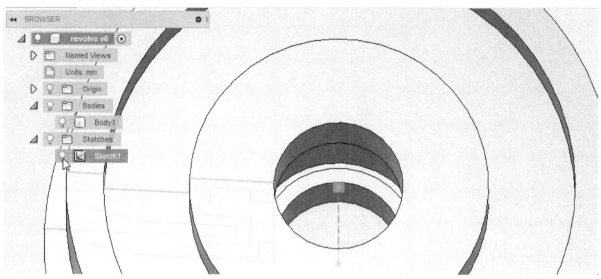

Figure-26. Sketch preview

- Click on the **Select** button for **Profile section** from the **REVOLVE** dialog box and then select the inner sketch for profile from the sketch.
- Click on the **Select** button for **Axis** from the **REVOLVE** dialog box and then select the same axis which was selected in first component, so that these two component will interlock with each other; refer to Figure-27.

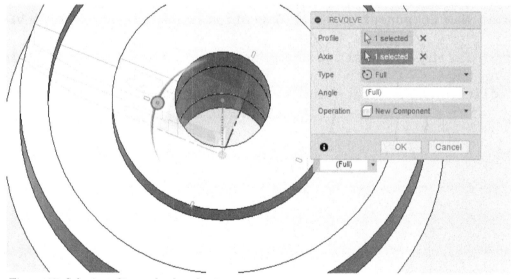

Figure-27. Selection of inner sketch and axis

- Click on **Full** option in **Type** drop-down and **New Component** option in **Operation** drop-down to create this component a separate part.
- On specifying the various parameters, click **OK** button in **REVOLVE** dialog box to complete the process.
- Here, new component is created successfully. To view the component, click on the component and move it away from the first component; refer to Figure-28.

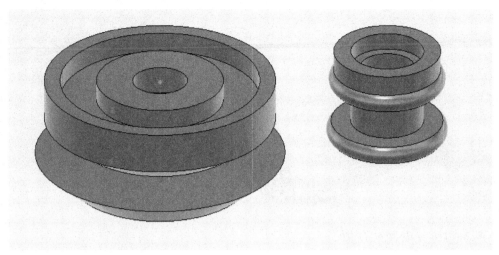

Figure-28. Preview of new component

Other tools of **Operation** drop-down were discussed earlier in **Extrude** tool.

Sweep

The **Sweep** tool is used to create a solid volume by moving a sketch along the selected path. In other words, If you move a sketch along a path then the volume that is covered by moving sketch boundary is called sweep tool. Note that to use this tool, you must have a sketch section and a path, then only the tool will be active. The steps to use this tool are discussed next.

• Click on **Sweep** tool form **Create** drop-down. The **SWEEP** dialog box will be displayed; refer to Figure-28.

Figure-29. Sweep tool

In this dialog box you, there are two tools in **Type** drop-down which are used to create sweep feature i.e. **Single Path** and **Path + Guide Rail**.

Single Path

The **Single Path** tool sweeps the selected profile in a single path.

• Click on the **Select** button of **Profile** section from the **SWEEP** dialog box and select the sketch for profile from the canvas screen.
• Click on the **Select** button of **Path** section from **SWEEP** dialog box and select the path from the canvas screen; refer to Figure-30.

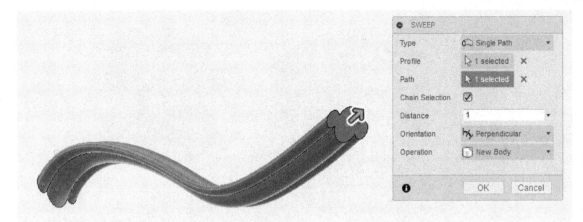

Figure-30. Selection of sketch and path

- Select the **Chain Selection** check box, if you want the edges of sweep are tangent to the selected path. Since in our case, there is single path so it wont affect the sweep feature.
- The distance in the **Distance** edit box is not the specific distance value of sweep but rather the percentage of total path distance. In this, **1** is the maximum value, which is the total path distance of sweep. Enter **0.5** value in the dimension box to create the sweep in only half of the total path. You can adjust the length of the sweep by entering the numerical value in distance dimension box or in the floating window. You can also adjust the length by moving the drag arrow.
- The **Orientation** drop-down of **SWEEP** dialog box is very important. It determine, how the profile is pulled along the sweep path.
- **Perpendicular** option is the default selection of **Orientation** drop-down. This option creates the sweep such that it is always perpendicular to the path. This allow more curvature on the path; refer to Figure-31.

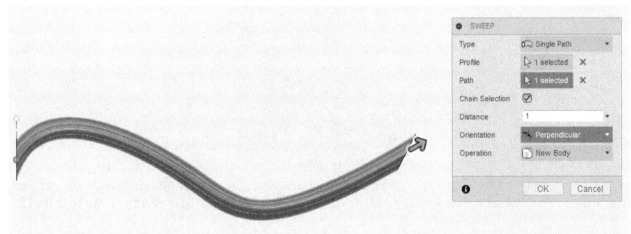

Figure-31. Perpendicular orientation

- **Parallel** option of **Orientation** drop-down, creates the profile path such that it is always parallel to the sketch plane; refer to Figure-32.

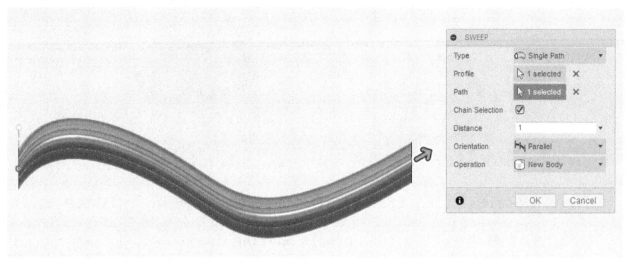

Figure-32. Parallel orientation

- After specifying the parameters, select **OK** button to complete the sweep feature.

Path + Guide Rail

This tool sweeps the selected sketch along the path and uses the guide rail to control the shape. The procedure to use this tool is discussed next.

- Click on **Path + Guide Rail** option of **Type** drop-down from **SWEEP** dialog box; refer to Figure-33.

Figure-33. Path + Guide Rail

- Click on the **Select** button for **Profile** from the **SWEEP** dialog box and then select the sketch for profile from the canvas screen.
- Click on the **Select** button for **Path** from **SWEEP** dialog box and then select the path from the canvas screen.
- Click on the **Select** button for **Guide Rail** from the **SWEEP** dialog box and then select the sketch for guide from the canvas screen; refer to Figure-34.

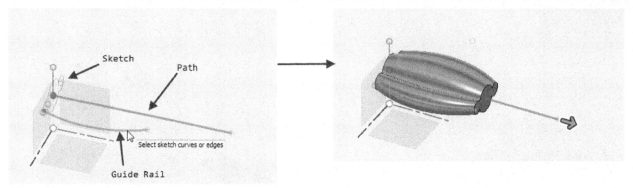

Figure-34. Selection of entities for path + guide rail

- Click on **Scale** option from **Profile Scaling** drop-down in **SWEEP** dialog box, if you want to scale the profile in both X and Y direction.
- Click on **Stretch** option from **profile Scaling** drop-down, if you want to scale the profile in X direction. It also stretch the profile between rails while maintaining height.
- Click on **None** option from **Profile Scaling** drop-down, if you don't want to scale the profile along sweep. It uses guide rail as a orientation guide; refer to Figure-35.

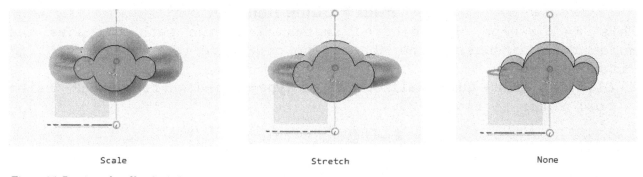

Figure-35. Preview of profile orientation

- The tools of **Operation** drop-down werewe discussed earlier in this book.

CONSTRUCTION GEOMETRY

Till now we have created all the features on default planes but sometimes we need to create features that cannot be created on default planes. In such cases, we create some construction geometries. Construction Geometry are the references for other features. Some of the well known entities that come under datum feature are:

- Planes
- Axes
- Points
- Coordinate Systems
- Curves
- Sketches

You have learned about sketches earlier in the book. Now, we will work on other features. All the tools for construction geometry features are available on the **Construct** drop-down; refer to Figure-36.

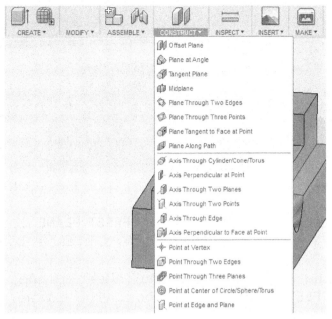

Figure-36. Construct tab

The tools of **Construct** drop-down are discussed next.

Offset Plane

The **Offset Plane** tool is used to create a plane with some specific distance from a reference. The Procedure to use this tool is discussed next.

* Click on the **Offset Plane** tool of **Construct** drop-down from **Toolbar**; refer to Figure-37. The **OFFSET PLANE** dialog box will be displayed; refer to Figure-38.

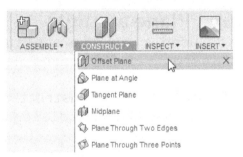

Figure-37. Offset Plane tool

*Figure-38. OFFSET PLANE
dialog tool*

* Plane selection of OFFSET PLANE dialog box is selected by default. Click on the plane of model as a reference for offset plane; as shown in Figure-39. The preview for offset plane will be displayed.

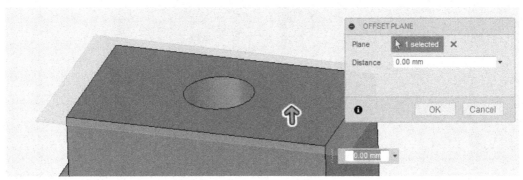

Figure-39. Selection of plane for Offset Plane

- Click on the **Distance** edit box from **OFFSET PLANE** dialog box and enter the distance. You can also set the distance by moving the drag handle.
- After specifying the distance for offset plane, click on the **OK** button from **OFFSET PLANE** dialog box. The plane will be created; refer to Figure-40.

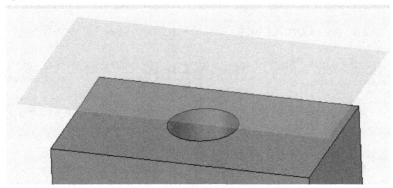

Figure-40. Preview of plane

Plane at Angle

The **Plane at Angle** tool is used to create a plane at a specified angle. The procedure to use this tool is discussed next.

- Click on the **Plane at Angle** tool of **Construct** drop-down from **Toolbar**; refer to Figure-41. The **PLANE AT ANGLE** dialog box will be displayed; refer to Figure-42.

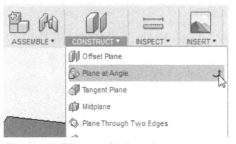

Figure-41. Plane at Angle tool

Figure-42. PLANE AT ANGLE
dialog box

- **Line** selection of **PLANE AT ANGLE** dialog box is selected by default. Click on the line from the model as reference for creating a plane. The updated **PLANE AT ANGLE** dialog box will be displayed; refer to Figure-43.

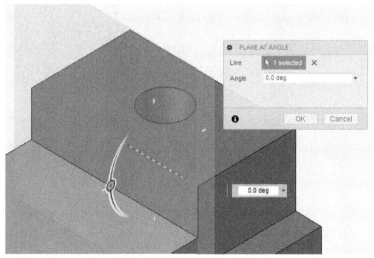

Figure-43. Updated PLANE AT ANGLE dialog box

- Click in the **Angle** edit box and specify the angle of plane with respect to the selected line. You can also set the angle by rotating the drag handle.
- After specifying the angle for plane, click on the **OK** button from **PLANE AT ANGLE** dialog box. The Plane will be created; refer to Figure-44.

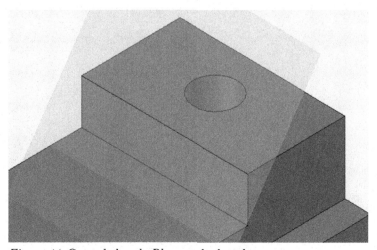

Figure-44. Created plane by Plane at Angle tool

Tangent Plane

The **Tangent Plane** tool is used to create a plane along the tangent of a body. The procedure to use this tool is discussed next.

- Click on the **Tangent Plane** tool of **Construct** drop-down from **Toolbar**; refer to Figure-45. The **TANGENT PLANE** dialog box will be displayed; refer to Figure-46.

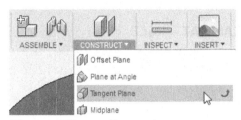

Figure-45. Tangent Pane tool

Figure-46. TANGENT PLANE dialog box

- The **Face** selection of **TANGENT PLANE** dialog box is selected by default. Click on the round face of model to create plane. The updated dialog box will be displayed; refer to Figure-47.

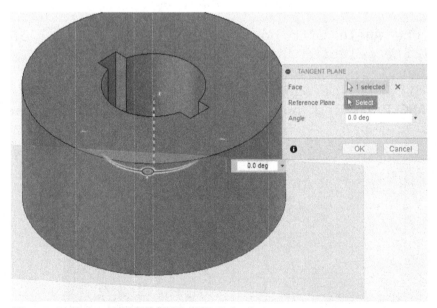

Figure-47. Update TANGENT PLANE dialog box

- Selection of **Reference Plane** is optional. If you want to select the reference plane then, click on the **Select** button of **Reference Plane** section and click on face of model as reference. The plane will be created and preview will be displayed.

- Click in the **Angle** edit box and specify the angle of plane with respect to the selected face. You can also set the angle by moving the drag handle.
- After specifying the parameters, click on the **OK** button from **TANGENT PLANE** dialog box to create a tangent plane.

Midplane

The **Midplane** tool is used to create a plane at the midpoint of two faces. The procedure to use this tool is discussed next.

- Click on the **Midplane** tool of **Construct** drop-down from **Modify** tab; refer to Figure-48. The tool will be activated.

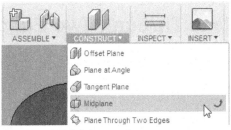

Figure-48. Midplane tool

- Now, you need to select two plane or faces to create a mid plane; refer to Figure-49.

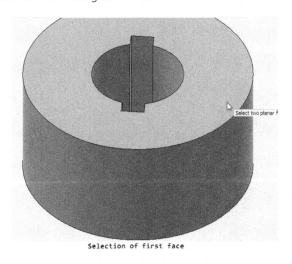
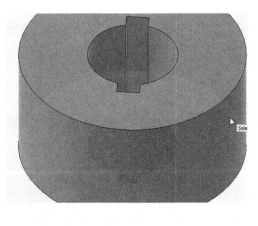

Selection of first face   Selection of second face

Figure-49. Selection of face to create midplane

- After selecting the second plane or face, a preview of plane will be displayed; refer to Figure-50. The plane will be created.

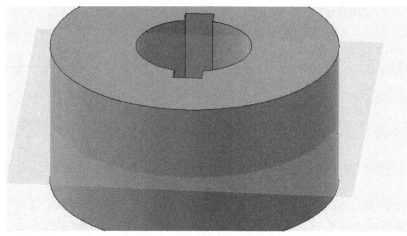

Figure-50. Midplane

Plane Through Two Edges

The **Plane Through Two Edges** tool is used to create a plane with the reference as two linear edges. The procedure to use this tool is discussed next.

- Click on the **Plane Through Two Edges** tool of **Construct** drop-down from **Toolbar**; refer to Figure-51. The tool will be activated.

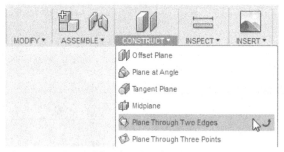

Figure-51. Plane Through Two Edges

- Click on the two edges of model to create plane; refer to Figure-52.

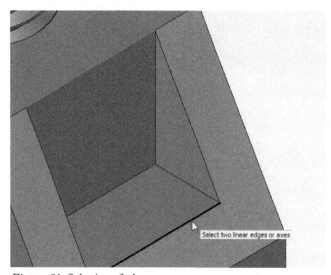

Figure-52. Selection of edges

- On selecting the second edge, the plane will be created; refer to Figure-53

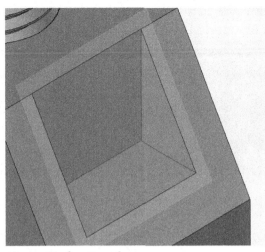

Figure-53. Creating plane

Plane Through Three Points

The **Plane Through Three Points** tool is used to create a plane with the help of three points as reference. The procedure to use this tool is discussed next.

- Click on the **Plane Through Three Points** tool of Construct drop-doen from Toolbar; refer to Figure-54. The tool will be activated.

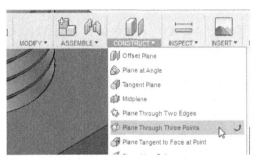

Figure-54. Plane Through Three Points tool

- Select three points from model as reference to create a plane; refer to Figure-55.

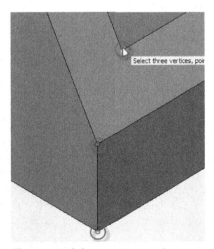

Figure-55. Selection of points for creating plane

- After selecting the third point, the plane will be created and displayed along with model; refer to Figure-56.

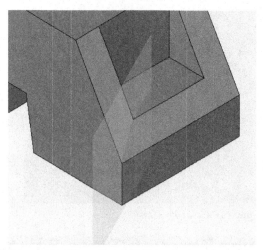

Figure-56. Plane Created after selecting three points

Plane Tangent to Face at Point

The **Plane Tangent to Face at Point** tool is used to create a plane which is tangent to a face and aligned to a point. The procedure to use this tool is discussed next.

- Click on the **Plane Tangent to Face at Point** tool of **Construct** drop-down from **Toolbar**; refer to Figure-57. The tool will be activated.

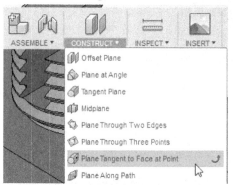

Figure-57. Plane Tangent to Face at Point tool

- Click on the model to select a point and face; refer to Figure-58.

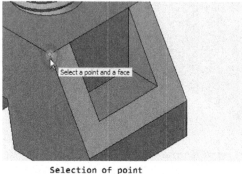

Selection of point

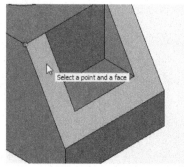

Selection of point

Figure-58. Selection of point and Face

- On selecting the second face or point, the plane will be created; refer to Figure-59.

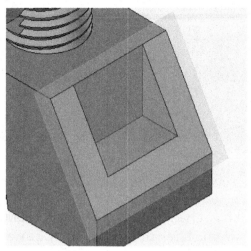

Figure-59. Plane created using point and face

Plane Along Path

The **Plane Along Path** tool is used to create a plane along the selected path. The procedure to use this tool is discussed next.

- Click on the **Plane Along Path** tool of Construct drop-down from Toolbar; refer to Figure-60. The **PLANE ALONG PATH** dialog box will be displayed; refer to Figure-61.

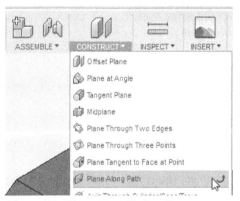

Figure-60. Plane Along Path tool

Figure-61. PLANE ALONG PATH dialog box

- **Path** selection of **PLANE ALONG PATH** dialog box is selected by default. Click on the path from model as reference for plane; refer to Figure-62.

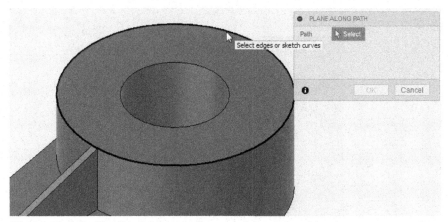

Figure-62. Selection of path

- On selecting the path, the updated **PLANE ALONG PATH** dialog box will be displayed along with the preview of plane; refer to Figure-63.

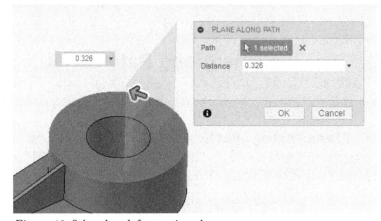

Figure-63. Selected path for creating plane

- Click in the **Distance** edit box and specify the distance of plane between 0 to 1. You can also adjust the distance by moving the drag handle.
- After specifying the parameters, click on the **OK** button from **PLANE ALONG PATH** dialog box. The plane will be created.

Till now we have created plane through different references. In next section we will create axis through different references.

Axis Through Cylinder/Cone/Torus

The **Axis Through Cylinder/Cone/Torus** tool is used to create axis through Cylinder, Cone, or Torus body. The procedure to use this tool is discussed next.

- Click on the **Axis Through Cylinder/Cone/Torus** tool of Construct drop-down from Toolbar; refer to Figure-64. The tool will be activated.

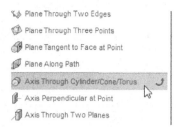

Figure-64. Axis Through Cylin-
der/Cone/Torus tool

- Click on Cylinder, Cone, or Torus from model, the axis will be created; refer to Figure-65.

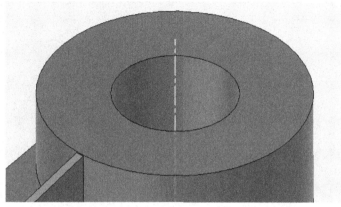

Figure-65. Creating axis of cylinder

Axis Perpendicular at Point

The **Axis Perpendicular at Point** tool is used to create an axis perpendicular to the selected point. The procedure to use this tool is discussed next.

- Click on the **Axis Perpendicular at Point** tool of **Construct** drop-down from **Toolbar**; refer to Figure-66. The tool will be activated.

Figure-66. Axis Perpendicular
at Point dialog box

- You need to select the specific point to create the perpendicular axis as shown in Figure-67.

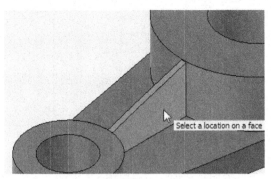

Figure-67. Selection of point for axis

- After selecting the point on any face, the axis will be created; refer to Figure-68.

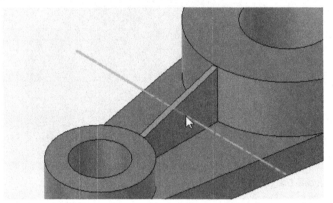

Figure-68. Perpendicular axis created from the selected point

Axis Through Two Planes

The **Axis Through Two Planes** tool is used to create an axis with the intersection of two planes or planer faces. The procedure to use this tool is discussed next.

- Click on the **Axis Through Two Planes** tool of **Construct** drop-down from **Toolbar**; refer to Figure-69. The **AXIS THROUGH TWO PLANES** dialog box will be displayed; refer to Figure-70.

Figure-69. Axis Through two Planes tool

Figure-70. AXIS THROUGH
TWO PLANES dialog box

- Planes selection of **AXIS THROUGH TWO PLANES** dialog box is selected by default. You need to select two planes to create an axis; refer to Figure-71.

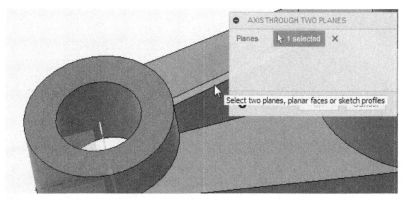

Figure-71. Selection of plane for creating axis

- On selecting the second plane, the axis will be created; refer to Figure-72.

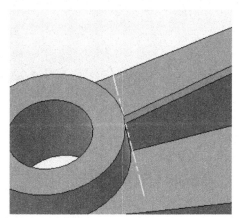

Figure-72. Axis created by selecting two
planes

Axis Through Two Points

The **Axis Through Two Points** tool is used to create an axis through two reference points. The procedure to use this tool is discussed next.

- Click on the **Axis Through Two Points** tool of **Construct** drop-down from **Toolbar**; refer to Figure-73. The tool will be activated.

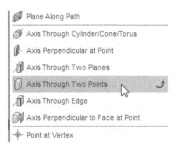

Figure-73. Axis Through Two Points

- Select the two points from model to create an axis; refer to Figure-74.

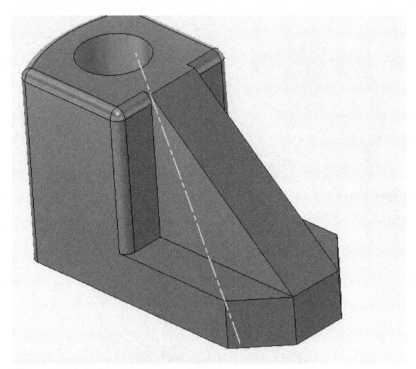

Figure-74. Axis through two selected points

Axis Through Edge

The **Axis Through Edge** tool is used to create an axis on the selected edge. The procedure to use this tool is discussed next.

- Click on the **Axis Through Edge** tool of **Construct** drop-down from **Toolbar**; refer to Figure-75. The tool will be activated.

Figure-75. Aaxis Through Edges

- Click on the **Edge** as reference to create the axis; refer to Figure-76.

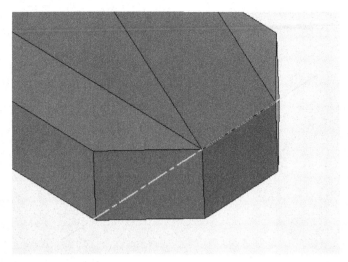

Figure-76. Creating axis through edge

Axis Perpendicular to Face at Point

The **Axis Perpendicular to Face at Point** tool is used to create a plane by using a face and a point. The procedure to use this point is discussed next.

- Click on the **Axis Perpendicular to Face at Point** tool of **Construct** drop-down from **Toolbar**; refer to Figure-77. The tool will be activated.

Figure-77. Axis Perpendicular to Face at point tool

- Click on the model to select the point and face as a reference for creating plane. On selecting the face, the axis will be created; refer to Figure-78.

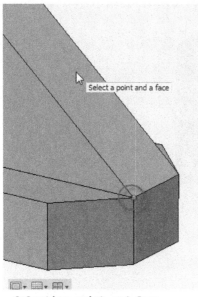

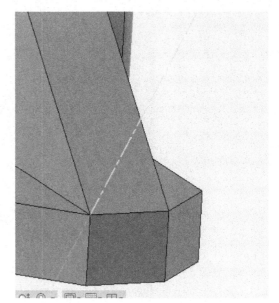

Selecting point and face   Created axis

Figure-78. Axis created by selecting face and point

Till now, we have learned to created plane and axis and in next topics we will learn the procedure to create a vertex.

Point at Vertex

The **Point at Vertex** tool is used to create a point by selecting a vertex from model. The procedure to use this tool is discussed next.

- Click on the **Point at Vertex** tool of **Construct** drop-down from **Toolbar**; refer to Figure-79. The tool will be activated.

Figure-79. The Point at Vertex tool

- Click on the vertex of model to create a constructional point; refer to Figure-80.

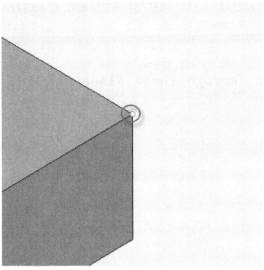

Figure-80. Creating point on vertex

Point Through Two Edges

The **Point Through Two Edges** tool is used to create a point using two edges as reference. The procedure to use this tool is discussed next.

- Click on the **Point Through Two Edges** tool of **Construct** drop-down from **Toolbar**; refer to Figure-81. The tool is activated.

Figure-81. Point Through Two Edges tool

- Click on the model to select two edges to create a point; refer to Figure-82.

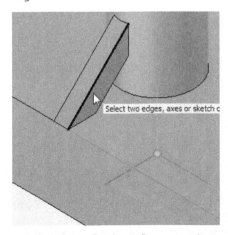

Selection of edges for creating points

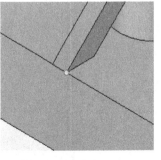

Created point

Figure-82. Creating point by selecting two edges

Point Through Three Planes

The **Point Through Three Planes** tool is used to create a point using three planes as references. The procedure to use this tool is discussed next.

- Click on the **Point Through Three Planes** tool of **Construct** drop-down from **Toolbar**; refer to Figure-83. The tool will be activated.

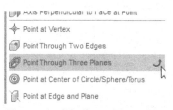

Figure-83. The Point Through Three Planes tool

- You need to select the three planes from the model as shown in Figure-83. On selecting the third plane, the constructional point will be created; refer to Figure-85.

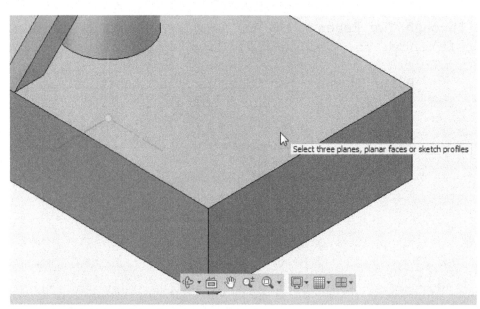

Figure-84. Selection of three planes

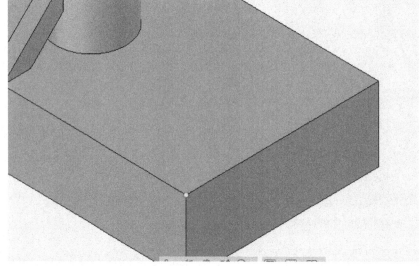

Figure-85. Point created by selecting three planes

Point at Center of Circle/Sphere/Torus

The **Point at Center of Circle/Sphere/Torus** tool is used to create a point at the center of selected circle/sphere/torus. The procedure to use this is discussed next.

• Click on the **Point at Center of Circle/Sphere/Torus** tool of **Construct** drop-down from **Toolbar**; refer to . The tool will be activated.

Figure-86. Point at Center of Circle/Sphere/Torus tool

• You need to click on circle, sphere, or torus to create the center point; refer to Figure-87.

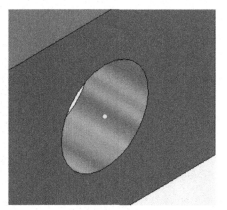

Figure-87. Creating Center Point

Point at Edge and Plane

The **Point at Edge and Plane** tool is used to create a point with the help of edge and plane as references. The procedure to use this tool is discussed next.

• Click on the **Point at Edge and Plane** tool of **Construct** drop-down from **Toolbar**; refer to Figure-88. The tool will be activated.

Figure-88. Point at Edges and Plane

• You need to select the edge and a plane to create a intersecting point. Select the edge and plane as shown in Figure-89. The point will be created; refer to Figure-90.

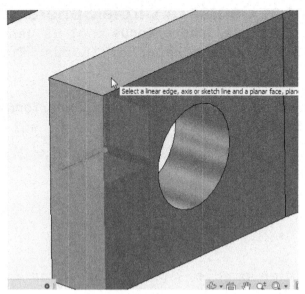

Figure-89. Selection of edge and plane

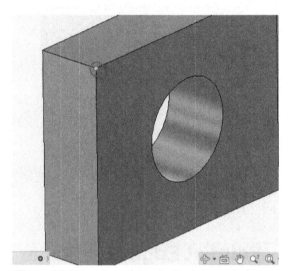

Figure-90. Point created on selecting edge and plane

SELF ASSESSMENT

Q1. Discuss the procedure of cutting material from a body.

Q2. Which tool is used to combine two bodies?

Q3. Discuss the use of Sweep tool with example?

Q4. Discuss the procedure of creating plane using Offset Plane tool with example.

Q5. Create a plane tangent to the face of model.

Q6. Create a plane using three points as reference.

Q7. Create a plane using edge of a model as reference.

Q8. Create a point at the center of cylinder.

FOR STUDENTS NOTES

Chapter 4

Advanced 3D Modeling

Topics Covered

The major topics covered in this chapter are:

- *Loft Tool*
- *Web Tool*
- *Multiple Holes Tool*
- *Torus Tool*
- *Rectangular Pattern*
- *Thicken Tool*
- *Create Mesh Tool*

INTRODUCTION

In the last chapter we have learned the procedure to use many tools and applying the constructional geometry. In this chapter we will use the constructional geometry tools to learn the advanced 3D modeling tool.

LOFT

The **Loft** tool is used to create a solid volume joining two or more sketches created on different planes; refer to Figure-1.

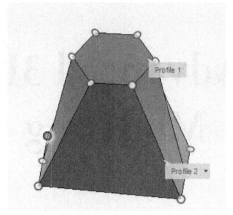

Figure-1. Lofted feature

- Click on the **Loft** tool from **Create** drop-down. The **LOFT** dialog box will be displayed as shown in Figure-2.

Figure-2. LOFT dialog box

- By default, **Profiles** selection box is active and you are asked to select sketches for loft feature.
- Click one by one on the sketches created at different planes. Note that you need to select the sketches in order by which they can be joined to each other successively. The preview of loft will be displayed ; refer to Figure-3.

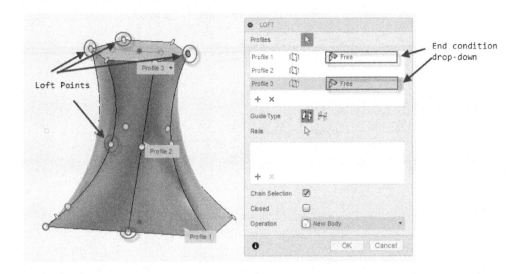

Figure-3. Preview of lofted feature

- To make adjustments to the loft, you need to move the loft points along the profile edge or adding a **Takeoff Weight** to the profile.
- To add **Takeoff Weight,** you need to select the end condition from the **End condition** drop-down.
- Click the **End condition** drop-down of **Profile 1** from **Profiles** selection box.
- End condition Controls the transition away from the start and end profiles of the loft. The end conditions available depends on the type of geometry selected for the profile. There are various tool in End condition drop-down; refer to Figure-4.

- **Free** No end condition applied.

- **Direction** Applies an angle measured off the sketch plane. Available when the loft profile is a 2D sketch.

- **Tangent** Applies a G1 condition off the loft profile. Available when the loft profile is the edge or face of a body.

- **Smooth** Applies a G2 condition off the loft profile. Available when the loft profile is the edge or face of a body.

- **Sharp** Transitions to a sharp point. Available when the profile is a sketch point or construction point.

- **Point Tangent** Applies tangency at the point to create a dome shape transition. Available when the profile is a sketch point or construction point.

Figure-4. End condition

- Click on **Direction** option from the **End Condition** drop-down. The updated **LOFT** dialog box will be displayed; refer to Figure-5.

Figure-5. Updated loft dialog box

- The **Takeoff Weight** and **Takeoff Angle** are displayed in the updated **LOFT** dialog box. These two tools causes the loft to bulge around the selected profile.
- The setting of **Takeoff Weight** and **Takeoff Angle** can be adjusted by entering the numerical value in their respective box or by moving the drag handle from the drawing area.
- Enter the value as **4** in **Takeoff Weight** edit box and specify the outward angle in **Takeoff Angle** edit box by **20** degree; refer to Figure-6.

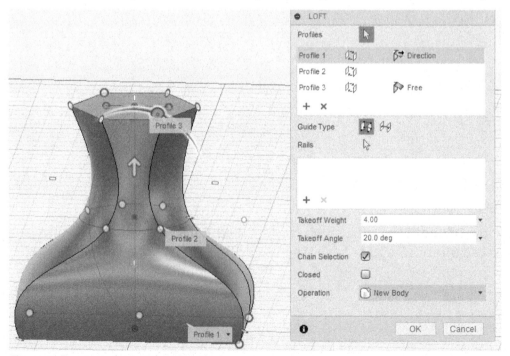

Figure-6. Preview after entering the value

- Similarly, click on **Direction** option of **End Condition** drop-down from **Profile 3**.

- Enter the value as **3** in **Takeoff Weight** edit box and specify the outward angle in **Takeoff Angle** edit box by **15** degree.
- Select the **Chain Selection** check box, if you want to select the adjacent edges together and included as one profile. **Clear** the check box if you want to select the adjacent edges individually.
- Click **OK** button of **LOFT** dialog box to finish the Loft feature.

Guide Type

- There are two different tools in this section which are used to create loft i.e. **Rails** and **Centerline**
- **Rails** are 2D or 3D curves that affect the loft shape between sections. To refine the shape of loft, You need to add number of rails. These rails must intersect each section and terminate on or beyond the first and last sections. **Rails** must be tangent continuous.
- A **Centerline** tool is a type of rail to which the loft sections are held normal, which causes behavior like a sweep path. Centerline lofts maintain a more consistent transition between the cross-sectional areas of selected loft sections. Center lines follow the same criteria as rails, except they need not intersect the sections, and only one can be selected.
- To select the particular rails for the loft, click the **Rails** or **Centerline** button in **Guide Type** section; refer to Figure-7.

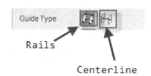

Figure-7. Rails and Cenrterline tools

- In our case we have created a sketch of tent like structure using a rectangle and four arcs on different planes; refer to Figure-8.

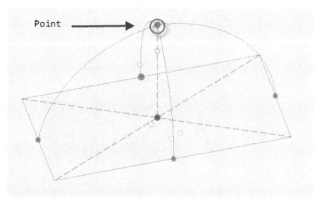

Figure-8. Sketch for using rails tool

- Now, click on the **Stop Sketch** button to exit the sketch.
- Select the **Loft** tool from **Create** drop-down. The **LOFT** dialog box will be displayed.
- Select the rectangle and point of the sketch in **Profiles** section of **LOFT** dialog box. The preview of model will be shown.
- Click on the **Arrow** selection button of **Rails** from **Guide Type** section box, to select the rails for the loft.
- Select the four rails of the model; refer to Figure-9.

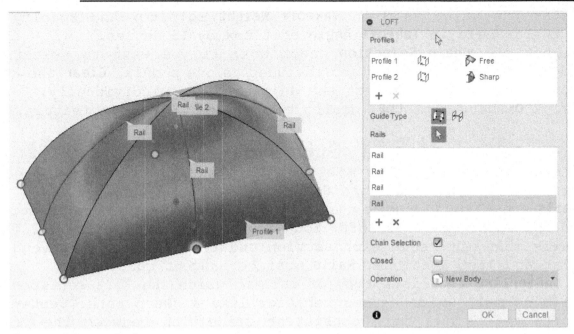

Figure-9. Selection of rails

- Click **OK** of **LOFT** dialog box to complete the process of making loft.
- To use the **Centerline** guide type, you need to create a sketch of ring like structure with the help of circle and rectangle on different planes; refer to Figure-10.

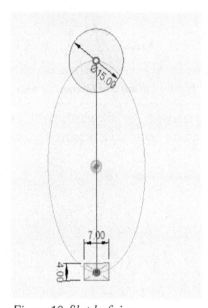

Figure-10. Sketch of ring

- Now, click on the **Stop Sketch** button to exit the sketch.
- Select the **Loft** tool from **Create** drop-down. The **LOFT** dialog box will be displayed.
- Select the circle and rectangle as profile in **Profiles** section of **LOFT** dialog box.
- Since these two profiles are not in the same plane to generate loft. So, Click on **Centerline** from **Guide Type** section and select the edge of larger circle as centerline in **Rails** section. The preview will be displayed; refer to Figure-11.

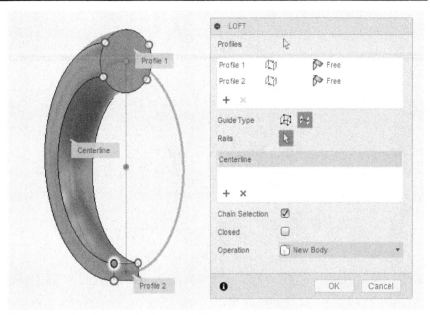

Figure-11. Preview on selecting centerline

- This centerline connects the both profile in both direction. So, check the **Closed** check box to complete the loft process.
- After specifying the parameters, the preview of loft will be displayed; refer to Figure-12.

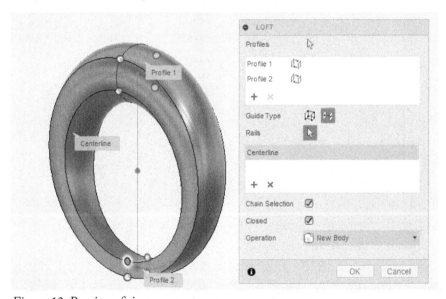

Figure-12. Preview of ring

- Click **OK** button of **LOFT** dialog box to complete the process.
- The tools of **Operation** drop-down were discussed earlier.

RIB

The **Rib** tool is used to create a support between two solid model. The procedure to use this tool is discussed next.

- Click on **Rib** tool from **Create** drop-down or right-click on the screen canvas and select **Rib** of **Create** cascading menu from **Marking menu**; refer to Figure-13. The **RIB** dialog box will be displayed; refer to Figure-14.

Figure-13. Selecting Rib tool

Figure-14. Rib dialog box

- Note that the sketch should be created in such a way that its projection is within solid faces of the model; refer to Figure-15.

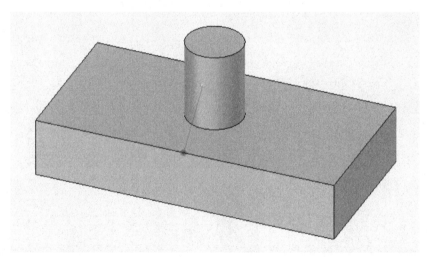

Figure-15. Sketch for rib

- The **Curve** section of **RIB** dialog box is selected by default. Select the sketch joining the two solid model to create rib. The updated **RIB** dialog box will be displayed; refer to Figure-16.

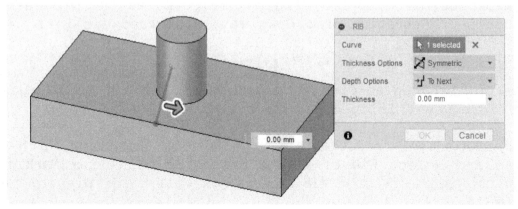

Figure-16. Updated Rib dialog box

- Click on **Symmetric** option from **Thickness Options** drop-down to set the thickness of rib on both side. Click on **One Direction** option from **Thickness Options** drop-down to apply the thickness of rib in only one direction or one side.
- Enter the value of thickness as **4** in **Thickness** edit box or floating window.
- Click on **To Next** in **Depth Options** drop-down from **RIB** dialog box to, create the rib up to next surface or model; refer to Figure-17.

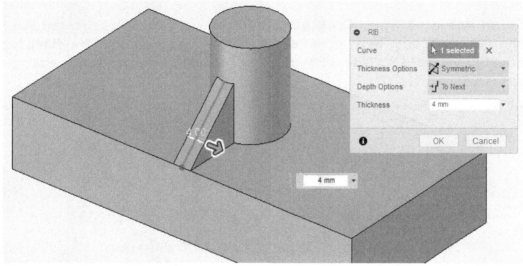

Figure-17. To next tool

- Click on the **Depth** tool from **Depth Options** drop-down to specify the depth of rib.
- Enter the numerical value of depth in **Depth** edit box or adjust the depth by moving the **Drag arrow.**
- You can also the enter the value of depth in minus(-) to reverse the direction of depth in **Depth** edit box; refer to Figure-18.

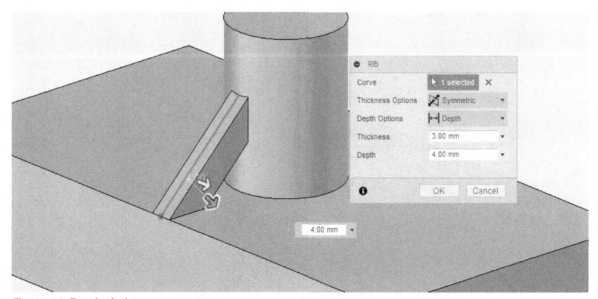

Figure-18. Depth of rib

- After specifying the parameters, click **OK** button of **RIB** dialog box complete the process.

WEB

The **Web** tool works like **Rib** tool, which is used to create geometry from sketch curves that intersect with pre-existing bodies. The **Web** tool uses multiple curves to create several merged elements at once. The procedure to use this tool is discussed next.

- Click on **Web** tool from **Create** drop-down or right-click on the screen canvas and select **Web** of **Create** cascading menu from **Marking menu**; refer to Figure-19. The **WEB** dialog box will be displayed; refer to Figure-20.

Figure-19. Web tool

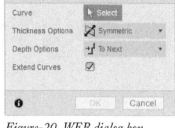

Figure-20. WEB dialog box

- In **WEB** dialog box you need to select the sketch for web tool. **Curve** option is selected by default. Select the sketch in Curve section; refer to Figure-21.

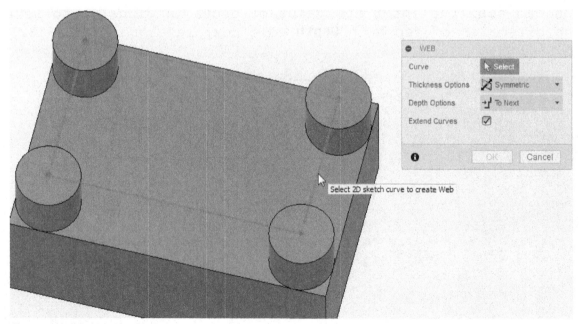

Figure-21. Selecting the sketch

- Enter the value as **1** in the **Thickness** edit box or floating window. You can also set the thickness by adjusting or moving the drag handle.
- The options of **Thickness options** and **Depth Options** were discussed earlier.

- After specifying the parameters, the preview will be displayed; refer to Figure-22.

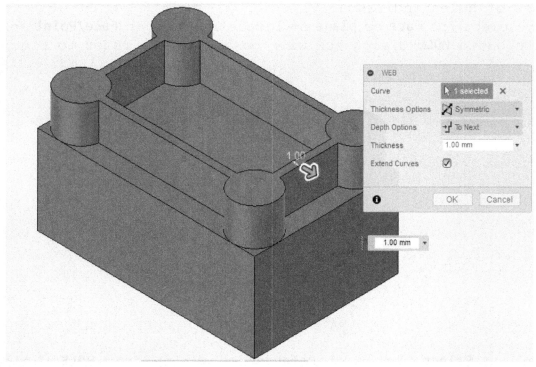

Figure-22. Preview of web tool

- Click **OK** button from **RIB** dialog box to complete the process.

HOLE

The **Hole** tool is used to create holes that comply with the real matching tools. With the help of **Hole** tool, three types of holes are created. The procedure to use this tool is discussed next.

- Click on **Hole** tool from **Create** drop-down or right-click on the screen canvas and select **Hole** of **Create** cascading menu from **Marking menu**; refer to Figure-23. The **HOLE** dialog box will be displayed refer to Figure-24.

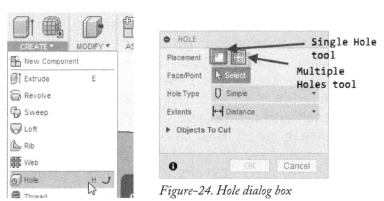

Figure-23. Hole tool

Figure-24. Hole dialog box

- There are two types of tools in **Placement** section which are discussed next.

Single Hole

- Click on **Single Hole** tool from **Placement** section to create various types of holes individually.
- Now, select the **face** or **plane** to locate the hole in **Face/Point** section.
- The updated **HOLE** dialog box will be displayed; refer to Figure-25.

Figure-25. Updated Hole dialog box

- Click on **Select** button of **References** section from **HOLE** dialog box and select the edge to add reference; refer to Figure-26.

Figure-26. Add Reference

- Enter the value as **25** in the floating window.
- Click on **Simple** in **Hole Type** drop-down from **HOLE** dialog box.
- Click on **Counterbore** in **Hole Type** drop-down to create a counterbore hole and specify the dimension in their respective edit box.
- Click on **Countersunk** in **Hole Type** drop-down to create a hole and specify the dimension in their respective edit box.
- Click in the **Depth** edit box and enter **7** as the depth of hole.
- Click in the **Diameter** edit box and enter **5** as the diameter of hole.
- Click in the **Tip Angle** edit box and enter **100** as the tip angle of hole.
- Click on **Distance** option from **Extent** drop-down to specify the distance of hole in the model. Click on **To** option from **Extent** drop-down to set the length of hole up to next model or face. Click on **All** option from **Extent** drop-down to create a hole passing through all the features of models; refer to Figure-27.

Figure-27. Extent drop - down

- Click on **Flip Direction** option to reverse the direction of hole.
- After specifying the parameters, click **OK** button of **HOLE** dialog box to complete the process; refer to Figure-28.

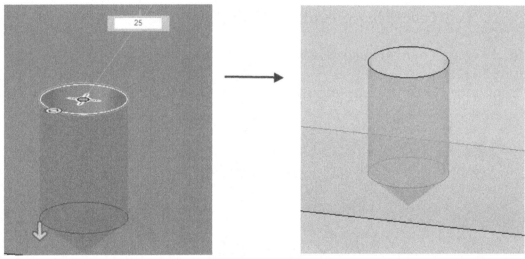

Figure-28. Preview of hole

Multiple Holes

The **Multiple Holes** tool is used to create various similar hole at a time. The procedure to use this tool is discussed next.

- Click on the **Multiple tool** button in **Placement** section from **HOLE** dialog box.
- **Face/Point** section is selected by default. Select the face/point for the position of holes; refer to Figure-29.

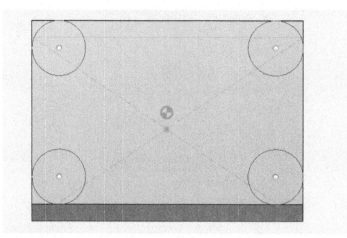

Figure-29. Sketch for multiple holes

- Click on **Counterbore** in **Hole Type** drop-down menu.
- Click in the **Depth** edit box and enter **10** as the depth of hole.
- Click in the **Diameter** edit box and enter **4** as the diameter of hole.
- Click in the **Tip Angle** edit box and enter **118** as the tip angle of hole.
- Click in the **Counterbore Depth** edit box and enter **3** as the depth of counterbore.
- Click in the **Counterbore Diameter** edit box and enter **5** as the depth of counterbore.
- The tools of **Extent** drop-down were discussed in Single Hole option.
- After specifying the parameters, click **OK** button from **HOLE** dialog box to finish the process. The preview will be displayed; refer to Figure-30.

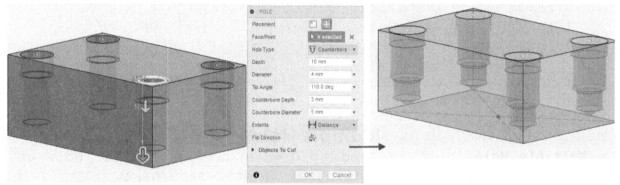

Figure-30. Preview of multiple holes

THREAD

The **Thread** tool is used to cut helical thread on cylindrical faces. Using this tool, you can save the custom threads in library. The Procedure to use this tool is discussed next.

- Click on the **Thread** tool from the **Create** drop-down in the **Toolbar**; refer to Figure-31. The **THREAD** dialog box will be displayed; refer to Figure-31.

Figure-31. Thread tool

Figure-32. THREAD dialog box

- **Faces** section of **THREAD** dialog box is selected by default. Select the round faces for thread. You can select multiple faces at once by pressing the **CTRL** key and click on the face.
- Select the **Modeled** check box to create the model feature of thread otherwise, it will be created as cosmetic feature.
- On selecting **Modeled check** box, a preview of thread will be displayed; refer to Figure-33.
- Clear the **Full Length** check box to specify the length of thread on the selected face. Otherwise select the check box to create thread on the entire face.
- Click on the **Thread Type** drop-down to select the type of thread profile; refer to Figure-34.

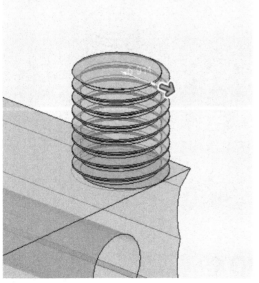

Figure-33. Preview on selecting Modeled check box

Figure-34. Thead drop - down

- Click in the **Size** edit box and enter **14** as the size of thread. You can also adjust the size by moving the **Drag Handle**.
- Click on the **Designation** drop-down and select the designation of thread ; refer to Figure-35.

- Click on the **Class** drop-down and select the class of thread.
- Click on **Direction** drop-down to and select the direction for thread; refer to Figure-36.

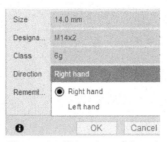

Figure-36. Direction of thread

Figure-35. Designation

- Select the **Remember Size** check box to remember these thread setting for the next time, when thread tool is invoked.
- Click **OK** button from **THREAD** dialog box to create the thread; refer to Figure-37.

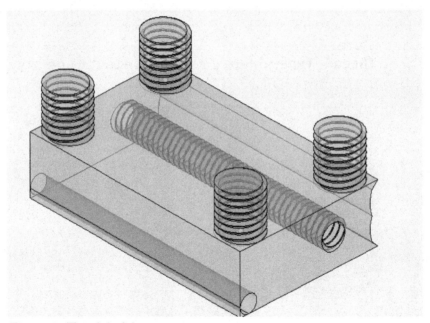

Figure-37. Threaded solid

BOX

The **Box** tool is used to create a rectangular body on a work plane. The procedure to use this tool is discussed next.

- Click on **Box** tool in **Create** drop-down from **Toolbar**; refer to Figure-38.

Figure-38. Box

- You are asked to select the plane for creating the sketch for box. Select **Top** plane.
- Click on the screen to specify the first point of sketch.
- Move the mouse to create the rectangle. You can also enter the numerical value of **Length** and **Width** in their specific edit box; refer to Figure-39.
- Enter the value as **40** and **70** in their respective edit box. You can toggle between these two edit box by pressing **TAB** key.

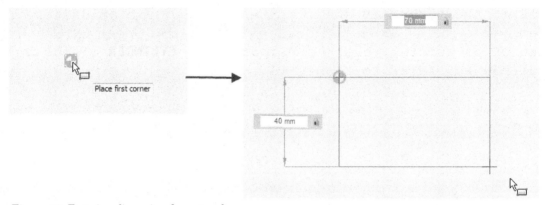

Figure-39. Entering dimension for rectangle

- Press **Enter** key or click on the screen to complete the sketch of rectangle.
- A **BOX** dialog box will be displayed; refer to Figure-40.

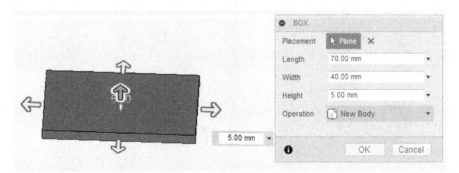

Figure-40. Box dialog box

- The value in **Length** and **Width** were same as mentioned earlier in the sketch.
- If you want to enter the new values of **Length** and **Width** then, click in the respective edit box and specify the desired values.
- Click in the **Height** edit box and enter **10** as the height of box.
- The tools of **Operation** section were discussed earlier.
- Click **OK** button from the **BOX** dialog box to create the box; refer to Figure-40.

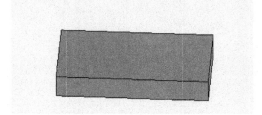

Figure-41. Preview of box

CYLINDER

This tool is used to create a cylindrical body by adding depth to a circular region. The procedure to use this tool is discussed next.\

- Click on **Cylinder** tool in **Create** drop-down from **Toolbar**; refer to Figure-42.
- You are asked to select the plane for creating the sketch. Select **Top** plane from the canvas screen.
- Click on the screen to specify the center point of sketch.
- Enter the value as **30** in the floating window and press **ENTER** to create a sketch of circle; refer to Figure-43. A **CYLINDER** dialog box will be displayed; refer to Figure-44.

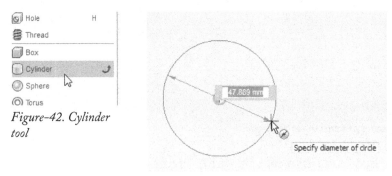

Figure-42. Cylinder tool

Figure-43. Specify diameter of circle

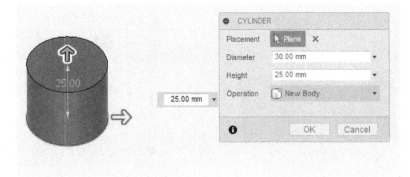

Figure-44. CYLINDER dialog box

- If you want to edit the diameter of cylinder then, click on the **Diameter** edit box of **CYLINDER** dialog box and specify the desired value.
- Click on the **Height** edit box and enter **25** as the height of cylinder.
- Click **OK** button from the **CYLINDER** dialog box to create the cylinder; refer to Figure-45.

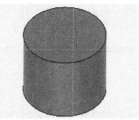

Figure-45. Preview of cylinder

SPHERE

The **Sphere** tool is used to create a T-spline, solid, or surface sphere body on a work plane or face. The procedure to use this tool is discussed next.

• Click on the **Sphere** tool in **Create** drop-down from **Toolbar**; refer to Figure-46.
• You are asked to select the plane for creating the sketch for circle. Select **Top** plane to create the sketch.
• Click on the screen to specify the point of sphere. The **SPHERE** dialog box will be displayed; refer to Figure-47.

Figure-46. Sphere tool

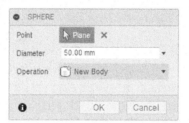

Figure-47. SPHERE dialog box

• Click in the **Diameter** edit box and enter **50** as diameter of sphere.
• The tools of **Operation** drop-down were discussed earlier.
• Click **OK** button of **SPHERE** dialog box to finish the process; refer to Figure-48.

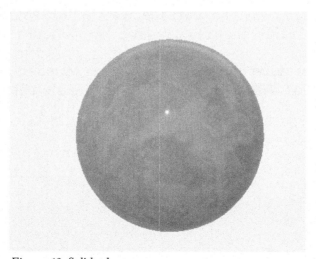

Figure-48. Solid sphere

TORUS

The **Torus** tool is used to create T-spline, solid, or surface torus body on a work plane or face. The procedure to use this tool is discussed next.

- Click on **Torus** tool in **Create** drop-down from **Toolbar**; refer to Figure-49.
- You are asked to select the plane for creating the sketch for circle. Select **Top** plane.
- Click on the screen to specify the center point of sketch. Enter **50** as the diameter of torus in floating window.
- Press **Enter** or click on the screen to create the sketch. A **TORUS** dialog box will be displayed; refer to Figure-50.

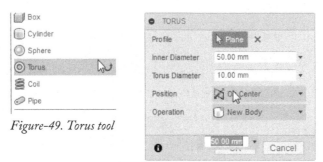

Figure-49. Torus tool

Figure-50. Torus dialog box

- If you want to edit the inner diameter of circle then, click in the **Inner Diameter** edit box and enter the desired value.
- Click in the **Torus Diameter** edit box and enter **20** as the diameter of torus. You can also adjust the diameter of torus by moving the drag handle.
- Click **On Center** button of **Position** drop-down from **TORUS** dialog box, if you want to position the torus section center point on the inner circle. Click on the **Inside** button of **Position** drop-down, if you want to position the torus section tangent to the inside of the inner circle. Click on the **Outside** button of **Position** drop-down, if you want to position the torus section tangent to the outside of the inner circle.
- The tools of **Operation** drop-down were discussed earlier.
- Click **OK** button of **TORUS** dialog box to finish the process; refer to Figure-51.

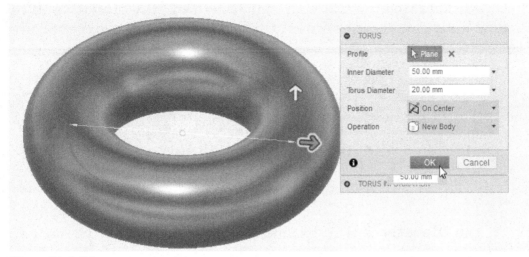

Figure-51. Solid torus

COIL

The **Coil** tool is used for creating helix-based geometry. The procedure to use this tool is discussed next.

- Click on **Coil** tool of **Create** drop-down from **Toolbar**; refer to Figure-52.
- Select **Top** plane when you are asked to select the plane for creating the sketch.
- Click on the screen to specify the center point of sketch. Enter **60** as the diameter of coil in the **Diameter** edit box. A **COIL** dialog box will be displayed; refer to Figure-53.

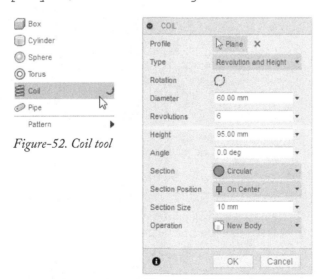

Figure-52. Coil tool

Figure-53. COIL dialog box

- Click on **Resolution and Height** from **Type** drop-down, if you want to create the coil using Resolution and height of coil. Click on **Resolution and Pitch** from **Type** drop-down, if you want to create the coil using resolution and pitch. Click on **Height and Pitch** from **Type** drop-down to create the coil using height and pitch. Click on **Spiral** from **Type** drop-down to create a spiral coil. In our case we are selecting the **Resolution and Height** button; refer to Figure-54.

- Click on the **Rotation** button to rotate the coil in **clockwise** direction. Click again to rotate the coil in **counterclockwise** direction; refer to Figure-55.

Figure-54. Type drop-down

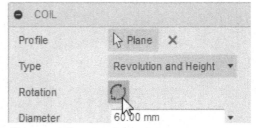

Figure-55. Rotation button

- Click in the **Diameter** edit box and enter the desired value to change the diameter of coil.
- Click in the **Revolutions** edit box and enter **6** as the revolution of coil.
- Click in the **Height** edit box and enter **80** as the height of coil. You can also set the height of coil by moving the drag handle.
- Click in the **Angle** edit box and specify the desired angle for coil.
- Click on **Circular** option of **Section** drop-down from **COIL** dialog box, if you want to set the circular profile of coil. Click on **Square** option from **Section** drop-down, if you want to set the square profile of coil. Click on **Triangular(External)** option of **Section** drop-down, if you want to set the external triangle profile of coil. Click on **Triangular(Internal)** option of **Section** drop-down, if you want to set the internal triangle profile of coil; refer to Figure-56

Figure-56. Section drop-down

- Click on **On Center** button of **Section Position** drop-down from **COIL** dialog box, if you want to positions the coil section center point on the inner circle. Click on the **Inside** button of **Section Position** drop-down, if you want to position the coil section tangent to the inside of the inner circle. Click on the **Outside** section of **Section Position** drop-down, if you want to position the coil section tangent to the outside of the inner circle.
- Click in the **Section Size** edit box and enter **15** as the size of section.
- The options of **Operations** drop-down were discussed earlier.
- Click **OK** button of **COIL** dialog box to finish the process of creating a coil; refer to Figure-57.

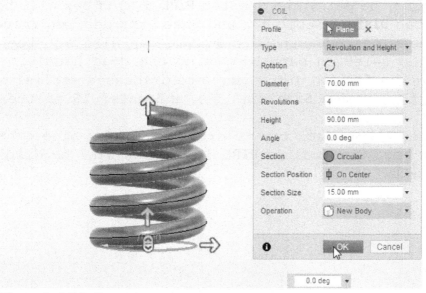

Figure-57. Soild coil

PIPE

The **Pipe** tool is used to create a solid or surface body that follows a selected path. The procedure to use this tool is discussed next.

- Click on **Pipe** tool of **Create** drop-down from **Toolbar;** refer to Figure-58. A **PIPE** dialog box will be displayed; refer to Figure-59.

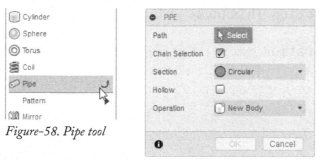

Figure-58. Pipe tool

Figure-59. PIPE dialog box

- You are asked to select the path for pipe. **Path** selection option is selected by default. Select the sketch or create a sketch for the path of the pipe; refer to Figure-60.

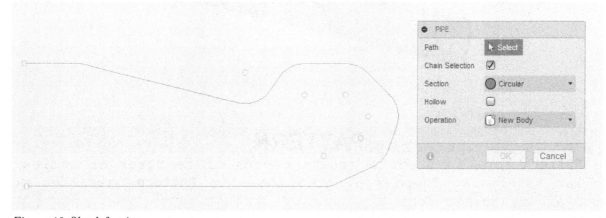

Figure-60. Sketch for pipe

- On selecting the path, the updated **PIPE** dialog box will be displayed. Click in the **Distance** edit box and enter the desired value. The value of distance lies between 0 to 1. You can also set the distance of pipe on the selected path by moving the drag handle.
- The options of **Section** drop-down were discussed earlier in **Coil** tool.
- Click in the **Section Size** edit box and enter **15** as the diameter of pipe.
- Select the **Hollow** Check box to create a hollow pipe otherwise clear the check box. The updated **PIPE** dialog box will be displayed; refer to Figure-61.

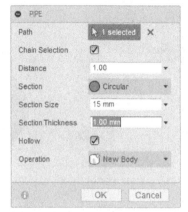

Figure-61. Updated PIPE dialog box

- Click in the **Section Thickness** edit box and enter **1** as thickness of pipe.
- The tools of **Operation** drop-down were discussed earlier in this book.
- Click **OK** button of **PIPE** dialog box to complete the process of creating the pipe; refer to Figure-62.

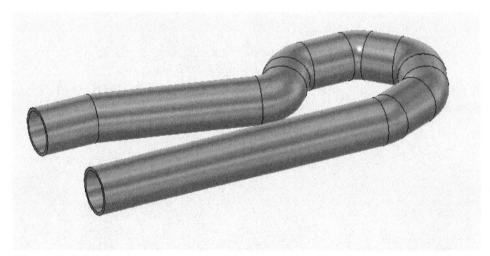

Figure-62. Solid pipe

PATTERN

The **Pattern** tool is used to create copies of features or bodies at regular intervals. There are three tools in **Pattern** cascading menu which are discussed next.

Rectangular Pattern

The **Rectangular Pattern** tool is used to create copies of objects in one or two directions. The procedure to use this tool are discussed next.

- Click on **Rectangular Pattern** of **Pattern** cascading menu from **Create** drop-down; refer to Figure-63. A **RECTANGULAR PATTERN** dialog box will be displayed; refer to Figure-64.

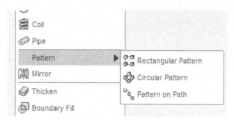

Figure-63. Rectangular Pattern tool

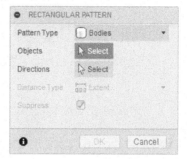

Figure-64. RECTANGULAR PATTERN dialog box

- Click on **Faces** from **Pattern Type** drop-down, if you want to select the faces for pattern. Click on **Bodies** from **Pattern Type** drop-down, if you want to select bodies for creating copies. Click on **Features** from **Pattern Type** drop-down, if you want to select features for creating copies. Click on **Component** from **Pattern Type** drop-down, if you want to select component for creating copies; refer to Figure-65.

Figure-65. Pattern Type drop-own

- **Bodies** option from **Pattern Type** drop-down, is selected by default. Select the body for creating copies; refer to Figure-66.

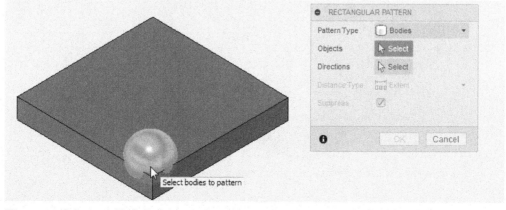

Figure-66. Selection of body for pattern

- Click on **Select** button of **Directions** option and select the **Z axis** for creating the copies; refer to Figure-67.

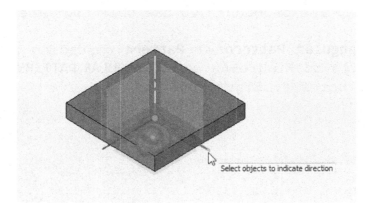

Figure-67. Selection of direction

- Click on **Extent** option from **Distance Type** drop-down, if you want to specify the distance between two entity. Click on **Spacing** from **Distance Type** drop-down, to set the total distance for the pattern. In our case we are selecting **Extent** option.
- Check the **Suppress** check box, if you want to suppress the particular pattern of the entity otherwise clear the check box.
- Click in the **Quantity** edit box and enter **2** as the quantity of copy.
- Click in **Distance** edit box and enter **-55** as the distance between two copies. You can also set the quantity by moving the drag handle.
- Click on **One Direction** option from **Direction Type** drop-down, if you want to create the copy in one direction. Click on **Symmetric** option from **Direction Type** drop-down to create the copies in both direction.
- Similarly, enter the value in **Quantity** and **Distance** edit box of other direction as discussed above.
- Click on **One Direction** from **Direction Type** drop-down.
- After specifying the parameter, click **OK** button from **RECTANGULAR PATTERN** dialog box to finish the process; refer to Figure-68.

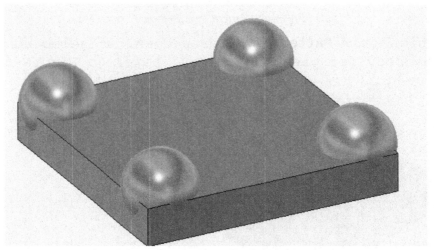

Figure-68. Solid model created after rectangular pattern tool

Circular Pattern

The **Circular Pattern** tool is used to create copies of selected objects around a selected axis. The procedure to use this tool is discussed next.

* Click on **Circular Pattern** tool from **Pattern** cascading menu; refer to Figure-69. A **CIRCULAR PATTERN** dialog box will be displayed; refer to Figure-70.

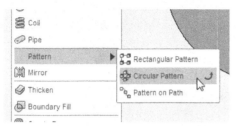

Figure-69. Circular Pattern tool

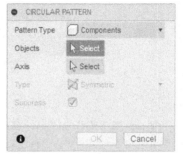

Figure-70. CIRCULAR PATTERN dialog box

* Click on **Components** option from **Pattern Type** drop-down. Other options of **Pattern Type** drop-down were discussed in **Rectangular Pattern** tool.
* **Objects** selection is selected by default. Select the component to create copies; refer to Figure-71.

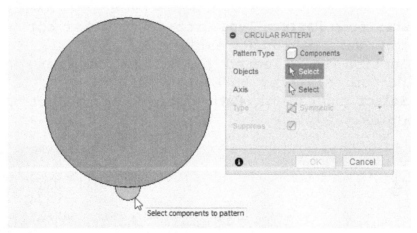

Figure-71. Select components to create copies

* Click on **Select** option of **Axis** section and select the **Y** axis from the canvas screen.
* Click on **Full** option from **Type** drop-down, if you want to create the copies of component which covers 360 degree path. Click on **Angle** from **Type** drop-down, if you want to create the copies in a specified angle. Click on **Symmetric** option from **Type** drop-down, if you want to cover a symmetric pattern.
* Check the **Suppress** check box, if you want to suppress the particular pattern of the entity otherwise clear the check box.
* Click in the **Quantity** edit box and enter **7** as the number of copies of selected component.
* After specifying the parameters, click **OK** button of **CIRCULAR PATTERN** dialog box to finish the process; refer to Figure-72.

Figure-72. Solid model after Circular Pattern tool

Pattern on Path

The **Pattern on Path** tool is used to copy the selected objects along a selected path rather than following a circular or rectangular path. The procedure to use this tool is discussed next.

• Click on **Pattern on Path** of **Pattern** cascading menu from **Create** drop-down. A **PATTERN ON PATH** dialog box will be displayed; refer to Figure-73.
• Click on **Faces** from **Pattern Type** drop-down. Other options of **Pattern Type** drop-down were discussed earlier.
• **Objects** selection is selected by default. Select the faces for creating the copies; refer to Figure-74.
• Click on **Select** button of **Path** option and select the path for creating copies.

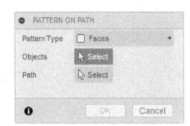

Figure-73. PATTERN ON PATH dialog box

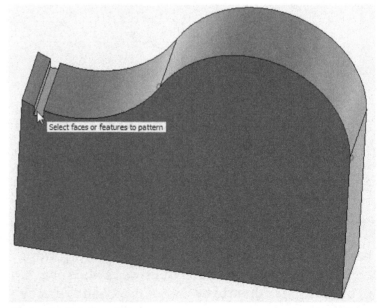

Figure-74. Selection of faces

- On selecting path, the updated **PATTERN ON PATH** dialog box will be displayed; refer to Figure-75.
- Check the **Suppress** check box, if you want to suppress the particular pattern of the entity otherwise clear the check box.
- Click in the **Quantity** option edit box and enter **10** as the quantity of copies.
- Click in the **Distance** option edit box and enter **100** as the total distance.
- The **Start Point** option is used to line up with the objects. It is shown in the screen canvas as a **white point** and determines the starting point of the selected path. It can be set to a specific value between zero and one from the Start Point edit box, but it is easier to adjust it manually by moving the start point; refer to Figure-76.

Figure-76. Start point

Figure-75. Updated PATTERN ON PATH dialog box

- The options of **Distance Type** and **Direction** drop-down were discussed earlier in this chapter.
- Click on **Path Direction** option of **Orientation** drop-down, if you want to rotate each instance relative to the pattern path. Click on **Identical** option of **Orientation** drop-down to keep the instances in the same position as the original object.
- Click **OK** button from **PATTERN ON PATH** dialog box to complete the process of creating the copies; refer to Figure-77.

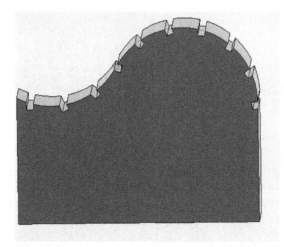

Figure-77. Solid model after Pattern On Path tool

MIRROR

The **Mirror** tool is used to create mirrors of selected objects to the opposite side of a selected face or plane. The procedure to use this tool is discussed next.

- Click on **Mirror** tool of **Create** drop-down from **Toolbar**; refer to Figure-78. A **MIRROR** dialog box will be displayed; refer to Figure-79.

Figure-78. Mirror tool

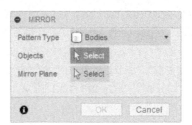

Figure-79. MIRROR dialog box

- Click on **Bodies** options from **Pattern Type** drop-down. The other options of Pattern Type section were discussed earlier.
- Click on **Select** button of **Mirror Plane** option and select the plane to create a mirror copy; refer to Figure-80. You can also select any face or plane of the model to create a mirror copy.

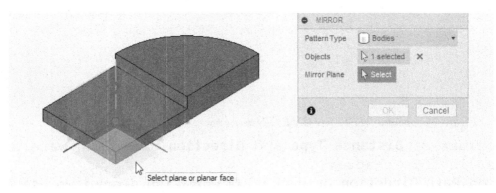

Figure-80. Selection of plane for Mirror

- Click **OK** button of **MIRROR** dialog box to finish the process of creating mirror; refer to Figure-81.

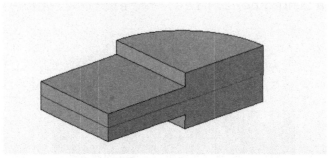

Figure-81. Solid model after applying Mirror tool

THICKEN

The **Thicken** tool is used to add thickness to surfaces to make them solid. The procedure to use this tool is discussed next.

- Click on **Thicken** tool of **Create** drop-down from **Toolbar**; refer to Figure-82. A **THICKEN** dialog box will be displayed; refer to Figure-83.

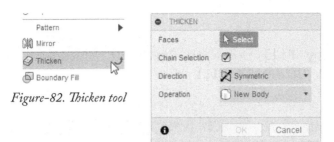

Figure-82. Thicken tool

Figure-83. THICKEN dialog box

- **Faces** selection is selected by default. Select the faces to add thickness; refer to Figure-84.

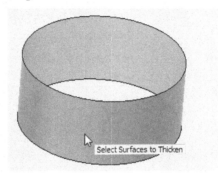

Figure-84. Selection of faces to add thickness

- Click in the **Thickness** edit box and enter **4** as the thickness of model.
- Click on **One Side** option from **Direction** drop-down to apply the thickness on one side. Click on **Symmetric** option from **Direction** drop-down to apply the thickness to both direction or side. In our case we are selecting **Symmetric** option.
- The options of **Operation** drop-down were discussed earlier in this chapter.
- Click **OK** button of **THICKEN** dialog box to finish the process; refer to Figure-85.

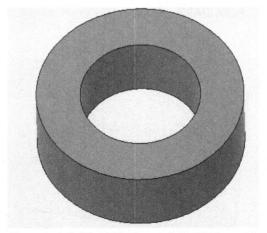

Figure-85. Solid model after applying Thicken tool

BOUNDARY FILL

The **Boundary Fill** tool is used to fill combines solids, work planes, and surfaces to create cells. The procedure to use this tool are discussed next.

* Click on **Boundary Fill** tool of **Create** drop-down from **Toolbar**; refer to Figure-86. A **BOUNDARY FILL** dialog box will be displayed; refer to Figure-87.

Figure-86. Boundary Fill tool

Figure-87. BOUNDARY FILL dialog box

* **Select Tools** selection is selected by default. Select the body for boundary fill; refer to Figure-88.

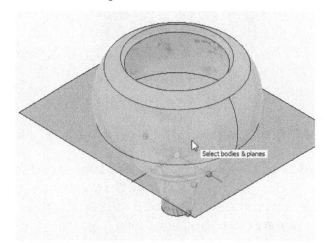

Figure-88. Selection for boundary fill

* Click on **Select** button of **Select Cells** and select the desired cells for boundary cells.
* The tools of **Operation** section were discussed earlier in this chapter.
* Click **OK** button of **BOUNDARY FILL** dialog box to finish the process; refer to Figure-89.

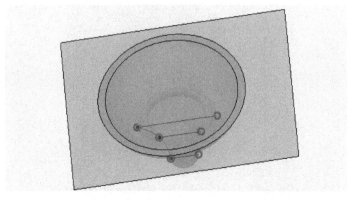

Figure-89. Preview after applying Boundary Fill tool

CREATE FORM

The **Create Form** tool adds a form operation to the **Timeline**. There is no dialog box for create form. The procedure to use this tool is discussed next.

- Click on **Create Form** of **Create** drop-down from **Toolbar**; refer to Figure-90. A **Sculpt Environment** box will be displayed; refer to Figure-91.

Figure-90. Create Form tool

Figure-91. Sculpt Environment box

- Click **OK** button to enter in Sculpt Environment.
- Now, create or modify the model as desired.
- Click **Finish Form** button from **Toolbar** to exit the Sculpt Environment and to return to **Model Workspace**; refer to Figure-92.

Figure-92. Finish Form tool

CREATE BASE FEATURE

The **Create Base Feature** tool inserts a base feature operation in the **Timeline**. Any operations performed on the design while in a base feature are not recorded in the **Timeline**.

- Click on **Create Base Feature** tool of **Create** drop-down from **Toolbar**; refer to Figure-93.
- Do the desired operation that you do not want to record in **Timeline**.
- Click on **FINISH BASE FEATURE** button from **Toolbar** to exit this tool; refer to Figure-94.

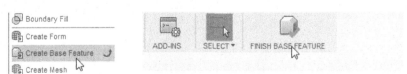

Figure-93. Create
Base Feature tool

Figure-94. FINIS BASE FEATURE button

CREATE MESH

The **Create Mesh** tool is used to create a mesh of the current model. The procedure to use this tool is discussed next.

- Click on **Create Mesh** tool of **Create** drop-down from **Toolbar**; refer to Figure-95. Mesh Workspace will be opened. If not, then check the **Mesh Workspace** check box of **Preview** section from **Preferences** tool under user account; refer to Figure-96.

Figure-95. Create Mesh tool

Figure-96. Preferences tool

- Create and modify the desired operation or command in **Mesh Workspace.**
- On defining the various parameters for mesh, click on **FINISH MESH** button from **Toolbar** to exit **Mesh Workspace** and to return to **Model Workspace**; refer to Figure-97.

Figure-97. FINISH MESH button

SELF ASSESSMENT

Q1. Which of the Following is not the default plane available in Autodesk Fusion 360?

a. Front Plane
b. Right Plane
c. Top Plane
d. Left Plane

Q2. Discuss the procedure of creating multiple holes with example.

Q3. Which of the following tool is not available in **Create** drop-down of Autodesk Fusion 360?

a. Thread
b. Box
C. Torus
d. Curve

Q4. **Thicken** tool is used to add thickness and length. (T/F)

Q5. The **Pattern On Path** tool is available in the **Create** drop-down. (T/F)

Q6. Discuss the use of **Pipe** tool with examples.

Q7. Discuss the use of **Web** tool with examples.

Q8. Which tool is used to create a solid model connecting two sketch from different planes?

Q9. Discuss the use of **Pattern On Path** tool with examples.

Q10. The --------- tool is used to cut helical thread on cylindrical faces.

FOR STUDENTS NOTES

Chapter 5

Practical and Practice

The major topics covered in this chapter are:

- *Practical*
- *Practice*

PRACTICAL 1

Sketch the model as shown in Figure-1. The views of the model with dimension are given in Figure-2.

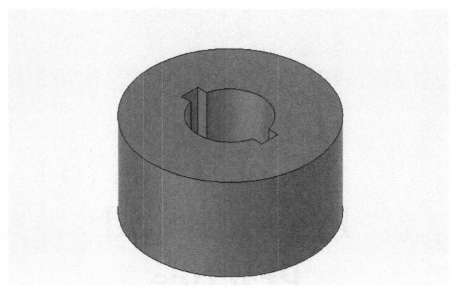

Figure-1. Model of Practical 1

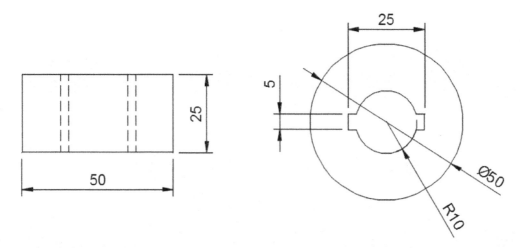

Figure-2. Views of practical 1

Before we start working on Practical 1, it is important to understand two terms; first angle projection and third angle projection. These are the standards of placing views in the engineering drawings. The views placed in the above figure are using third angle projection. In first angle projection, the top view of model is placed below the front view and right side view is placed below the front view. You will learn more about projection in chapter related to drafting.

Creating the sketch

- Click on **Center Diameter Circle** tool of **Circle** cascading menu from **Sketch** drop-down in **Toolbar**. You will asked to select the plane for creating sketch.
- Select **Top Plane** from the canvas screen.
- Click at the **Origin** to place the center point of circle. Enter **50** in floating window as diameter of circle and press **ENTER** key.
- Sketch another circle of radius **10** taking center point as origin.
- Select **Center Rectangle** tool of **Rectangle** cascading menu from **Sketch** drop-down in **Toolbar**.
- Click at **Origin** as center point for rectangle and enter **5** as width and **25** as length in floating window to create a rectangle; refer to Figure-3.

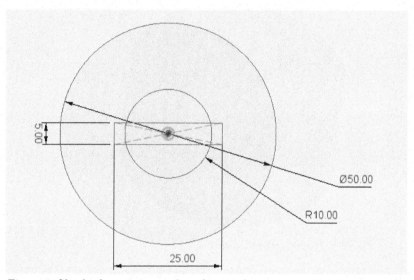

Figure-3. Sketch after creating circle and rectangle

- Select **Trim** tool of **Sketch** drop-down from **Toolbar** and trim the entities in such a way that the sketch is displayed as shown in Figure-4.

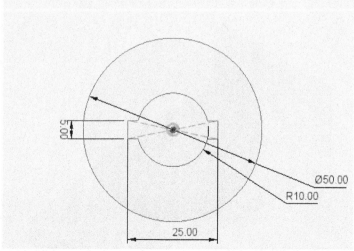

Figure-4. Sketch after trimmig

- Click on **Stop Sketch** button from **Toolbar** to exit the sketch mode.

Creating Extrude

- Click on **Extrude** tool of **Create** drop-down from **Toolbar** or you can also press **E** key from keyboard to select **Extrude** tool. The **EXTRUDE** dialog box will be displayed.
- Profile selection is active by default. Select the recently created sketch profile. The updated **EXTRUDE** dialog box will be displayed.
- Specify the height of extrusion as **25** in **Distance** edit box or floating window. The preview of extrusion will be displayed; refer to Figure-5.
- Click **OK** button from **EXTRUDE** dialog box to complete the process of extrusion.

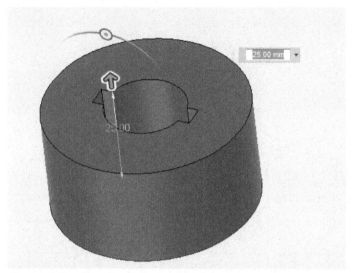

Figure-5. Preview of extrude feature

PRACTICAL 2

Create the model as shown in Figure-6. The dimensions and views are given in Figure-7.

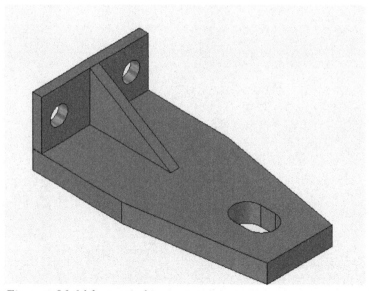

Figure-6. Model for practical 2

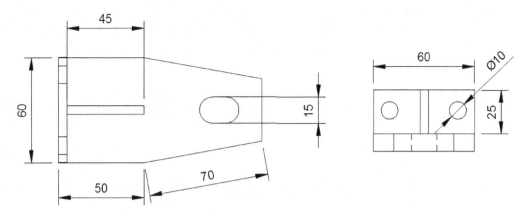

Figure-7. Practical 2 drawing

Creating sketch for model

- Start a new sketch in **Top plane.**
- Select **Centre Rectangle** tool of **Rectangle** cascading menu from **Sketch** drop-down.
- Click on the origin to specify the centre point of rectangle and enter **60** and **50** in floating window as dimensions of rectangle.
- Select **Line** tool from **Sketch** drop-down and create a line of length as **70** and angle of **80** degree as shown in Figure-8.

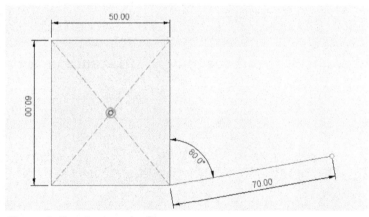

Figure-8. Creating angular line

- Select **Mirror** tool of **Sketch** drop-down from **Toolbar** and create a mirror copy of recently created line about **Front** Plane.
- Create a line with the help of **Line** tool joining the two earlier created lines as shown in Figure-9.

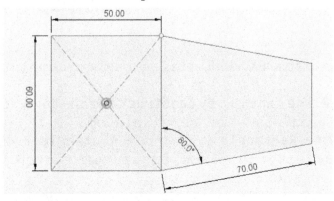

Figure-9. Sketch after applying mirror

- Select **Center Point Slot** tool of **Slot** cascading menu from **Sketch** drop-down and create a slot of diameter as **15** and length as **8**; refer to Figure-10.

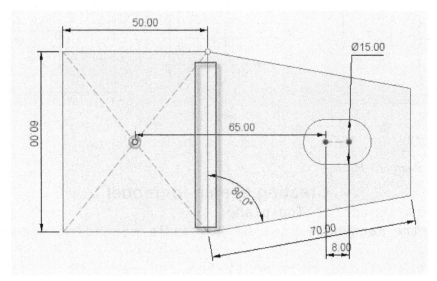

Figure-10. Creating slot

- Select **Extrude** tool from **Create** drop-down. The **EXTRUDE** dialog box is displayed.
- Profile selection is active by default. Select the recently created sketch to extrude.
- Enter **10** as height of extrusion in **Distance** edit box of **EXTRUDE** dialog box; refer to Figure-11.

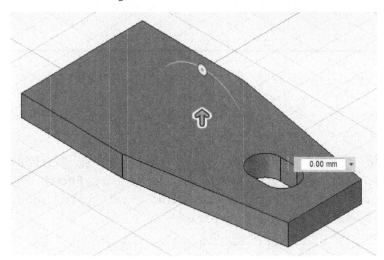

Figure-11. Preview of extrusion1

- Click **OK** button from **EXTRUDE** dialog box to complete the process of extrusion.
- Select **Offset Plane** tool of **Construct** drop-down from **Toolbar**. The **OFFSET PLANE** dialog box will be displayed.
- Select the **Centre Rectangle** tool from **Rectangle** cascading menu. You are asked to select the plane to create sketch. Select the side face of model as shown in Figure-12.
- Create a rectangle of dimensions **25** and **60** as shown in Figure-12.

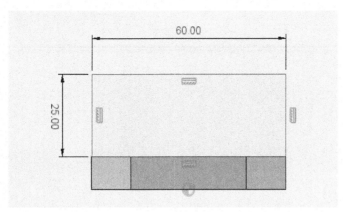

Figure-12. Creating a rectangle

- Select **Extrude** tool from **Create** drop-down. The **EXTRUDE** dialog box will be displayed.
- Select the recently created sketch in **Profiles** selection and enter **5** as height of extrusion.
- Select **Join** from **Operation** drop-down of **EXTRUDE** dialog box to join the two extrusions.
- Click **OK** button from **EXTRUDE** dialog box to finish this process.

Creating Cylinder cut

- Select **Cylinder** tool of **Create** drop-down from **Toolbar**. The **CYLINDER** dialog box will be displayed.
- Select the recently created extrusion face as reference to create a cylinder cut and place the cylinder as displayed in Figure-13.

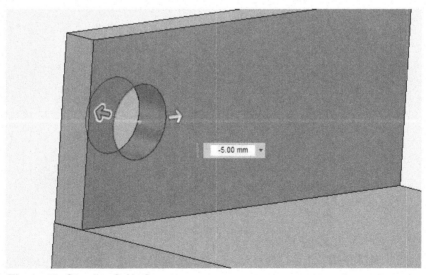

Figure-13. Creating Cykinder

- Click in the **Diameter** edit box and enter **10** as diameter of cylinder.
- Click in the **Height** edit box and enter **-5** as the height of the cylinder.
- Select **Cut** from **Operation** drop-down to cut the material of cylinder from solid; refer to Figure-14.

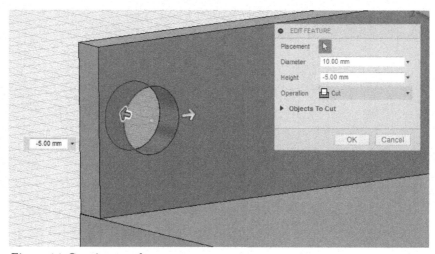

Figure-14. Creating round cut

- Click **OK** button from **CYLINDER** dialog box to finish the process.
- Select **Mirror** tool of **Create** drop-down from **Toolbar**. The **MIRROR** dialog box will be displayed.
- Select **Faces** of **Pattern Type** drop-down from **MIRROR** dialog box.
- Click on **Select** button of **Objects** section and select the face of recently created cylinder.
- Click on **Select** button of **Mirror Plane** section and select the **Front Plane**; refer to Figure-15.

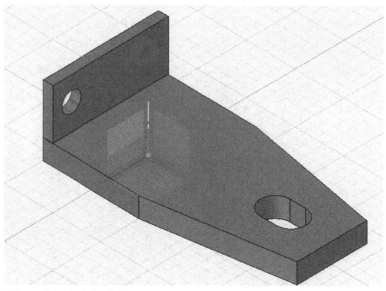

Figure-15. Applying Mirror

- Click **OK** from **MIRROR** dialog box to complete the process of creating mirror feature.

Creating Rib

- Select **Line** tool from **Sketch** drop-down. You will be asked to select the plane for creating sketch. Select **Front Plane**.
- Create a line as shown in Figure-16 with end points coincident to the vertices of model.

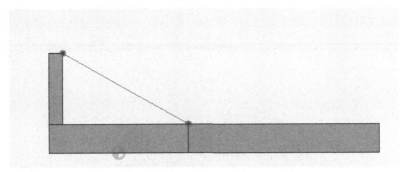

Figure-16. Creating sketch for Rib

- Select **Rib** tool from **Create** drop-down. The **RIB** dialog box will be displayed.
- Select the recently created line for **Curve** section.
- Select **Symmetric** option form **Thickness Options** drop-down.
- Select **To Next** Option from **Depth Options** drop-down.
- Click in **Thickness** edit box and enter **5** as thickness of rib; refer to Figure-17.

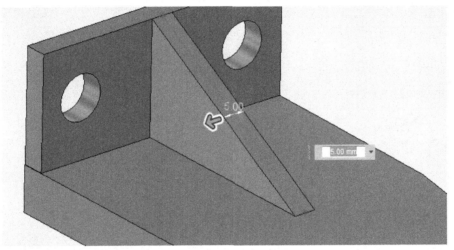

Figure-17. Creating Rib

- Click **OK** button of **RIB** dialog box to finish the process. The model will be displayed as shown in Figure-18.

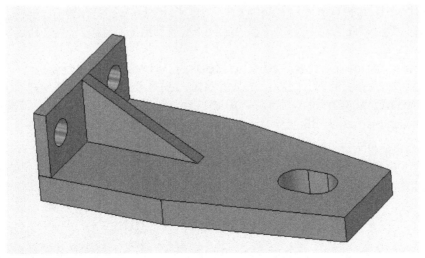

Figure-18. Final model

PRACTICAL 3

Create the model as shown in Figure-19. The dimensions of the model are given in Figure-20.

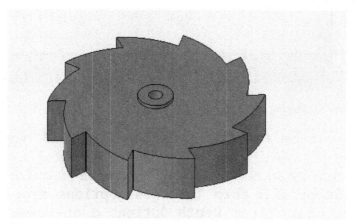

Figure-19. Model for practical 3

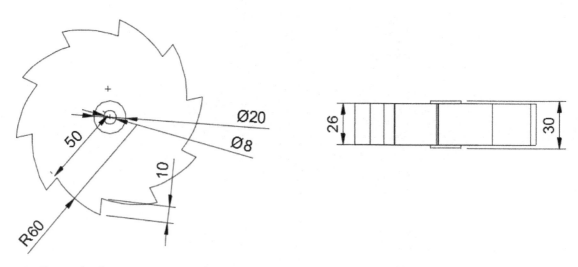

Figure-20. Practical 3 drawing views

Creating sketch for model

- Select **Center Diameter Circle** tool of **Circle** cascading menu from **Sketch** drop-down. You are asked to select the plane. Select **Top Plane** to create sketch.
- Create three circle of **8, 20** and **100** as diameter dimension of circles.
- Press **L** key to select **Line** tool and create a line of **10** as length.
- Select **3-Point Arc** tool from **Arc** cascading menu and create an arc of **60** as radius and **35** as length as shown in Figure-21.

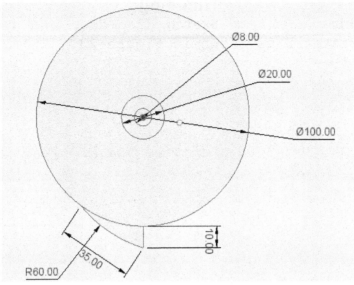

Figure-21. Creating arc

Creating Circular Pattern

- Select **Circular Pattern** from **Sketch** drop-down. The **CIRCULAR PATTERN** dialog box will be displayed.
- Objects selection is selected by default. Select the recently created line and arc for circular pattern.
- Click on **Select** button of **Center Point** section and select the major circle to define its center as center point for pattern; refer to Figure-22 .

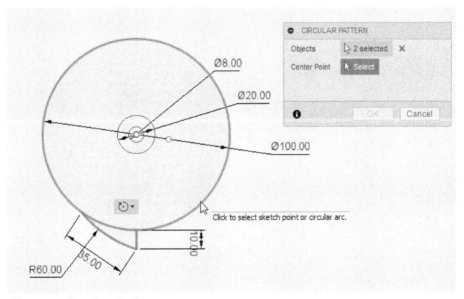

Figure-22. Creating circular pattern

- The updated **CIRCULAR PATTERN** dialog box will be displayed. Click in **Quantity** edit box and enter **10** as quantity of circular pattern.
- Click **OK** button from **CIRCULAR PATTERN** dialog box to complete the process.

Trimming

- Select the **Trim** tool from **Sketch** drop-down and remove the extra entities from sketch as shown in Figure-23.

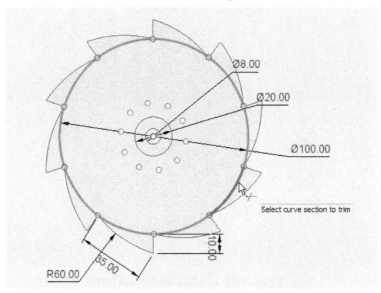

Figure-23. Trimming the extra entity

- After removing the extra entities from sketch, the sketch will be displayed as shown in Figure-24.

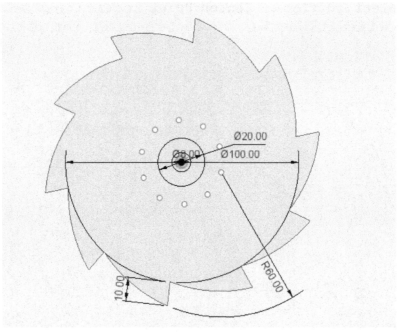

Figure-24. After removing extra entities

Creating Extrude

- Select **Extrude** tool from **Create** drop-down or press **E** key. The **EXTRUDE** dialog box will be displayed.
- Profile selection is active by default. Select the inner circle area to extrude.
- Select **Symmetric** button of **Direction** drop-down from **EXTRUDE** dialog box.

- Click in **Distance** edit box of **EXTRUDE** dialog box and enter **15** as height of extrusion; refer to Figure-25.

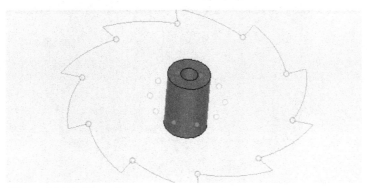

Figure-25. 1st Extrusion

- Click **OK** button of **EXTRUDE** dialog box to finish this extrusion.
- Right-click on the screen and select **Repeat Extrude** button from **Marking Menu**. The **EXTRUDE** dialog box will be displayed.
- Select the other area to extrude under **profile** section.
- Select **Symmetric** button of **Direction** drop-down from **EXTRUDE** dialog box.
- Click in **Distance** edit box of **EXTRUDE** dialog box and enter **13** as height of extrusion; refer to Figure-26.

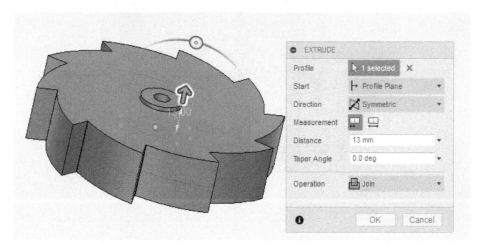

Figure-26. Second extrusion

- Click **OK** button from **EXTRUDE** dialog box to complete the process of extrusion. The model will be displayed as shown in Figure-27.

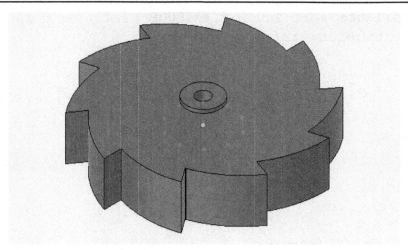

Figure-27. Final model for Practical 3

PRACTICE 1

Create the model as shown in Figure-28. The dimensions are given in Figure-29.

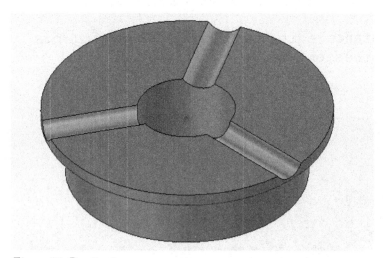

Figure-28. Practice 1

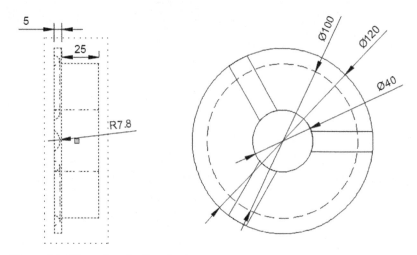

Figure-29. Dimensions for Practice 1

PRACTICE 2

Create the model as shown in Figure-30. The dimensions are given in Figure-31.

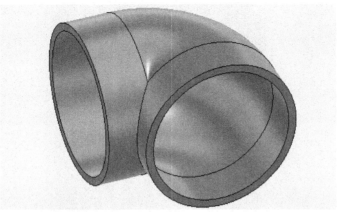

Figure-30. Practice 2

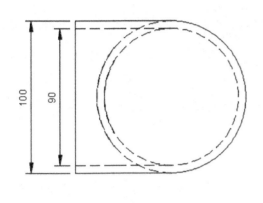

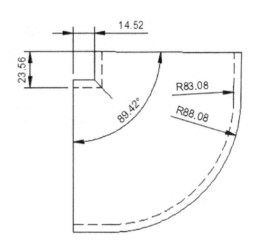

Figure-31. Dimensions for Practice 2

PRACTICE 3

Create the model as shown in Figure-32. The dimensions are given in Figure-33.

Figure-32. Practice 3

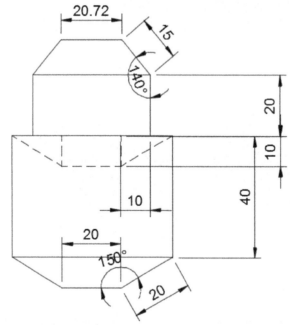

Figure-33. Dimension for practice 3

PRACTICE 4

Create the model as shown in Figure-34. The dimensions are given in Figure-35.

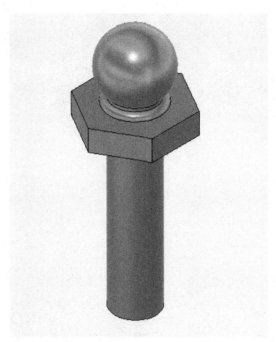

Figure-34. Practice 4

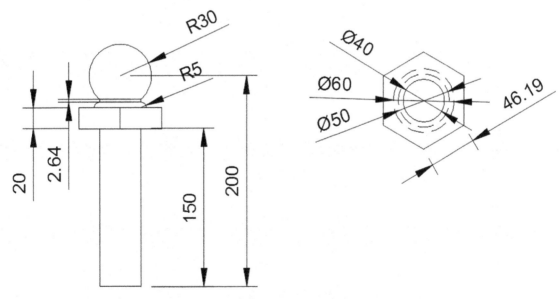

Figure-35. Dimensions for practice 4

PRACTICE 5

Create the model as shown in Figure-36. The dimensions are given in Figure-37.

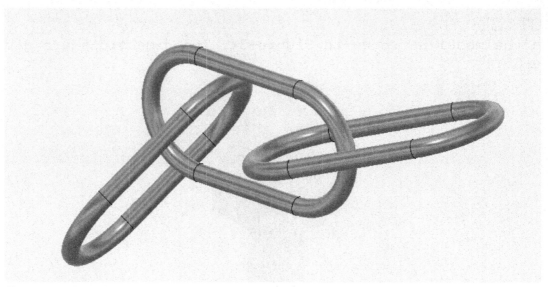

Figure-36. Practice 5

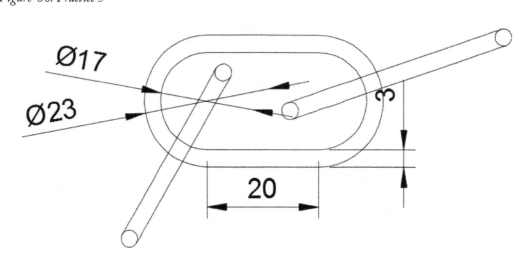

Figure-37. Dimensions for Practice 5

PRACTICE 6

Create the model as shown in Figure-38. The dimensions are given in Figure-39.

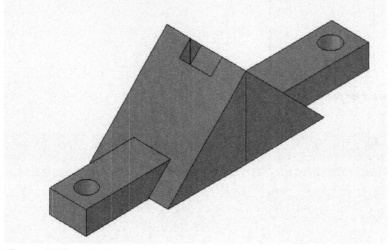

Figure-38. Practice 6

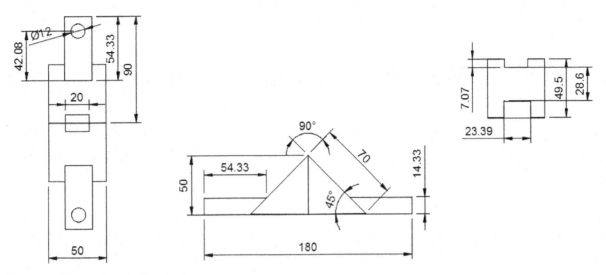

Figure-39. Dimensions for Practice 6

PRACTICE 7

Create the model as shown in Figure-40. The dimensions are given in Figure-41.

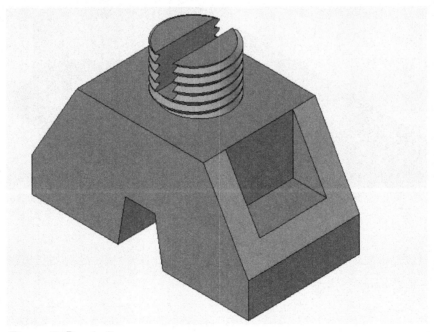

Figure-40. Practice 7

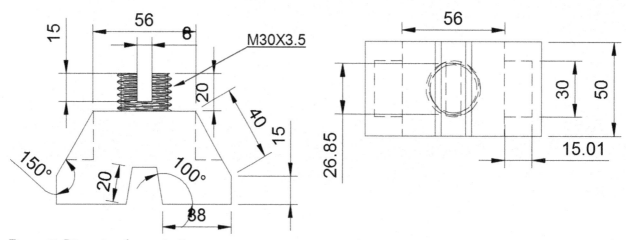

Figure-41. Dimensions for practice 7

PRACTICE 8

Create the model as shown in Figure-42. The dimensions are given in Figure-43.

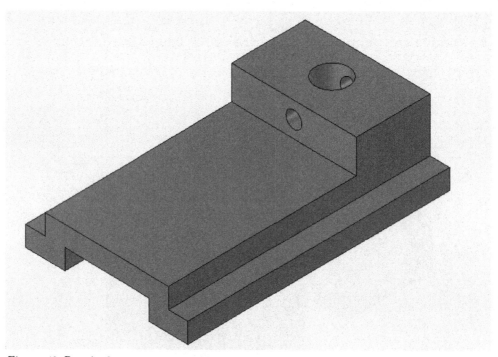

Figure-42. Practice 8

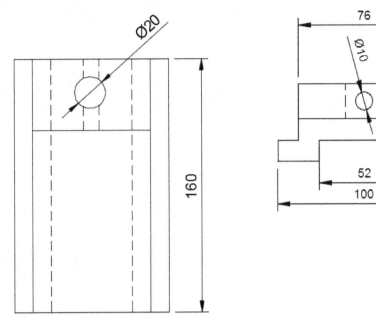

Figure-43. Dimensions for practice 8

PRACTICE 9

Create the model as shown in Figure-44. The dimensions are given in Figure-45.

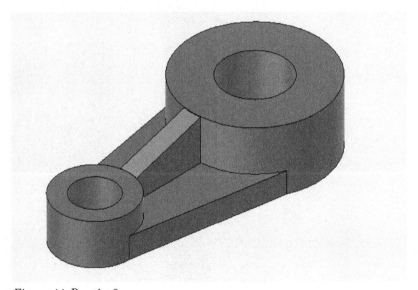

Figure-44. Practice 9

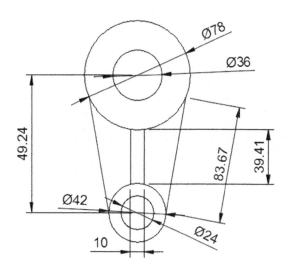

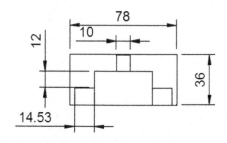

Figure-45. Dimensions for Practice 9

PRACTICE 10

Create the model as shown in Figure-46. The dimensions are given in Figure-47.

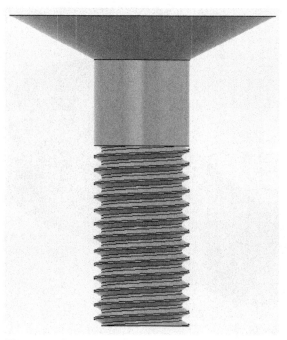

Figure-46. Practice 10

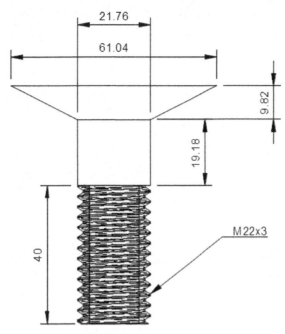

Figure-47. Dimensions for Practice 10

FOR STUDENT NOTES

Chapter 6

Solid Editing

Topics Covered

The major topics covered in this chapter are:

- *Press Pull Tool*
- *Fillet Tool*
- *Shell Tool*
- *Combine Tool*
- *Split Body Tool*
- *Align Tool*
- *Physical Material*
- *Appearance*

SOLID EDITING TOOLS

In the previous chapters, we have learned to create solid models and remove materials from them. In this chapter, we will learn to edit the models. Like the other tools, Autodesk Fusion 360 has packed all the editing tools into one drop-down. These tools are available in **Modify** drop-down from **Toolbar**; refer to Figure-1. The tools in this drop-down are discussed next.

Figure-1. Modify drop-down

PRESS PULL

The **Press Pull** tool is used to modify the body geometry. The options displayed to perform press pull operations depends on the selected geometry. The procedure to use this tool is discussed next.

- Select **Press Pull** tool of **Modify** drop-down from **Toolbar**; refer to Figure-2. The **PRESS PULL** dialog box will be displayed; refer to Figure-3.

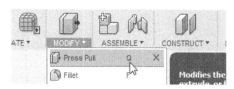

Figure-2. Press Pull tool

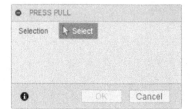

Figure-3. PRESS PULL dialog box

- Select a face, if you want to offset or move the face. The **OFFSET FACES** dialog box will be displayed. The **Offset Faces** dialog box will be discussed later.

- Select an edge, if you want to fillet the edge. The **FILLET** dialog box will be displayed. The **Fillet** tool will be discussed later in this unit.

- Select a sketch, if you want to extrude the profile. The **EXTRUDE** dialog box will be displayed. This tool is discussed earlier in this book.

Offset Faces

The **Offset Faces** tool is used to offset or move the face. The procedure to use this tool is discussed next.

- Select **Press Pull** tool from **Modify** drop-down. The **PRESS PULL** dialog box will be displayed. You can also select **Press Pull** tool by pressing **Q** key.

- Click on the face of the model to offset. The **OFFSET FACES** dialog will be displayed; refer to Figure-4.

- Select **Automatic** button from **Offset Type** drop-down, if you do not want to update the sketch on applying the offset tool. This action is not added to timeline.

- Select **Modify Existing Feature** button from **Offset Type** drop-down, if you want to update the sketch on applying Offset tool. This action is not added to timeline.

- Select **New Offset** button from **Offset Type** drop-down, if you want to create a new face. This action is added to timeline.

- Click in **Distance** edit box of **OFFSET FACES** dialog box or floating window from screen and enter the desired value as the distance for offset. You can also move drag handle to set the distance. The preview will be displayed; refer to Figure-5.

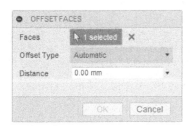

Figure-4. OFFSET FACE dialog box

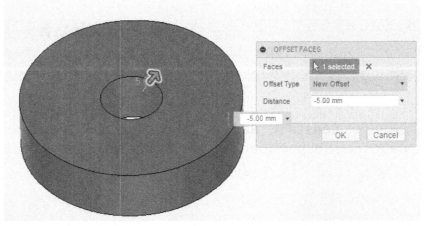

Figure-5. Preview on applying offset faces

- Click **OK** button from **OFFSET FACES** dialog box to complete the process.

FILLET

The **Fillet** tool is used to apply round at the edges. This tool works in the same way as the **Sketch Fillet** do. It is recommended that, apply the fillets after creating all the features required in the model because fillets can increase the processing time during modifications. The procedure to use this tool is given next.

- Select **Fillet** tool of **Modify** drop-down from **Toolbar**; refer to Figure-6. The **FILLET** dialog box will be displayed; refer to Figure-7.

Figure-6. *Fillet tool*

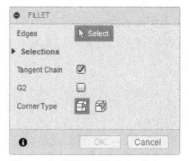

Figure-7. *FILLET dialog box*

- **Edges** selection is active by default. Click on the edge from the model to apply fillet. The updated **FILLET** dialog box will be displayed; refer to Figure-8.

Figure-8. *The updated FILLET dialog box*

- The options of the updated **FILLET** dialog box are discussed next.

Type drop-down

The options of **Type** drop-down are used to specify the type of round to be created. These options are discussed next.

Constant Radius

- Select **Constant Radius** button from **Type** drop-down in **Selections** section to constantly apply the fillet of fix size on the selected edge; refer to Figure-9.

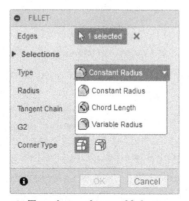

Figure-9. *Type drop - down of Selections section*

- Click in the **Radius** edit box of **Selections** section from **FILLET** dialog box and enter **10** as the radius of fillet. You can also move the drag-handle from canvas screen to set the radius of fillet.
- Select the **Tangent Chain** check box to include tangentially connected edges in selection.
- Select the **G2** check box, if you want to apply the G2 continuity to the selected fillet.
- Click **OK** button from **FILLET** dialog box to apply the fillet.

Chord Length

- Select **Chord Length** button of **Type** drop-down from **FILLET** dialog box, if you want to specify the chord length to control the size for the fillet.
- Click in the **Chord Length** edit box and enter **15** as the length for fillet. You can also set this length by moving the drag handle. The preview will be displayed; refer to Figure-10.

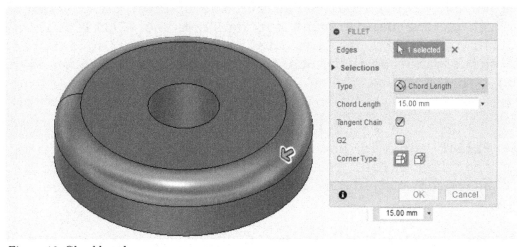

Figure-10. Chord length

- The other parameters of **Cord Length** button were discussed in last topic.

Variable Radius

- Select **Variable Radius** of **Type** drop-down from **FILLET** dialog box, if you want to specify radii at selected points along the edge. The updated **FILLET** dialog box will be displayed.
- Click on the edge to create points for applying fillet and specify radius for each point in the dialog box; refer to Figure-11.

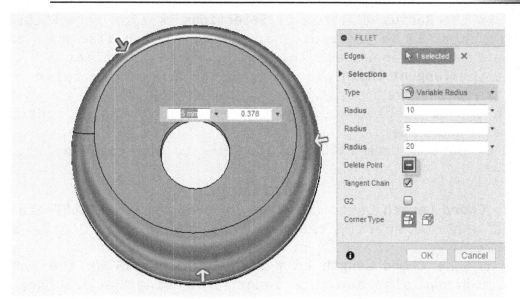

Figure-11. Variable Radius

- To delete the created point, select the specific point from the edge and click on the **Delete Point** button from the **FILLET** dialog box.
- The other parameters of this dialog were discussed earlier.
- After specifying the parameters, click on the **OK** button from **FILLET** dialog box to create the fillet.
-

RULE FILLET

The **Rule Fillet** tool is used to add fillets or rounds to a design based on specified rules. The edges are determined by specified rules rather than selection in the canvas. The procedure to use this tool is discussed next.

- Select **Rule Fillet** tool of **Modify** drop-down from **Toolbar**; refer to Figure-12. The **RULE FILLET** dialog box is displayed; refer to Figure-13.

Figure-12. Rule Fillet tool

Figure-13. RULE FILLET DIALOG BOX

- **Input Features/Faces** selection is active by default. Select the required feature/face from model to apply rule fillet.
- Click in the **Radius** edit box and enter the desired value as the radius for fillet. You can also move the drag handle to set the radius.
- Select **Against Features** options from **Scope Options** drop-down if you want to add the fillets at the intersection of the input faces/ features and the against features.

- Select **All Edges** option from **Scope Options** drop-down, if you want to apply the fillet at all edges.
- Click **OK** button from **RULE FILLET** dialog box to finish this process; refer to Figure-14.

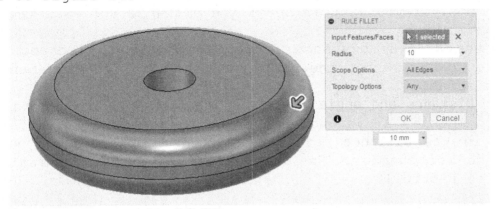

Figure-14. Preview on applying Rule Fillet

CHAMFER

The **Chamfer** tool is used to bevel the sharp edges of the model. The procedure to create chamfer by using this tool is given next.

- Click on the **Chamfer** tool of **Modify** drop-down from **Toolbar**; refer to Figure-15. The **CHAMFER** dialog box will be displayed; refer to Figure-16.

Figure-15. Chamfer tool

Figure-16. CHAMFER dialog box

- **Edges** selection is active by default. Click on the edge of model to apply chamfer. The updated **CHAMFER** dialog box will be displayed; refer to Figure-17.

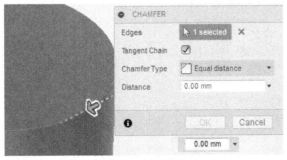

Figure-17. Updated CHAMFER dialog box

- Select the **Tangent Chain** check box to include tangentially connected edges in selection.

- Select **Equal Distance** option of **Chamfer Type** drop-down from **CHAMFER** dialog box, if you want to specify a equal distance for both sides of the chamfer.
- Select **Two Distances** option of **Chamfer Type** drop-down from **CHAMFER** dialog box, if you want to specify a distance for each face of the chamfer.
- Enter the specific distance for each face in the respective edit box. You can also set the distance by moving the drag handle from canvas scree.
- Select **Distance and angle** option of **Chamfer Type** drop-down, if you want to specify a distance and angle to create the chamfer.
- Click in the **Distance** edit box and enter the desired value for distance for chamfer. The preview of chamfer will be displayed; refer to Figure-18.

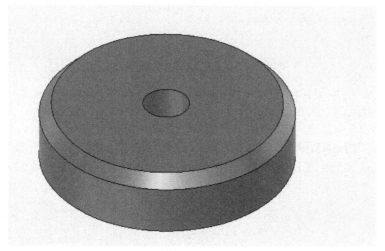

Figure-18. Preview of chamfer

- Click **OK** button of **CHAMFER** dialog box to finish the process.

SHELL

The **Shell** tool is used to make a solid part hollow and remove one or more faces. The procedure to use this tool is given next.

- Select **Shell** tool of **Modify** drop-down from **Toolbar**; refer to Figure-18. The **SHELL** dialog box will be displayed; refer to Figure-19.

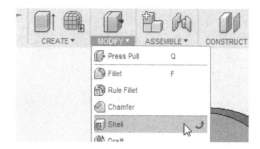

Figure-19. SHELL dialog box

- **Faces/Body** selection is active by default. Click on the Face/Body of the solid to apply shell. The updated **SHELL** dialog box will be displayed; refer to Figure-20.

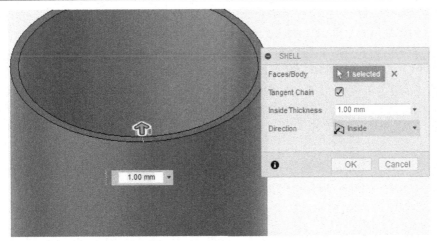

Figure-20. The updated SHELL dialog box

- Select the **Tangent Chain** check box to include tangentially connected edges in selection.
- Click in the **Inside Thickness** edit box and enter the desired value of shell thickness. The preview of shell will be displayed; refer to Figure-21.

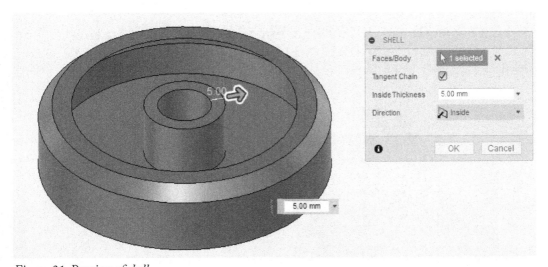

Figure-21. Preview of shell

- Select **Inside** option from **Direction** drop-down, if you want to offset the faces toward the interior of the part.
- Select **Outside** option from **Direction** drop-down, if you want to offset the faces towards the exterior of the part.
- Select **Both** option from **Direction** drop-down, if you want to offset the faces towards the interior and the exterior of the part by equal amount. Click in the specific thickness edit box and enter the specify thickness for shell.
- Click **OK** button from **SHELL** dialog box to complete the process.

DRAFT

The **Draft** tool is used to apply taper to the faces of a solid model. This tool is mainly useful when you are designing components for molding or casting. Draft tool applies taper on the face and this taper allows easy and safe ejection of parts from dies. The procedure to use this tool is discussed next.

- Select **Draft** tool of **Modify** drop-down from **Toolbar**; refer to Figure-22. The **DRAFT** dialog box will be displayed; refer to Figure-23.

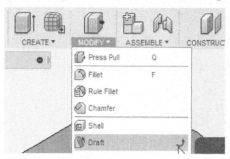

Figure-22. Draft tool

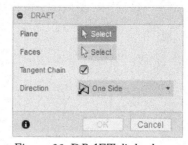

Figure-23. DRAFT dialog box

- **Plane** selection is active by default. Click on the face or plane of the model which you want to use as fixes reference.
- Click on **Select** button of **Faces** and select the faces to draft.
- Click on the **Flip Direction** button from **DRAFT** dialog box to flip the direction of draft.
- Click in the **Angle** edit box and enter the desired value of angle for draft. The preview of draft will be displayed; refer to Figure-24. You can also set the angle of draft by moving the drag handle.

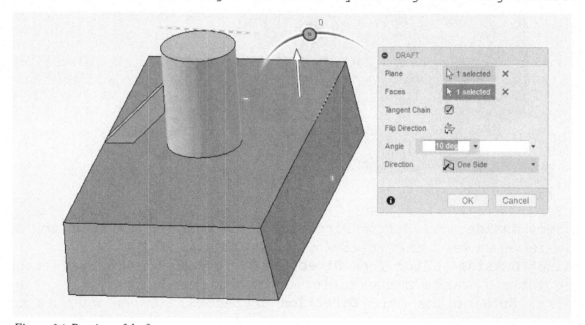

Figure-24. Preview of draft

- The options of **Direction** drop-down were discussed earlier in this book.
- Click **OK** button from **DRAFT** dialog box to finish the process.

SCALE

The **Scale** tool is used to enlarges or reduces selected bodies, sketches, or components based on specified scale factor. The procedure to use this tool is discussed next.

- Select **Scale** tool of **Modify** drop-down from **Toolbar;** refer to Figure-25. The **SCALE** dialog box will be displayed; refer to Figure-26.

Figure-25. Scale tool

Figure-26. SCALE dialog box

- **Entities** selection is active by default. Click on the entity to select.
- On selecting the entity for scale, point is automatically selected in **Point** section. If you want to use a different point as scale reference then click on the **Select** button for **Point** option in the dialog box and select the desired point.
- Select **Uniform** option from **Scale Type** drop-down, if you want to scale the selected entity uniformly.
- Select **Non-Uniform** option from **Scale Type** drop-down, if you want to scale the entity non-uniformly.
- Click in the **Scale Factor** edit box and enter the desired value for scale. The preview will be displayed; refer to Figure-27.
- Click on **OK** button of **SCALE** dialog box to complete the process.

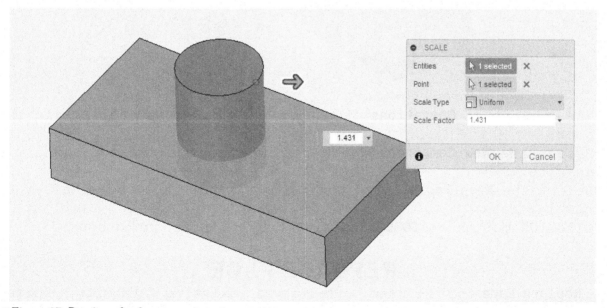

Figure-27. Preview of scale

COMBINE

The **Combine** tool is used to cut, join, and intersect the selected bodies. The procedure to use this tool is discussed next.

- Select **Combine** tool of **Modify** drop-down from **Toolbar**; refer to Figure-28. The **COMBINE** dialog box will be displayed; refer to Figure-29.

Figure-28. Combine tool

Figure-29. Combine dialog box

- **Target Body** is active by default. Click on the component from model as a target body.
- Click on other component to select in **Tool Bodies** section. The preview of selection will be displayed; refer to Figure-30.

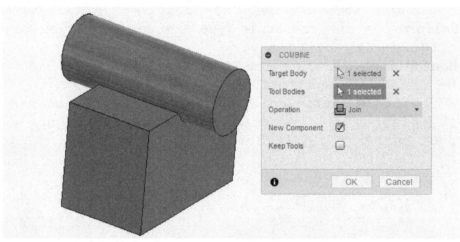

Figure-30. Preview of Combine tool

- The options of **Operations** drop-down were discussed earlier in this book.
- Select the **New Component** check box to create a new component with results.
- Select the **Keep Tools** check box to retain the tool bodies after the combine result.
- Click **OK** button of **COMBINE** dialog box to finish the process.

REPLACE FACE

The **Replace Face** tool is used to replace an existing face with a surface or work plane. The procedure to use this tool is discussed next.

- Select **Replace Face** tool of **Modify** drop-down from **Toolbar**; refer to Figure-31. The **REPLACE Face** tool dialog box will be displayed; refer to Figure-32.

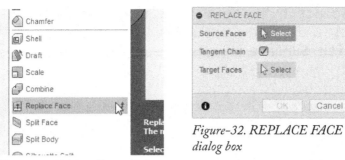

Figure-31. Replace Face tool

Figure-32. REPLACE FACE
dialog box

- **Source Faces** option selection is active by default. Click on the component face to select the face.
- Select the **Tangent Chain** check box to include tangentially connected edges in selection.
- Click **Select** button of **Target Faces** options and click on the face or plane to apply this tool. The preview will be displayed; refer to Figure-33.

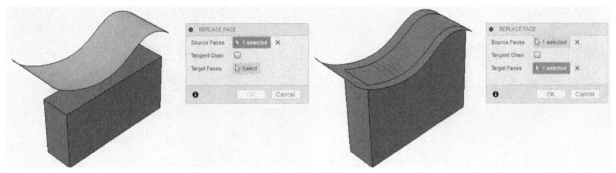

Figure-33. Preview on applying Replace Face tool

- Click **OK** button of **REPLACE FACE** dialog box to finish the process.

SPLIT FACE

The **Split Face** tool is used to divide faces on a surface or solid to add draft, delete an area, or create features. The procedure to use this tool is discussed next.

- Select **Split Face** tool of **Modify** drop-down from **Toolbar**; refer to Figure-34. The **SPLIT FACE** dialog box will be displayed; refer to Figure-35.

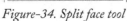

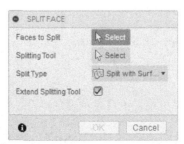

Figure-34. Split face tool   Figure-35. SPLIT FACE dialog box

- **Faces To Split** option is active by default. Click on the model face to split. You can also select multiple faces by holding the **CTRL** key.

- Click on the **Select** button of **Splitting Tool** option and select the face which will split the other faces.
- Select **Split With Surface** option from **Split Type** drop-down, if you want to project the splitting tool face which you had select upon the splitting face.
- Select **Along Vector** option from **Split Type** drop-down, if you want to define the projection direction of splitting face by selecting an face, edge, or axis.
- Select **Closest Point** option from **Split Type** drop-down, if you want to define the direction of projection as the closest distance between the splitting body and target faces.
- Select the **Extend Splitting Tool** check box to extends the tool to completely intersect the target faces.
- After specifying the parameters, click on **OK** button from **SPLIT FACE** dialog box to finish the process; refer to Figure-36.

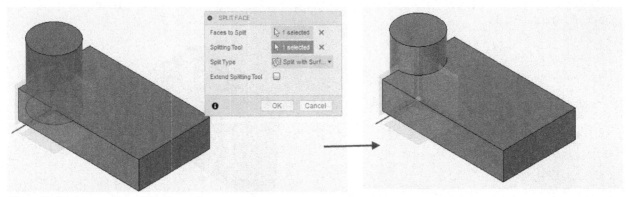

Figure-36. Preview on applying Split Faces tool

SPLIT BODY

The **Split Body** tool works similarly to **Split Face** tool but in this tool component or parts are selected in place of faces. The procedure to use this tool is discussed next.

- Select **Split Body** tool of **Modify** drop-down from **Toolbar**; refer to Figure-37. The **SPLIT BODY** dialog box will be displayed; refer to Figure-38.

Figure-37. Split Body tool

Figure-38. SPLIT BODY dialog box

- **Body To Split** selection is active by default. Click on the component to be splitted.
- Click on **Select** button of **Splitting Tool** option and select the body to split.
- Select the **Extend Splitting Tool** check box to extends the tool to completely intersect the target faces.

• After specifying the various parameters, click **OK** button of **SPLIT BODY** dialog box to complete the process; refer to Figure-39.

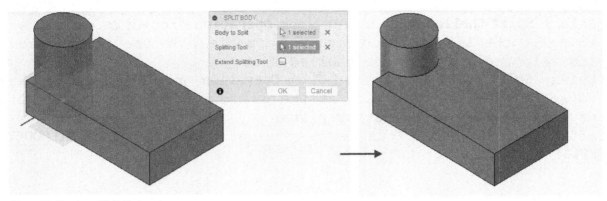

Figure-39. Preview of Split Body tool

SILHOUETTE SPLIT

The **Silhouette Split** tool is used to split selected body at the silhouette outline visible from the view direction. The procedure to use this tool is discussed next.

• Select **Silhouette Split** tool of **Modify** drop-down from **Toolbar**; refer to Figure-40. The **SILHOUETTE SPLIT** dialog box will be displayed; refer to Figure-41.

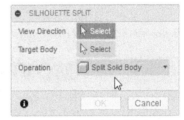

Figure-40. Silhouette Split tool

Figure-41. SILHOUETTE SPLIT dialog box

• **View Direction** selection option is active by default. Click on the plane or face to select the direction to split.
• Click on **Select** button of **Target Body** and select the body which you want to split. The preview of split will be displayed; refer to Figure-42.

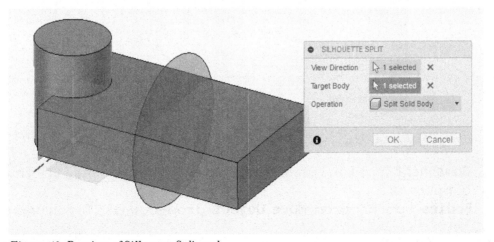

Figure-42. Preview of Silhouette Split tool

- Select **Split Faces Only** option from **Operation** drop-down, if you want to split the faces of the selected body but keeps it as a single body.
- Select **Split Shelled Body** option from **Operation** drop-down, if you want to split the selected body into two bodies. To use this option , the selected body must be shelled first.
- Select **Split Solid Body** option from **Operation** drop-down, if you want to splits the selected body at the parting line. The parting line of the selected body must be planar.
- After specify the various parameters, click **OK** button of **SILHOUETTE SPLIT** dialog box to finish the process; refer to Figure-43.

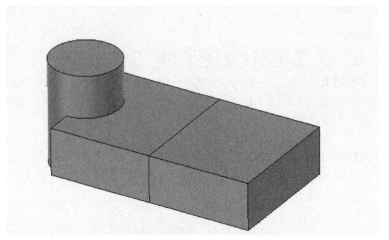

Figure-43. Solid model after applying Silhouette Split tool

MOVE/COPY

The **Move/Copy** tool is used to move a face, body, sketch curve, components, or sketch geometry. Using this tool you can also move the whole geometry. The procedure to use this tool is discussed next.

- Select **Move/Copy** tool of **Modify** drop-down from **Toolbar**; refer to Figure-44. The **MOVE/COPY** dialog box will displayed; refer to Figure-45.

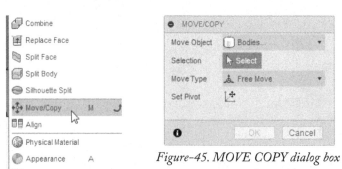

Figure-44. Move Copy tool

Figure-45. MOVE COPY dialog box

- Select **Component** option from **Move Object** drop-down, if you want to move a component.
- Select **Bodies** option from **Move Object** drop-down, if you want to move a body.

- Select **Faces** option from **Move Object** drop-down, if you want to move a face.
- Select **Sketch** Objects from **Move Object** drop-down, if you want to move the selected sketch.
- In our case, we are selecting the **Faces** option from **Move Object** drop-down.
- **Selection** section is active by default. Click on the face to be moved. The updated **MOVE/COPY** dialog box will be displayed; refer to Figure-46.

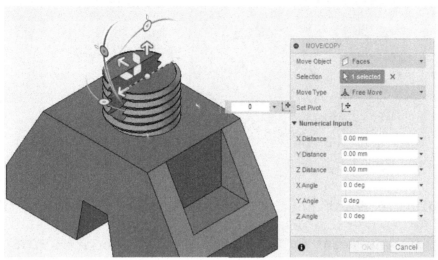

Figure-46. The Updated MOVE/COPY dialog box

- Select **Free Move** option from **Move Type** drop-down, if you want to move selected face freely
- On selecting the **Free Move** option, a Manipulator will be displayed on the selected face. Use the Manipulator to move the face or enter the value in respective field of **MOVE/COPY** dialog box.
- Select **Translate** option from **Move Type** drop-down, if you want to move the face along X, Y and Z directions.
- On selecting the **Translate** option, three arrows will be displayed on the face. Drag the desired arrow to move the face in respective direction. You can also enter the specific value of direction in the respective edit box from **MOVE/COPY** dialog box; refer to Figure-47.

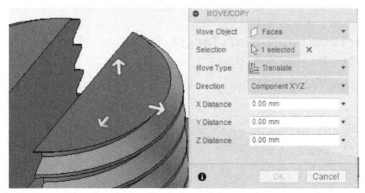

Figure-47. MOVE/COPY dialog box for translate option

- Select **Rotate** option from **Move Type** drop-down, if you want to rotate the face.

- On selecting the **Rotate** option, the **Axis** option will be displayed in the **MOVE/COPY** dialog box.
- Click on the **Select** button from **Axis** section of **MOVE/COPY** dialog box and select the axis for rotation of face. The drag handle will be displayed on the selected face.
- Move the drag handle as required or specify the value in the **Angle** edit box of **MOVE/COPY** dialog box; refer to Figure-48.

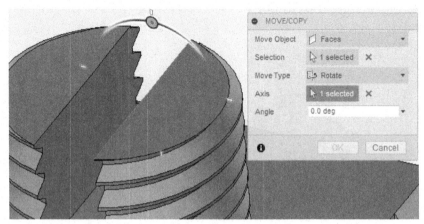

Figure-48. MOVE/COPY dialog box for Rotate option

- Select **Point to Point** option from **Move Type** drop-down, if you want to move the face from one point to another point.
- On selecting the **Point to Point** option, the updated **MOVE/COPY** dialog box will de displayed; refer to Figure-49.

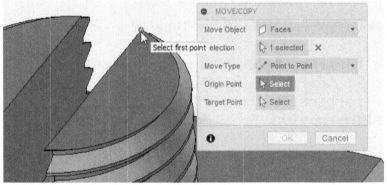

Figure-49. MOVE/COPY dialog box for Point to Point option

- Click on the **Select** button of **Origin Point** option from **MOVE/COPY** dialog box and select the first point of face from where you want to move the face.
- Click on the **Select** button of **Target Point** option from **MOVE/COPY** dialog box and select the finish point from model; refer to Figure-50.

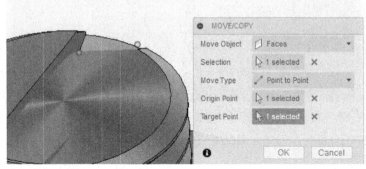

Figure-50. Selection of points

- After selecting the points, the preview will be displayed. If the preview is as required then, click on **OK** button from **MOVE/COPY** dialog box.
- Select **Point to Position** option from **Move Type** drop-down, if you want to move the face by selecting a point and a position from model.
- Select a point of the face. The updated **MOVE/COPY** dialog box will be displayed; refer to Figure-51.

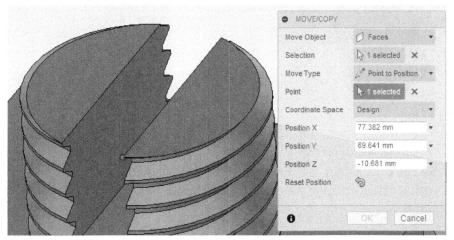

Figure-51. MOVE/COPY dialog box for Point to Position option

- Click on the **Position X** edit box and specify the value. Similarly enter the value in **Position Y** and **Position Z** edit boxes as required.
- After specifying the parameters click on the **OK** button from **MOVE/ COPY** to complete the process.

ALIGN

The **Align** tool is used to align the selected object to a location on another object. The procedure to use this tool is discussed next.

- Click on the **Align** tool of **Modify** drop-down from **Toolbar**; refer to Figure-52. The **ALIGN** dialog box will be displayed; refer to Figure-53.

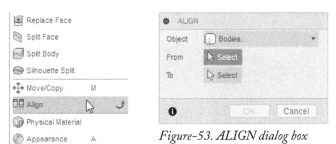

Figure-52. Align tool

Figure-53. ALIGN dialog box

- Select **Bodies** option of **Object** drop-down from **ALIGN** dialog box if you want to select bodies to align.
- Select **Components** option of **Object** drop-down from **ALIGN** dialog box if you want to select components to align.
- In our case, we are selecting **Bodies** option from **Object** drop-down.
- Click on **Select** button of **From** section of **ALIGN** dialog box and select the body; refer to Figure-54.

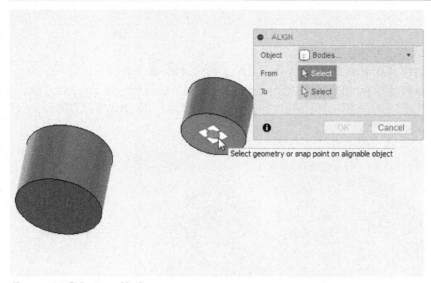

Figure-54. Selection of body

- Click on **Select** button of **From** section of **ALIGN** dialog box and select the face of other body as displayed in Figure-55.

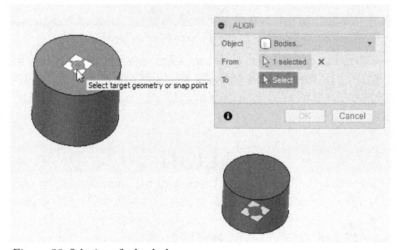

Figure-55. Sslection of other body

- On selecting the second body, the two selected body will align to each other at selected location; refer to Figure-56.

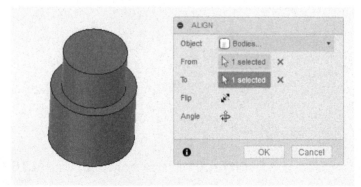

Figure-56. Preview of align

- Click on the **Flip** button, if you want to flip the aligned body.
- Click on **Angle** button if you want to rotate the aligned body.
- After specifying the parameters, click on the **OK** button from **ALIGN** dialog box to finish the process.

PHYSICAL MATERIAL

The **Physical Material** tool is used to apply physical and visual material to the component or body. The procedure to use this tool is discussed next.

- Click on the **Physical Material** tool of **Modify** drop-down from **Toolbar**; refer to Figure-58. The **PHYSICAL MATERIAL** dialog box will be displayed; refer to Figure-58.

Figure-57. Physical Material tool

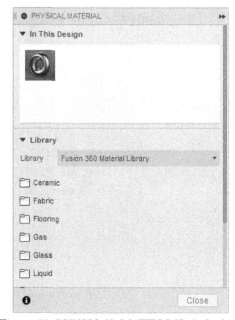

Figure-58. PHYSICAL MATERIAL dialog box

- Select the required option from **Library** drop-down.
- Select the required physical material from **Library** section of **PHYSICAL MATERIAL** dialog box.
- Now, drag the specific material from the Library section and drop it to the body or component. The physical material of body/component will be changed; refer to Figure-59.

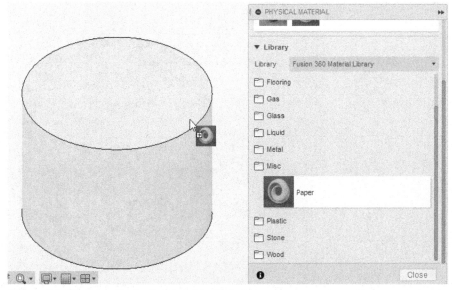

Figure-59. Applying physical material to the body

• After applying material, click on the **Close** button of **PHYSICAL MATERIAL** dialog box to apply changes.

APPEARANCE

The **Appearance** tool is used to change the appearance of the body. This tool is generally used to change the color of the body. The procedure to use this tool is discussed next.

• Click on the **Appearance** tool of **Modify** drop-down from **Toolbar**; refer to Figure-60. The **APPEARANCE** dialog box will be displayed; refer to Figure-61.

Figure-60. Appearance tool

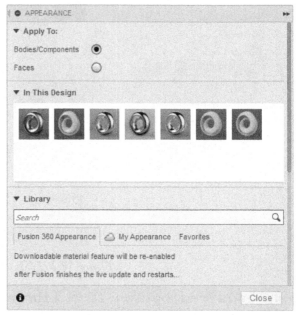

Figure-61. APPEARANCE dialog box

• Select the required radio button of **Apply To** section from **APPEARANCE** dialog box.
• Similarly, drag an appearance and drop it to the body or face; refer to Figure-62.

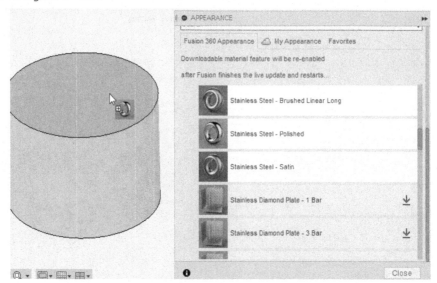

Figure-62. Applying apearance on face

• After applying the appearance, click on the **Close** button from **APPEARANCE** dialog box to finish the process.

MANAGE MATERIALS

The **Manage Materials** tool is used to define the material properties shown in **Physical Material** and **Appearance** tool. The procedure to use this tool is discussed next.

• Click on the **Manage Materials** tool of **Modify** drop-down from **Toolbar**; refer to Figure-63. The **Material Browser** dialog box will be displayed; refer to Figure-64.

Figure-63. Manage Materials tool

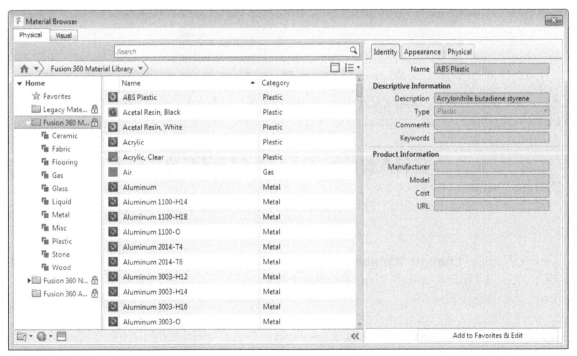

Figure-64. Material Browser dialog box

• Click on any material or appearance from list section to check or modify the property of selected material or appearance.
• Change the properties as required using the options displayed in the right area of **Material Browser** dialog box.

DELETE

The **Delete** tool is used to delete the selected object or component. The procedure to use this tool is discussed next.

- Click on the **Delete** tool of **Modify** drop-down from **Toolbar**; refer to Figure-65. The **DELETE** dialog box will be displayed; refer to Figure-66.

Figure-65. Delete Tool

Figure-66. DELETE dialog box

- The **Entities** selection is active by default. Select the entity to be delete from the existing model; refer to Figure-67.

Figure-67. Selection of enntity to delete

- After selecting the required entity, click on the **OK** button from **DELETE** dialog box to delete the selected entity.

CHANGE PARAMETERS

The **Change Parameters** tool is used to change the parameters and apply equations to the model. The procedure to use this tool is discussed next.

- Click on the **Change Parameters** tool of **Modify** drop-down from **Toolbar**; refer to Figure-69. The **Parameters** dialog box will be displayed; refer to Figure-69.

Figure-68. Change Parameters tool

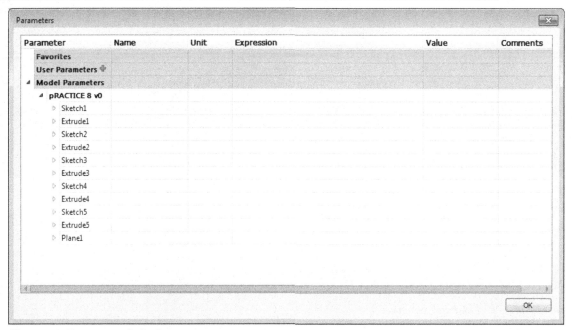

Figure-69. Parameters dialog box

- If you want to create some parameters which are used to be used later in equation then click on the **User Parameters** button from **Parameters** dialog box. The **Add User Parameters** dialog box will be displayed; refer to Figure-70.

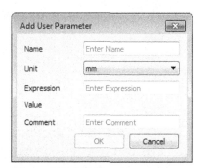

Figure-70. Add User Parameters
dialog box

- Click on the **Name** edit box and specify the name of parameter.
- Click on the **Unit** drop-down and select the required unit of the parameter.
- Click on the **Expression** edit box and enter the desired expression like d1/4. The **Value** will be displayed.
- Click in the **Comment** edit box and enter the required comment.
- After specifying the parameters, click on the **OK** button from **Add User Parameters** dialog box. The user defined parameter will be added in **Parameters** dialog box; refer to Figure-71.

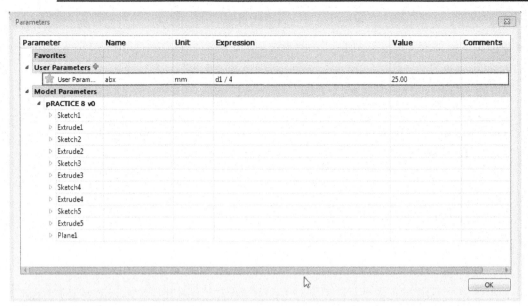

Figure-71. Added user parameter

- Click on the **User Parameter** node from **Parameters** dialog box, all the model feature will be displayed.
- Click on the specific model feature node of File name node, the parameters applied in that feature will be displayed; refer to Figure-72.

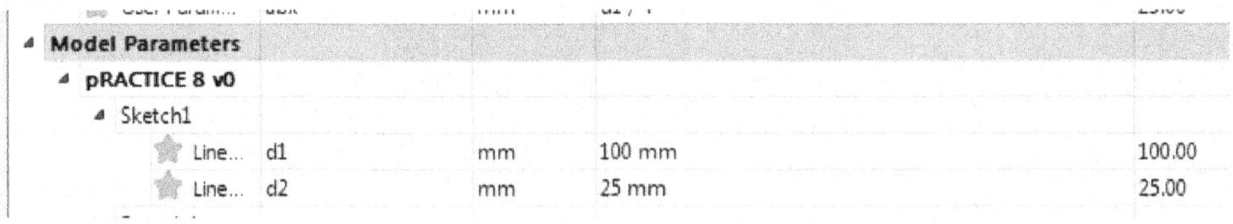

Figure-72. Parameters of sketch

- If you want to apply equation in parameters then click on the dimension parameter and enter the desired equation like d1/2; refer to Figure-73. Note that here d1 is dimension number applied automatically to the sketch dimension.

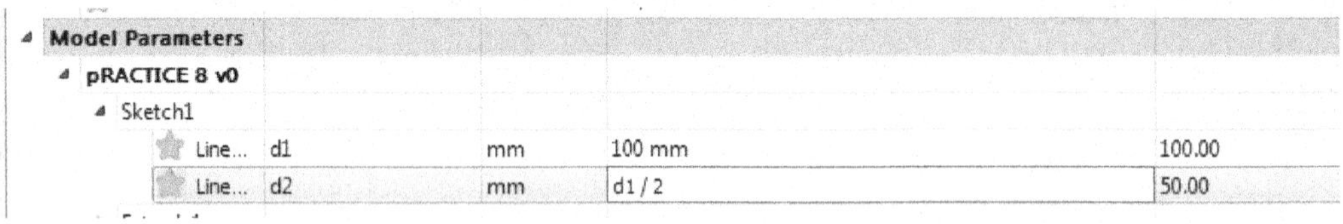

Figure-73. entering equation

- After specifying the parameters, click on the **OK** button from **Parameters** dialog box.
- Now, when you change the dimension of d1, the dimension of d2 will automatically change; refer to Figure-74.

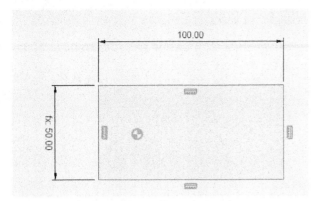 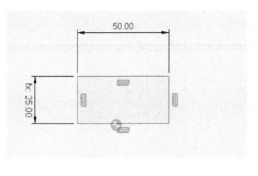

Figure-74. Applying Change Parameter tool

FOR STUDENTS NOTES

FOR STUDENTS NOTES

Chapter 7

Assembly Design

Topics Covered

The major topics covered in this chapter are:

- *New Component Tool*
- *Joint Tool*
- *Joint Origin*
- *Rigid Group*
- *Motion Links*
- *Motion Study*

INTRODUCTION

Most of the things you find around you in real-world are assembly of various components; the computer you are working on is an assembly, the automotive you may be driving or sitting in for travelling is an assembly, there are lots of examples. Till this point, we have learned to create solid parts. But most of the time, we need to assemble the parts to get some use of them. In this chapter, we will work on the Assembly environment of Autodesk Fusion 360 in which we will be assembling two or more components using some assembly constraints.

NEW COMPONENT

The **New Component** tool is used to create an empty component or creating the components from the selected body. Note that we need to convert all parts to components before performing their assembly. The procedure to use this tool us discussed next.

- Click on the **New Component** tool of **Assemble** drop-down from **Toolbar**; refer to Figure-1. The **NEW COMPONENT** dialog box will be displayed; refer to Figure-2.

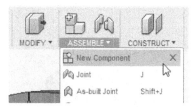

Figure-1. New Component tool

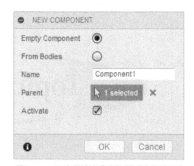

Figure-2. NEW COMPONENT dialog box

- Select the **Empty Component** radio button from **NEW COMPONENT** dialog box, if you want to create an empty component in the assembly structure.
- Select the **From Bodies** Radio button from **NEW COMPONENT** dialog box, if you want to create a new component from existing body.

Empty Component

In this topic we will discuss the procedure of creating empty component.

- Select the **Empty Component** radio button from the **NEW COMPONENT** dialog box if not selected by default.
- Click in the **Name** edit box of **NEW COMPONENT** dialog box and enter the name of component.
- Click on **Select** button of **Parent** option and select the parent body to create an empty component.
- Select the **Activate** check box from **New Component** dialog box to activate the new component. Note that all the modeling operations performed after activating the component will become child feature of that component.

- After specifying the parameters for empty component, click on the **OK** button from **NEW COMPONENT** dialog box. The component will be created; refer to Figure-3.

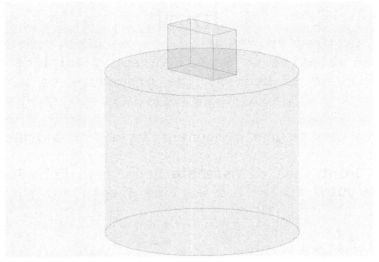

Figure-3. Empty component

From Bodies

The **From Bodies** radio button is used to create a component from the selected body. The procedure to use this tool is discussed next.

- Select the **From Bodies** radio button from **NEW COMPONENT** dialog box. The updated dialog box will be displayed; refer to Figure-4

Figure-4. Updated NEW COM-PONENT dialog box

- The **Select** button of **Bodies** option is active by default. Click on the component from model to create a new component; refer to Figure-5. You can also select multiple bodies by holding the **CTRL** key.

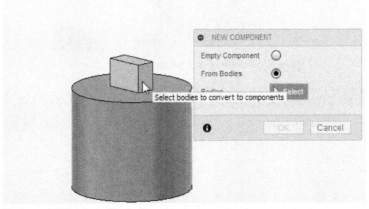

Figure-5. Selection of body for new component

- After selection the body, click on the **OK** button from **NEW COMPONENT** dialog box. The new component will be created and action will be displayed in **Timeline**.

JOINT

The **Joint** tool is used to define joints between components to align their occurrence relative to other. To apply **Joint** tool, you need to create the component of bodies. The procedure to use this tool is discussed next.

The procedure to create new component is discussed next.

- Click on the **Joint** tool of **Assemble** drop-down from **Toolbar**; refer to Figure-6. The **JOINT** dialog box will be displayed; refer to Figure-7.

Figure-6. Joint tool

Figure-7. JOINT dialog box

- **Select** button of **Component1** section is active by default. Click on the first component to assemble; refer to Figure-8. You will be asked to select a location on second component.

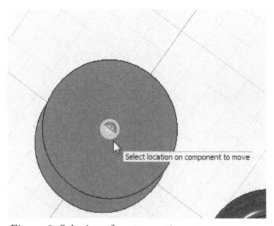

Figure-8. Selection of component1

- Click on the second component to select; refer to Figure-9.

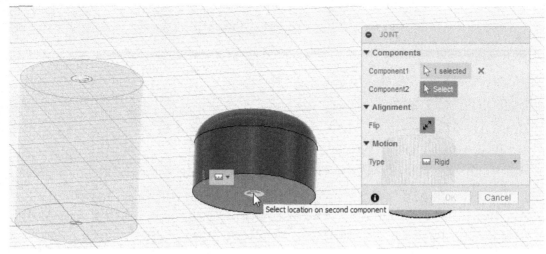

Figure-9. Selection of second component

- On selecting the second component, the selected two component will be joined at each other at the locations selected on them and the updated **JOINT** dialog box will be displayed; refer to Figure-10.

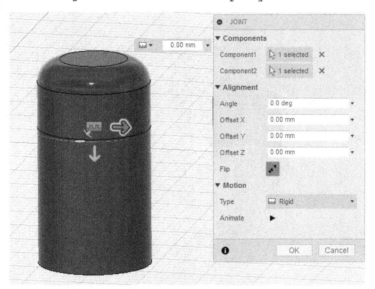

Figure-10. Updated JOINT dialog box

- Click in the **Angle** edit box of **JOINT** dialog box and specify the angle of rotation of component.
- If you want to specify the Offset distance of component then click in the **Offset X , Offset Y**, and **Offset Z** edit boxes as required and enter the value. You can also set the offset distance by moving the drag handle.
- Click on the **Flip** button from **JOINT** dialog box, if you want to flip the direction of second component.

Motion

There are seven motions in **JOINT** dialog box which we can apply to the component as required; refer to Figure-11. These motions are described next.

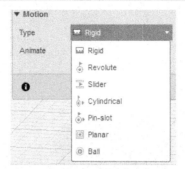

Figure-11. Motions for component

- Click on the **Rigid** option of **Type** drop-down from **JOINT** dialog box, if you want the joint of two component to be rigid to each other and there should be no motion between the two joined component.

- Click on the **Revolute** option of **Type** drop-down from **JOINT** dialog box, if you want to revolve the component about other at the joint location. On selecting the **Revolute** option a revolve drag handle will displayed. You can also revolve the component by moving the drag handle; refer to Figure-12.

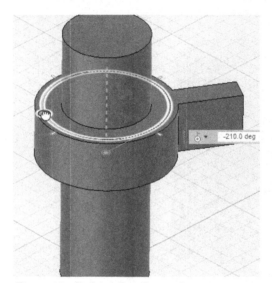

Figure-12. Applying Revolute option

- Click on the **Slider** option of **Type** drop-down from **JOINT** dialog box if you want to assemble the component in such a way that one component can slide over the other; refer to Figure-13.

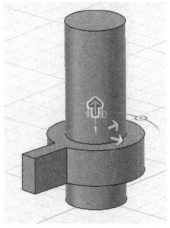
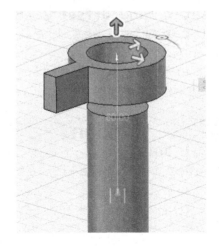
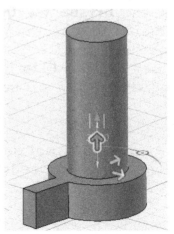

Figure-13. Slider option

- Click on the **Cylindrical** option of **Type** drop-down from **JOINT** dialog box, if you want to assemble the component in such a way that the component is free to move along the selected axis and free to rotate about the same selected axis; refer to Figure-14.

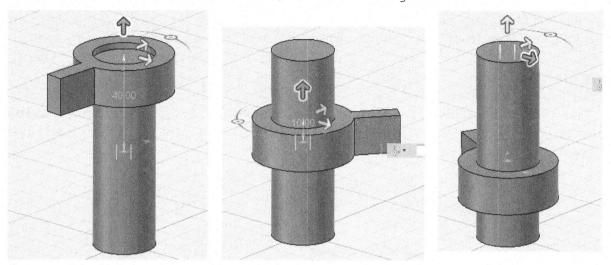

Figure-14. Cylindrical option

- Click on the **Pin-Slot** option of **Type** drop-down from **JOINT** dialog box, if you want revolve and slide the component together like in nut & bolt locking-unlocking.
- Click on the **Planer** option of **Type** drop-down from **JOINT** dialog box, if you want to assemble the component in such a way that the component can move along the selected plane in defined boundary.
- Click on the **Ball** option of **Type** drop-down from **JOINT** dialog box, if you want to assemble the component in such a way that the component is free to move 360 degree in 3D space pivoted to a point; refer to Figure-15.

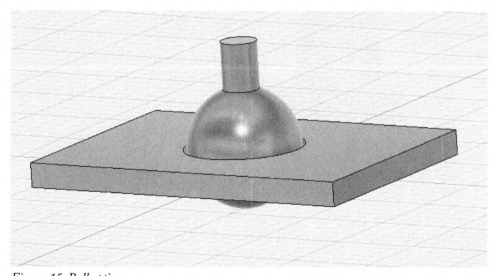

Figure-15. Ball option

- Click on the **Animate** button from **JOINT** dialog box to watch the allowed motion by animating **Component1** and click the button again to stop the animation.

After specifying the parameters for joint, click on the **OK** button from **JOINT** dialog box to complete the process.

AS-BUILT JOINT

The **As-built Joint** tool allows the joints to be applied to components in their current position as they're built, making it easy to join them in the right orientation. The procedure to use this tool is discussed next.

- Click on the **As-built Joint** tool of **Assemble** drop-down from **Toolbar**; refer to Figure-16. The **AS-BUILT JOINT** dialog box will be displayed; refer to Figure-17.

Figure-16. As-built Joint tool

Figure-17. AS-BUILT JOINT dialog box

- Select the components from the canvas.
- Select the required operation from **Type** drop-down. The updated **AS-BUILT JOINT** dialog box will be displayed; refer to Figure-18.

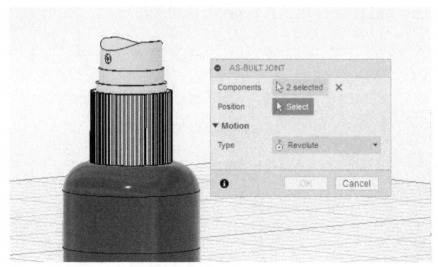

Figure-18. Updated AS-BUILT dialog box

- Select the desired references on the components based on the option selected in **Type** drop-down; refer to Figure-19.

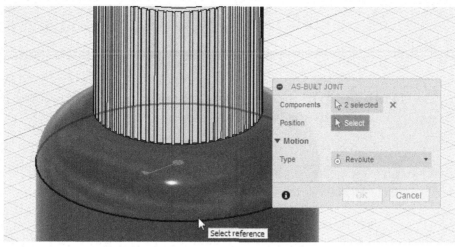

Figure-19. Selecting reference for revolve

- After specifying the parameters, click on the **OK** button from **AS-BUILT JOINT** dialog box to complete the process.
- If you want to rotate the then double-click on the flag displayed; refer to Figure-20. The drag handle will be displayed. Rotate the component as required; refer to Figure-21.

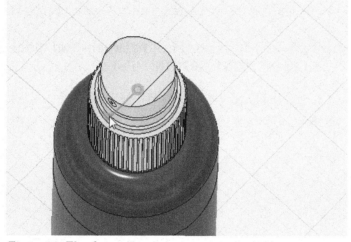

Figure-20. Flag for rotation

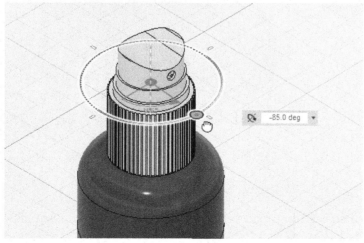

Figure-21. Rotating the component

JOINT ORIGIN

The **Joint Origin** tool is used to define a joint origin on a component. The procedure to use this tool is discussed next.

- Click on the **Joint Origin** tool of **Assemble** drop-down of **Toolbar**; refer to Figure-22. The **JOINT ORIGIN** dialog box will be displayed; refer to Figure-23.

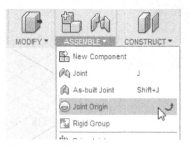

Figure-22. Joint Origin tool

Figure-23. JOINT ORIGIN dialog box

- Select **Simple** option of **Type** drop-down from **JOINT ORIGIN** dialog box, if you want to place the joint origin on a face.
- Select **Between Two Faces** option of **Type** drop-down from **JOINT ORIGIN** dialog box if you want to place the joint origin midway between the faces.
- In our case, we are selecting the **Between Two Faces** option of **Type** drop-down. The updated **JOINT ORIGIN** dialog box will be displayed as shown in Figure-24.
- The **Select** button of **Plane 1** option is active by default. Click on the face from model to select; refer to Figure-25.

Figure-24. The JOIN ORIGIN dialog box

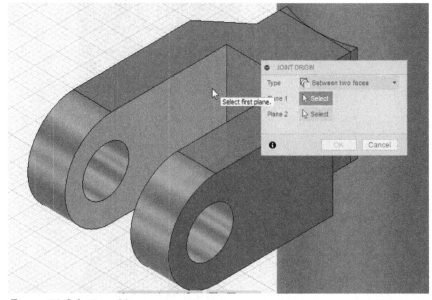

Figure-25. Selection of face

- Similarly, select the opposite face; refer to Figure-26. The updated **JOINT ORIGIN** dialog box will be displayed.

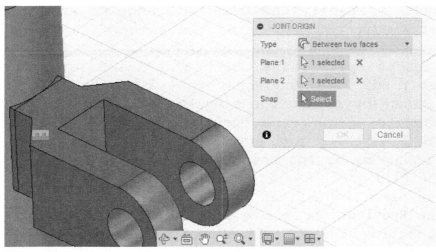

Figure-26. Selecting second face

- The **Select** button of **Snap** option is active by default. Click at desired location to define the snap point; refer to Figure-27.

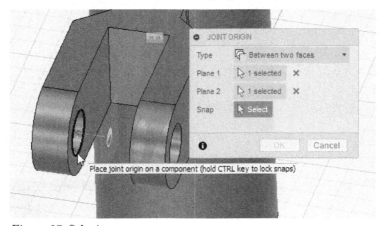

Figure-27. Selecting snap

- On selecting the snap point, the **JOINT ORIGIN** dialog box will be updated; refer to Figure-28.

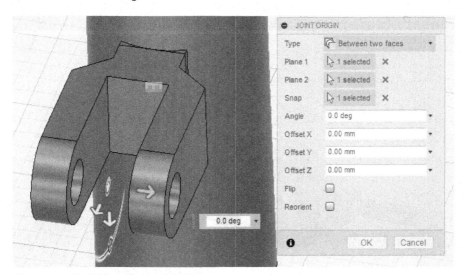

Figure-28. Specifying the location of snap

- If you want to specify the angle of snap then click in the **Angle** edit box of **JOINT ORIGIN** dialog box and specify the angle of rotation of Snap.

- If you want to specify the Offset distance of snap then click on the **Offset X** edit box and enter the value. You can also set the offset distance by moving the drag handle.
- If you want to specify the Offset distance of snap then click on the **Offset Y** edit box and enter the value. You can also set the offset distance by moving the drag handle.
- If you want to specify the Offset distance of snap then click on the **Offset Z** edit box and enter the value. You can also set the offset distance by moving the drag handle.
- Select the **Flip** check box from **JOINT ORIGIN** dialog box, if you want to flip the direction of snap.
- Select the **Reorient** check box from **JOINT ORIGIN** dialog box, if you want to position the axes of origin of geometry.
- After specifying the parameters, click on the **OK** button from **JOINT ORIGIN** dialog box to complete the process.
- Now, you are able to join the two component with the help of **Joint** tool; refer to Figure-29.

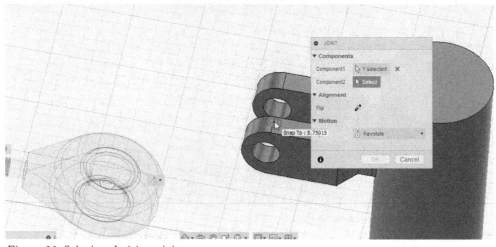

Figure-29. Selecting the joint origin

- Click on **OK** button from **JOINT** dialog box to create joint between two component; refer to Figure-30.

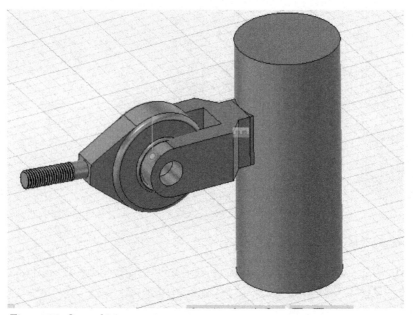

Figure-30. Created joint

RIGID GROUP

The **Rigid Group** tool is used to combine selected components into one group so that the created group act as one object. The procedure to use this tool is discussed next.

• Click on the **Rigid Group** tool of **Assemble** drop-down from **Toolbar**; refer to Figure-31. The **RIGID GROUP** dialog box will be displayed; refer to Figure-32.

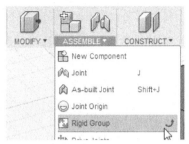

Figure-31. Rigid Group tool

Figure-32. RIGID GROUP dialog box

• The **Select** button of **Components** section is active by default. You need to select the components from the model to create a rigid group.
• Select the **Include Child Components** check box if you want to include the child components into the rigid group.
• Click on **OK** button from **RIGID GROUP** dialog box to complete the process.

DRIVE JOINTS

The **Drive Joints** tool is used to specifying the rotation angle or distance value for joint degrees of freedom. The procedure to use this tool is discussed next.

• Click on the **Drive Joints** tool of **Assemble** drop-down from **Toolbar**; refer to Figure-33. The **DRIVE JOINTS** dialog box will be displayed; refer to Figure-34.

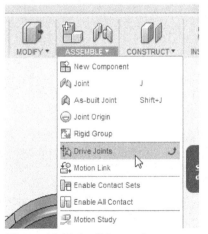

Figure-33. Drive Joints tool

Figure-34. DRIVE JOINTS dialog box

• The **Joint Input** option is active by default. Click on the joint from model. The updated **DRIVE JOINTS** dialog box will be displayed; refer to Figure-35.

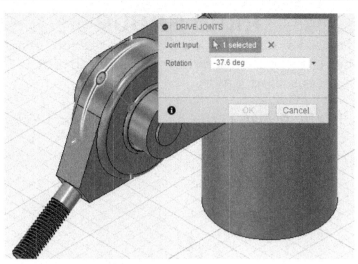

Figure-35. Updated DRIVE JOINTS dialog box

- Click in the **Rotation** edit box and specify total span of the angle of rotation. You can also rotation limits by moving the drag handle.
- After specifying the parameters, click on the **OK** button from **DRIVE JOINTS** dialog box.

MOTION LINK

The **Motion Link** tool is used to specifying the relationship between the degree of freedom of selected joints. The procedure to use this tool is discussed next.

- Click on the **Motion Link** tool of **Assemble** drop-down from **Toolbar**; refer to Figure-36. The **MOTION LINK** dialog box will be displayed; refer to Figure-37.

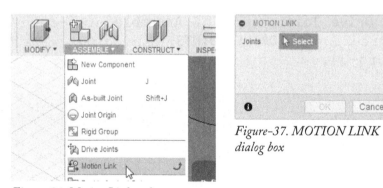

Figure-36. Motion Link tool

Figure-37. MOTION LINK dialog box

- **Joints** selection is active by default. Click on the model to select joints; refer to Figure-38. In our case, we are selecting the slider and revolve joints to create a motion of vice.

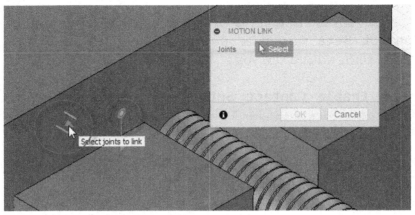

Figure-38. Selection of joints to link

- On selecting the two joints, the animation of vice will starts and the dialog box will be updated; refer to Figure-39.

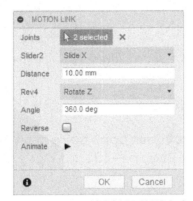

Figure-39. Updated MOTION LINK dialog box

- Click in the **Distance** edit box of **MOTION LINK** dialog box and specify the value of distance per revolution.
- Click in the **Angle** edit box of **MOTION LINK** dialog box and specify the value of angle of revolution.
- Select the **Reverse** check box of **MOTION LINK** dialog box, if you want to reverse the direction of revolution.
- Click on the Animate button to watch the animation of joints.
- After specifying the parameters, click on **OK** button of **MOTION LINK** dialog box to complete the process; refer to Figure-40.

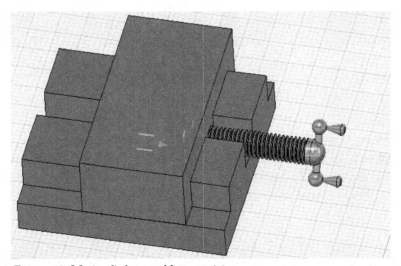

Figure-40. Motion link created between joints

ENABLE CONTACT SETS

The **Enable Contact Sets** tool is used to enable the physical contact conditions. The procedure to use this tool is discussed next.

*   Click on the **Enable Contact Sets** tool of **Assemble** drop-down from **Toolbar**; refer to Figure-41. The tool will be displayed in **Browser**; refer to Figure-42.

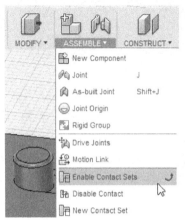

Figure-41. Enable Contact Sets tool

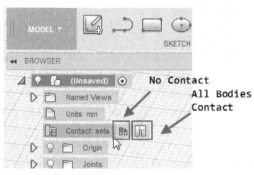

Figure-42. Contact sets

*   Click on **All Body Contact** button to activate the contact analysis between sets; refer to Figure-43.

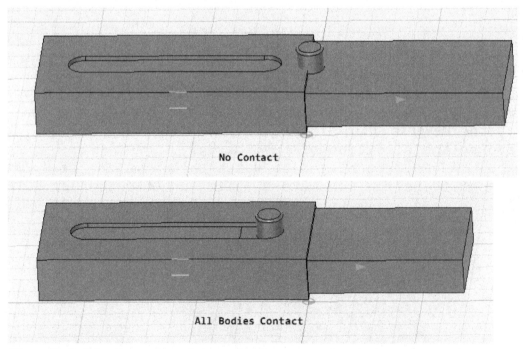

Figure-43. Applying Contact Sets tool

*   Click **No Contact** button to disable the contact between sets.
*   In the above figure, the slider joint is applied before applying **Contact Sets** tool between them.

ENABLE ALL CONTACT

The **Enable All Contact** tool is used to enable all contacts which were applied earlier in the part or model.

* To enable contact, click on the **Enable All Contact** tool of **Assemble** drop-down from **Toolbar**. All contacts will be enabled and the **Assemble** drop-down will be updated with two new tools which are discussed next.

DISABLE CONTACT

The **Disable Contact** tool is used to disable all contacts which were made earlier in the part or model.

* To disable contact, click on the **Disable Contact** tool of **Assemble** drop-down from **Toolbar**. All contacts will be disabled.

CREATE NEW CONTACT

The **Create New Contact** tool is used to create new contact set between two selected components. These contact sets prevent parts from passing through one another. The procedure to use this tool is discussed next.

* Click on the **Create New Contact** tool of **Assemble** drop-down from **Toolbar**; refer to Figure-44. The **CREATE NEW CONTACT** dialog box will be displayed; refer to Figure-45.

Figure-44. New Contact Set dialog box

Figure-45. The NEW CONTACT SET dialog box

* The **Select** button of **Bodies or Components** section is active by default. Select the two component or bodies to create sets and click on **OK** button from **NEW CONTACT SET** dialog box to complete the process.

MOTION STUDY

The **Motion Study** tool is used for performs kinematic motion analysis based on joints. The procedure to use this tool is discussed next.

* Click on the **Motion Study** tool of **Assemble** drop-down from **Toolbar**; refer to Figure-46. The **Motion Study** dialog box will be displayed; refer to Figure-47.

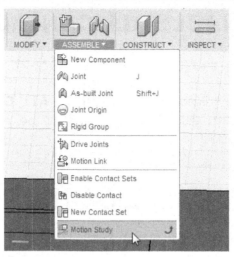

Figure-46. Motion Study tool

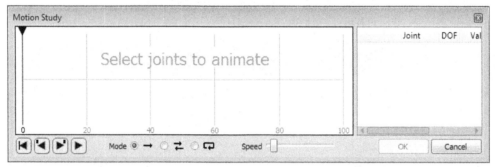

Figure-47. Motion Study dialog box

- You need to select the joint from part or model. Click to select the joint; refer to Figure-48.

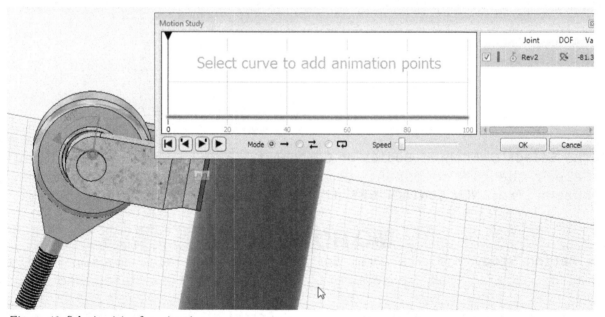

Figure-48. Selecting joint for animation

- After selecting joint, adjust the points of motion curve as required; refer to Figure-49.

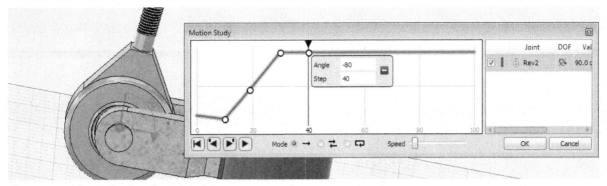

Figure-49. Adjusting points for animation

- Click in the **Angle** edit box and specify the required angle for animation point.
- After creating the animation points, set the speed of animation from **Speed** bar.
- Select **Play Once** radio button from **Mode** section to play the animation once.
- Select **Play Forward/Backward** radio button from **Mode** section to play the animation forward and then backward.
- Select **Play Loop** radio button from **Mode** section to play the animation in cycle.
- After specifying the parameters for animation, click on **Play** button to watch the animation of selected joint.
- If the animation is as expected then, click on **OK** button from **Motion Study** dialog box to exit the motion study of components.

PRACTICAL 1

Assemble the parts of handle as shown in Figure-51. The exploded view of assembly is displayed as shown in Figure-50.

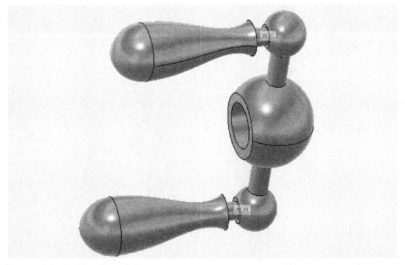

Figure-50. Handle assembled

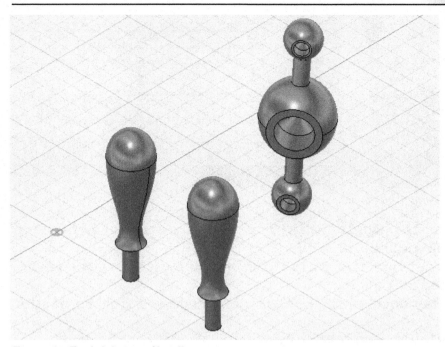

Figure-51. Exploded view of handle

All the part files can be downloaded from resource kit of Fusion 360 software, which is provided at **CADCAMCAEWORKS.COM.**

The steps to assemble these parts are given next.

Adding file into Fusion 360

- After downloading the resource kit from CADCAMCAEWORKS.COM, open the **Chapter 7** folder.
- Double click on **Handle v0.f3d** file. The file will open in **Autodesk Fusion 360** software; refer to Figure-52.

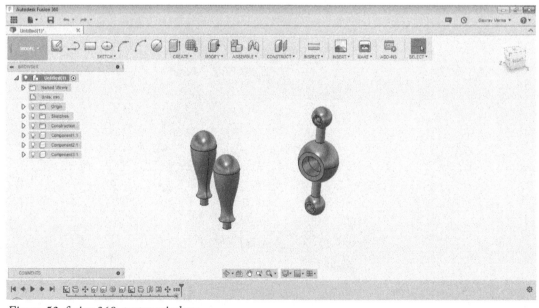

Figure-52. fusion 360 startup window

Assembling components

- Click on **New Component** tool of **Assemble** drop-down from **Toolbar**; refer to Figure-54. The **NEW COMPONENT** dialog box will be displayed; refer to Figure-54.

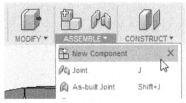

Figure-53. New Component tool

Figure-54. NEW COMPONENT dialog box

- Select the **From Bodies** radio button from **NEW COMPONENT** dialog box and select the three parts to make them component; refer to Figure-55

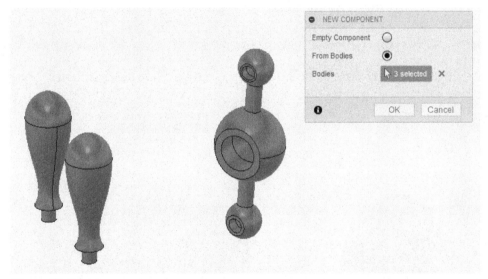

Figure-55. Selecting parts to create components

- After selecting the parts, click on **OK** button from **NEW COMPONENT** dialog box. The components will be created and added in **Timeline**.
- Click on **Joint** tool of **Assemble** drop-down from **Toolbar.** You can also activate the **Joint** tool by pressing **J** key from keyboard. The **Joint** dialog box will be displayed; refer to Figure-56.

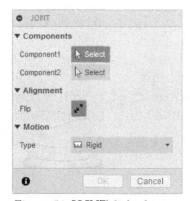

Figure-56. JOINT dialog box

- The **Select** button of **Component 1** selected by default. Click on the component to join; refer to Figure-57.

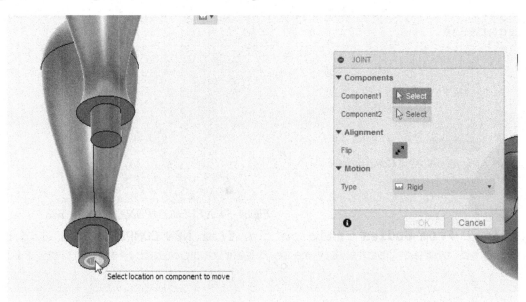

Figure-57. Selection of first component

- After selection of first component, select the second component as shown; refer to Figure-58.

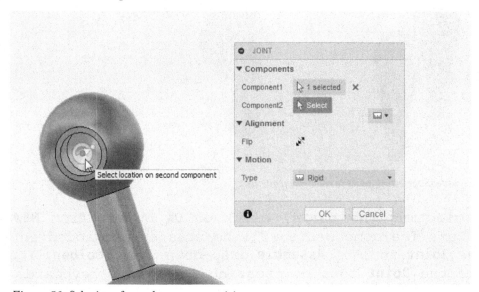

Figure-58. Selection of second component to join

- Select **Rigid** option from **Type** drop-down to create a rigid joint between two component.
- After specifying the parameters, click on **OK** button from **JOINT** dialog box to create a rigid joint.
- Similarly, create another rigid joint between other component.
- After joining these component the assembly will be displayed; as shown in Figure-59.

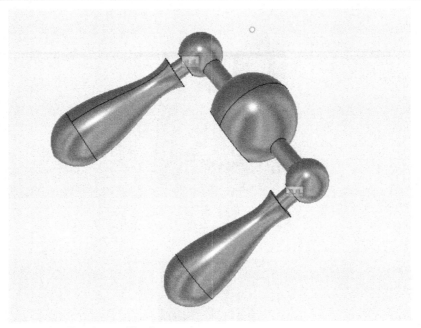

Figure-59. Final assembly of component

PRACTICE 1

Assemble the parts of handle as shown in Figure-60 and Figure-61. The exploded view of assembly is displayed as shown in Figure-62.

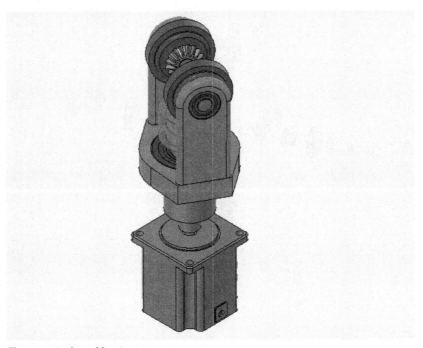

Figure-60. Assembly view 1

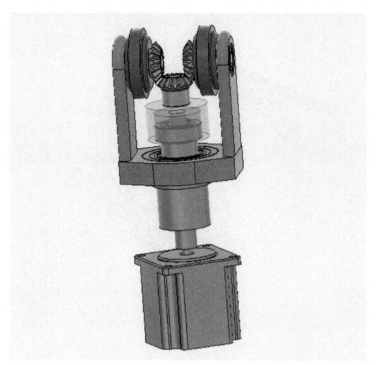

Figure-61. Assembly view 2

Figure-62. Exploded view of assembly

Chapter 8

Importing Files and Inspection

Topics Covered

The major topics covered in this chapter are:

- *Interference Tool*
- *Curvature Comb Analysis*
- *Data Analysis*
- *Section Analysis*
- *Decal tool*
- *Attached Canvas*
- *Insert SVG*
- *Inserting a Manufacturing Part*

INTRODUCTION

In the previous chapter, we have learned the procedure of joining and assembling component. In this chapter we will learn the procedure of inspection the part and importing files, in **Autodesk Fusion 360** software.

INSERTING FILE

In this section we will learn the procedure to insert the various types of files in **Autodesk Fusion 360**.

Decal

The **Decal** tool is used to place an image on the selected face/plane. The procedure to use this tool is discussed next.

- Click on the **Decal** tool of **Insert** drop-down from **Toolbar**; refer to Figure-1. The **DECAL** dialog box will be displayed; refer to Figure-2.

Figure-1. Decal tool

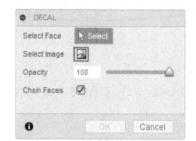

Figure-2. DECAL dialog box

- The **Select** button of **Select Face** section is active by default. Click on the face of model to select.
- Click on **Select Image** button from **DECAL** dialog box. The **Open** dialog box will be displayed; refer to Figure-3.
- Select the required image and click on **Open** button from **Open** dialog box. The updated **DECAL** dialog box will be displayed; refer to Figure-4.

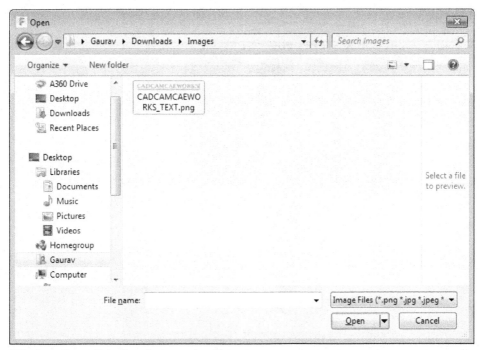

Figure-3. Open dialog box for importing file

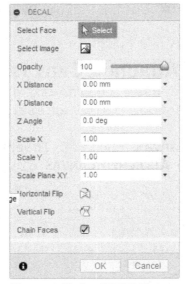

Figure-4. Updated DECAL
dialog box

- Set the image on the selected face with the help of drag handle displayed on image. You can also set the image by specifying the values in respective edit box.
- Click on the **Horizontal Flip** button to flip the image horizontally.
- Click on the **Vertical Flip** button to flip the image vertically.
- Select the **Chain Faces** check box to include tangent faces in selection.
- After specifying the parameters, the preview of model along with image will be displayed; refer to Figure-5.

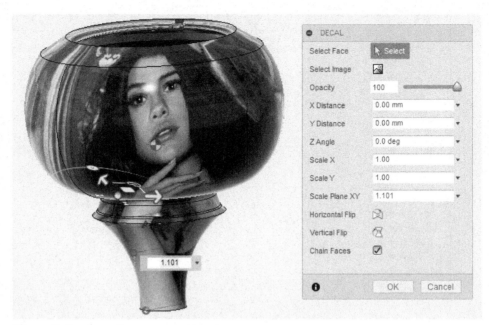

Figure-5. Preview of image on component

Yeah! You are right, I am a die hard fan of salena. If the preview is as desired then click on **OK** button from **DECAL** dialog box to complete the process.

Attached Canvas

The **Attached Canvas** tool is used to insert an image on a planer face. The procedure to use this tool is discussed next.

• Click on the **Attached Canvas** tool of **Insert** drop-down from **Toolbar**; refer to Figure-6 . The **ATTACHED CANVAS** dialog box will be displayed; refer to Figure-7.

Figure-6. Attached Canvas tool

*Figure-7. ATTACHED CAN-
VAS dialog box*

• The **Select** button of **Face** section is active by default. Click on

the face of model when you want to place the canvas.
- Click on **Select Image** button from **ATTACHED CANVAS** dialog box. The **Open** dialog box will be displayed; refer to Figure-8.
- Select the required image and click on **Open** button from **Open** dialog box. The updated **ATTACHED CANVAS** dialog box will be displayed; refer to Figure-9.

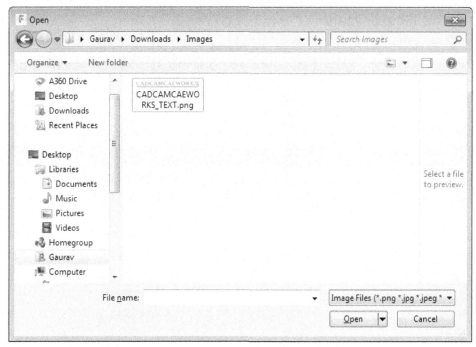

Figure-8. Open dialog box for importing file

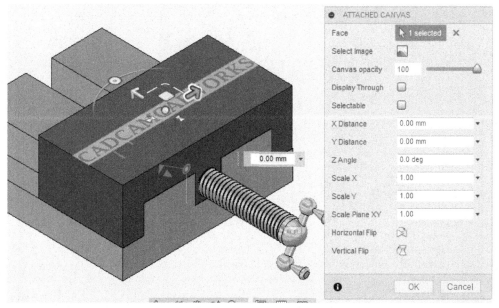

Figure-9. Updated ATTACHED CANVAS dialog box

- Click on **Canvas Opacity** edit box and specify the opacity of image. You an also set the opacity by moving the **Canvas Opacity** slider.
- Select the **Display Through** check box if you want to make the image visible through the component.
- Select the **Selectable** check box if you want the image to be selectable.
- Click in the **X Distance** edit box and specify a value of distance to

move in X direction.
- Click in the **Y Distance** edit box and specify a value of distance to move in Y direction
- Click in the **Z Angle** edit box and specify a value to rotation around z axis.
- Click in the **Scale X** edit box and specify a scale value in X direction.
- Click in the **Scale Y** edit box and specify a scale value in Y direction.
- Click in the **Scale Plane XY** edit box and specify the value of scale in the XY plane.
- Click on the **Horizontal Flip** button to mirror the image horizontally.
- Click on the **Vertical Flip** button to mirror the image vertically.
- After specifying the parameters, click on the **OK** button from **ATTACHED CANVAS** dialog box to complete the process.

Insert Mesh

The **Insert Mesh** tool is used to insert a mesh file to Autodesk Fusion 360 software. The procedure to use this tool is discussed next.

- Click on the **Insert Mesh** tool of **INSERT** drop-down from **Toolbar**; refer to Figure-10. The **Open** dialog box will be displayed; refer to Figure-11 .

Figure-10. Insert Mesh tool

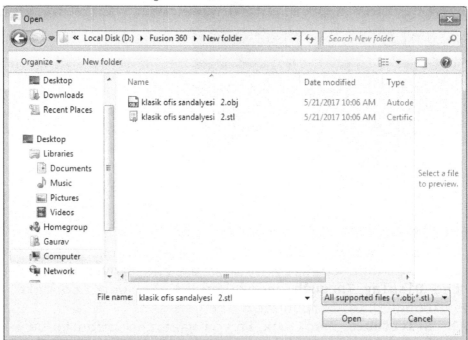

Figure-11. Open dialog box for inserting file

- Select the file and click on **Open** button of **Open** dialog box. The file will be added in application window along with **INSERT MESH** dialog box; refer to Figure-12.

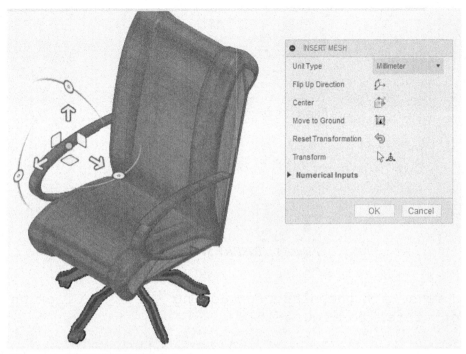

Figure-12. INSERT MESH dialog box

- Click on the **Unit Type** drop-down from **INSERT MESH** dialog box and select the required unit to scale the model.
- Click on **Flip Up Direction** button to flip the direction of model.
- Click on the **Center** button to drag the model to center or coordinate system.
- Click on **Move To Ground** button to move the model to ground plane.
- Click on **Reset Transformation** button to reset the model position to the original state.
- If you want to enter the numerical value to move the model then click on the **Numerical Inputs** section and enter the desired value.
- After specifying the parameters, click on the **OK** button from **INSERT MESH** dialog box to complete the process.

Insert SVG

The **Insert SVG** tool is used to insert an SVG file into sketch. The procedure to use this tool is discussed next.

- Click on the **Insert SVG** tool of **INSERT** drop-down from **Toolbar**; refer to Figure-13. The **INSERT SVG** dialog box will be displayed; refer to Figure-14.

Figure-13. Insert SVG tool

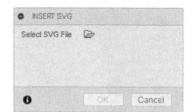

*Figure-14. INSERT SVG dialog
box*

- You need to select the plane for adding SVG file. Click on the plane
  to select; refer to Figure-15.

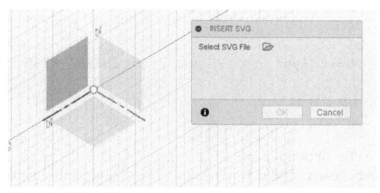

Figure-15. Seleting Face or plane

- On selecting the face or plane, the sketch will be enabled.
- Click on **Select SVG File** button from **INSERT SVG** dialog box the **Open**
  dialog box will be displayed.
- You need to select the required file from **Open** dialog box and click
  on **Open** button. The file will be added in application window along
  with updated **INSERT SVG** dialog box; refer to Figure-16 .

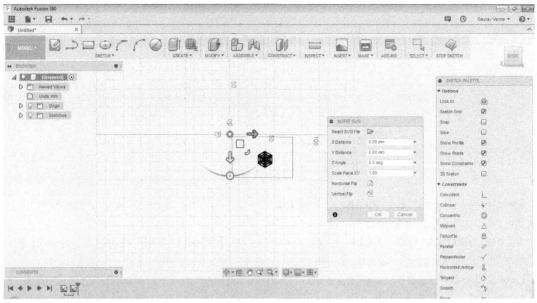

Figure-16. File along with updated INSERT SVG dialog box

- Click in the **X Distance** edit box and specify a value of distance to move in X direction.
- Click in the **Y Distance** edit box and specify a value of distance to move in Y direction.
- Click in the **Z Angle** edit box and specify a value to rotation around z axis.
- Click in the **Scale Plane XY** edit box and specify the value of scale in the XY plane.
- Click on the **Horizontal Flip** button to mirror the image horizontally.
- Click on the **Vertical Flip** button to mirror the image vertically.
- You can also use the drag handle to move or flip the model.
- After specifying the parameters, click on the **OK** button from **INSERT SVG** dialog box to complete the process.

Insert DXF

The **Insert DXF** tool is used to insert a DXF file into the active design. The procedure to use this tool is discussed next.

- Click on the **Insert DXF** tool of **INSERT** drop-down from **Toolbar**; refer to Figure-17 . The **INSERT DXF** dialog box will be displayed; refer to Figure-18.

Figure-17. Insert DXF tool

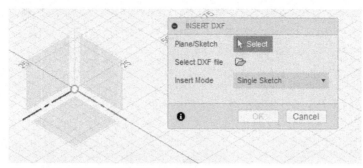

Figure-18. INSERT DXF dialog box

- The **Select** button of **Plane/Sketch** section is active by default. Click on the plane to select.
- Click on **Select DXF File** button from **INSERT DXF** dialog box the **Open** dialog box will be displayed.
- You need to select the file of .dxf format and click on **Open** button. The preview of DXF file will be displayed along with updated **INSERT DXF** dialog box; refer to Figure-19.

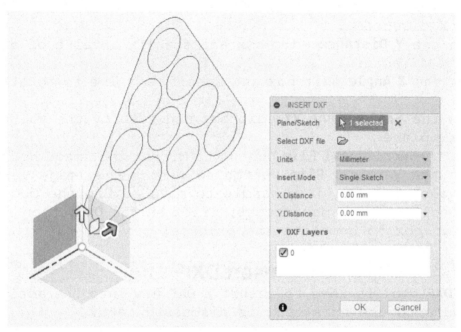

Figure-19. Updated INSERT DXF dialog box

- Click on the **Units** drop-down and select the required unit.
- Select **Single Sketch** option from **Insert Mode** drop-down to insert the layers of selected .DXF file into one sketch.
- Select **One Sketch Per Layer** option from **Insert Mode** drop-down to insert each sketch of DXF file on different layer.
- Click in the **X Distance** edit box and specify a value of distance to move in X direction.
- Click in the **Y Distance** edit box and specify a value of distance to move in Y direction.
- Select the DXF Layers check box to list the layers available in the selected DXF file. Select the layers to import.

- After specifying the parameters, click on the **OK** button from **INSERT DXF** dialog box to complete the process.

Insert a manufacturer part

The **Insert a manufacturer part** tool is used for inserting a part of manufacturer or supplier to the software through parts4cad. The procedure to use this tool is discussed next.

- Click on the **Insert a manufacturer part** tool of **INSERT** drop-down from **Toolbar**; refer to Figure-20. The AUTODESK Part Community window will be displayed; refer to Figure-21.

Figure-20. Insert a manufacturer part tool

Figure-21. AUTODESK FUSION 360 part community window

- Select the part from manufacturer or supplier and open the file in Autodesk Fusion 360.

INSPECT

Till now, we have discussed various tools to insert the file in Autodesk Fusion 360. In this section we will discuss about measurement and analysis tools available in Autodesk Fusion 360.

Measure

The **Measure** tool is used to measure selected geometry. The procedure to use this tool is discussed next.

- Click on the **Measure** tool of **Inspect** drop-down from **Toolbar**; refer to Figure-22. The **MEASURE** dialog box will be displayed; refer to Figure-23.

Figure-22. Measure tool

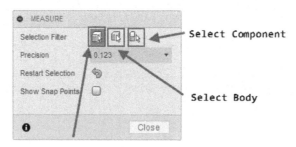

Figure-23. MEASURE dialog box

- Select the **Select Face/Edge/Vertex** button from **Selection Filter** option if you want to select face/edge/vertex for measurement.
- Select the **Select Body** button from **Selection Filter** option if you want to select body for measurement.
- Select the **Select Component** button from **Selection Filter** option if you want to select component for measurement.
- Click on the **Precision** drop-down and select the required precision option for displaying the number of decimal places in results.
- Click on the **Restart Selection** button to restart the selection process once you have performed measurement.
- In our case, we are selecting **Select Face/Edge/Vertex** button from **Selection Filter** option. Click on the face to measure; refer Figure-24.

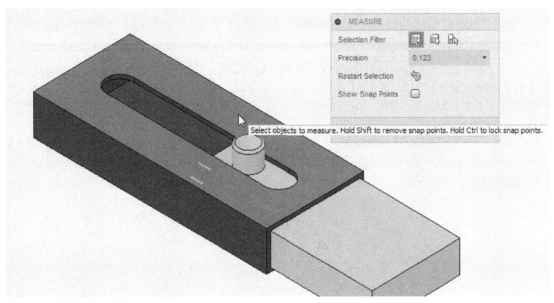

Figure-24. Selection of face for measurement

- On selecting the face, the updated **MEASURE** dialog box will be displayed with the measurement of selected face; refer to Figure-25.

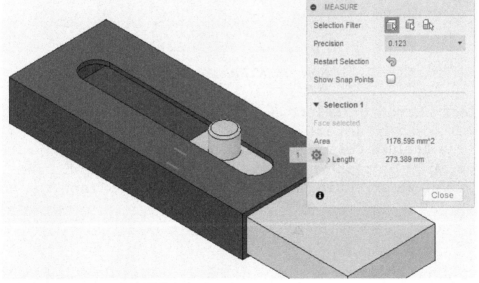

Figure-25. Updated MEASURE dialog box

- If you want to measure the distance between two faces then hold the **CTRL** key and select the other face.
- Hover the cursor on the respective measurement to copy the value to clipboard; refer to Figure-26.

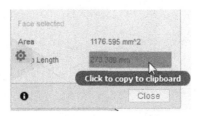

Figure-26. Coping measurement

- Select the **Show Snap points** check box to display snap points when you hover over your model with the **Measure** tool activated.

- Click on **Close** button of **MEASURE** dialog box to exit the **Measure** tool.

Interference

The **Interference** tool is used to display interference of selected components or bodies. The procedure to use this tool is discussed next.

- Click on the **Interference** tool of **Inspect** drop-down from **Toolbar**; refer to Figure-27. The **INTERFERENCE** dialog box will be displayed; refer to Figure-28.

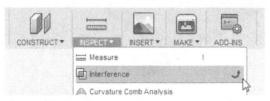

Figure-27. Interference tool

Figure-28. INTERFERENCE dialog box

- The **Select** button of **Select** section is active by default. Click on the model to check interference. You can also use window selection.
- Select the **Include Coincident faces** check box if you want to include the face in selection which are touching with each other.
- In our case, we are using window selection to select the component; refer to Figure-29.

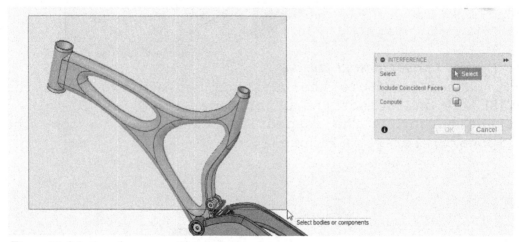

Figure-29. Selection of component for interference

- After selecting the components, click on **Compute** button to calculate the interference. The **Interference Results** dialog box will be displayed; refer to Figure-30.

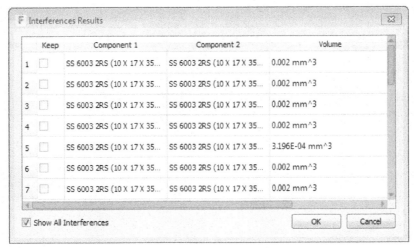
Figure-30. Interference Results dialog box

- Click on the **OK** button from **Interference Results** dialog box to exit the **Interference** tool.

Curvature Comb Analysis

The **Curvature Comb Analysis** tool is used to display a curvature comb on the spline or circular edges to determine the change in curvature. The procedure to use this tool is discussed next.

- Click on the **Curvature Comb Analysis** tool of **Inspect** drop-down from **Toolbar**; refer to Figure-31. The **CURVATURE ANALYSIS** dialog box will be displayed; refer to Figure-32.

Figure-31. Curvature Comb Analysis tool

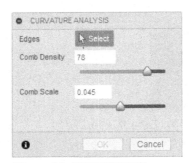
Figure-32. CURVATURE ANALYSIS dialog box

- **Edges** selection of **CURVATURE ANALYSIS** dialog box is active by default. Click on the geometry to select the edge. The preview of comb on edge will be displayed; refer to Figure-33.

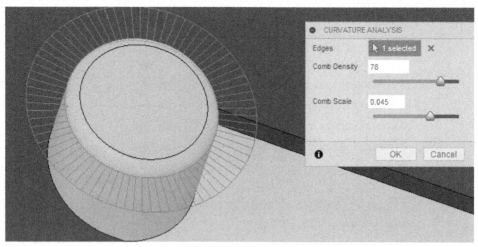

Figure-33. Preview of comb on circular edge

- Click in the **Comb Density** edit box and specify the value. You can also set the comb density by moving the **Comb Density** slider.
- Click in the **Comb Scale** edit box and specify the value. You can also set the comb scale by moving the **Comb Scale** slider.
- After analyzing, click on **OK** button from **CURVATURE ANALYSIS** dialog box to exit the tool.

Zebra Analysis

The **Zebra Analysis** tool is used to analyze continuity between the surfaces of component by displaying strips on model. The procedure to use this tool is discussed next.

- Click on the **Zebra Analysis** tool of **Inspect** drop-down from **Toolbar**; refer to Figure-34. The **ZEBRA ANALYSIS** dialog box will be displayed; refer to Figure-35.

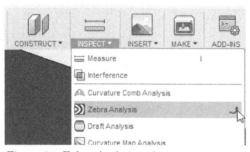

Figure-34. Zebra Analysis tool

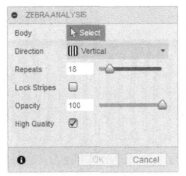

Figure-35. ZEBRA ANALYSIS dialog box

- The **Select** button of **Body** selection option is active by default. Click to select the body for analysis.
- On selecting the body, preview will be displayed; refer to Figure-36.

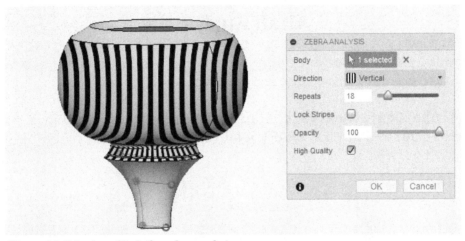

Figure-36. Selection of body for zebra analysis

- Select **Vertical** option from **Direction** drop-down for display of vertical lines on model.
- Select **Horizontal** option from **Direction** drop-down for display of horizontal lines on model.
- Set the number of strips for model from **Repeats** slider.
- Select the **Lock Strips** check box of **ZEBRA ANALYSIS** dialog box to freeze the stripes on body.
- Click on the **Opacity** edit box and enter the value for opacity. You can also set the value of opacity by moving the **Opacity** slider.
- Select the **High Quality** check box of **ZEBRA ANALYSIS** dialog box to increase the display quality of strips.
- After specifying the parameters, click on **OK** button from **ZEBRA ANALYSIS** dialog box. The strips will be displayed on model; refer to Figure-37.

Figure-37. Strips on model

Draft Analysis

The **Draft Analysis** tool is used to display a color gradient on the selected body to evaluate draft angle of various faces of the part. The procedure to use this tool is discussed next.

- Click on the **Draft Analysis** tool of **Inspect** drop-down from **Toolbar**; refer to Figure-38. The **DRAFT ANALYSIS** dialog box will be displayed; refer to Figure-39.

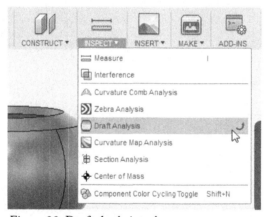

Figure-38. Draft Analysis tool

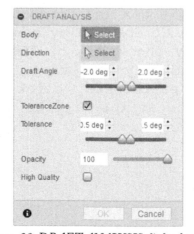

Figure-39. DRAFT ANALYSIS dialog box

- The **Select** button of **Body** section is active by default. Click on the body to perform analysis.
- Click on **Select** button of **Direction** option and select the direction for analysis. On selection, the preview will be displayed; refer to Figure-40.

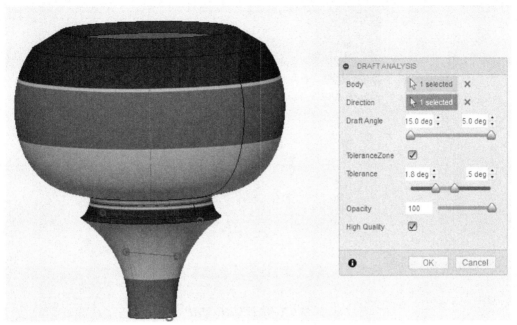

Figure-40. Preview for draft analysis

- Click in **Draft Angle** edit boxes and specify the range of draft angle for which you want to perform analysis. You can also set the draft angle limits by moving the **Draft angle** slider.
- Select the **ToleranceZone** check box if you want to displays a tolerance gradient on the results.
- Click on the **Tolerance** edit box and specify the start and stop angles tolerance. You can also set the tolerance angle by moving the slider.
- Click on the **Opacity** edit box and enter the value for opacity for draft colors. You can also set the value of opacity by moving the **Opacity** slider.
- Select the **High Quality** check box of **DRAFT ANALYSIS** dialog box to increase the display quality of gradient.
- After specifying the parameters for analysis, click on **OK** button from **DRAFT ANALYSIS** dialog box to complete the analysis.

Curvature Map Analysis

The **Curvature Map Analysis** tool is used to displays a color gradient on the faces of selected bodies to evaluate the amount of curvature. The procedure to use this tool is discussed next.

- Click on the **Curvature Map Analysis** tool of **Inspect** drop-down from **Toolbar**; refer to Figure-41. The **CURVATURE MAP ANALYSIS** dialog box will be displayed; refer to Figure-42.

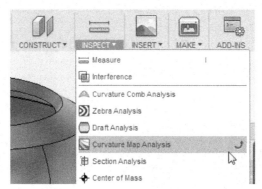

Figure-41. Curvature Map Analysis tool

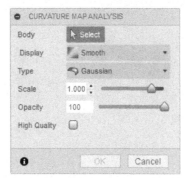

Figure-42. CURVATURE MAP ANALYSIS dialog box

- The **Select** button of **Body** option is active by default. Click on the body to select for analysis.
- Select **Smooth** option from **Display** drop-down if you want to display a blend of two colors.
- Select **Bands** option from **Display** drop-down if you want to display no transition between colors.
- Select **Gaussian** option from **Type** drop-down if you want to display the gradient as the product of curvature in the U direction and curvature in the V direction.
- Select **Principle Minimum** option from **Type** drop-down if you want to display the gradient highlighting areas of lowest curvature.
- Select **Principle Maximum** option from **Type** drop-down if you want to display the gradient highlighting areas with the highest curvature.
- Click in the **Scale** edit box of **CURVATURE MAP ANALYSIS** dialog box and enter the value. You can also set the value of scale by moving the **Scale** slider.

- Click on the **Opacity** edit box and enter the value for opacity. You can also set the value of opacity by moving the **Opacity** slider.
- Select the **High Quality** check box of **CURVATURE MAP ANALYSIS** dialog box to increase the display quality of gradient.
- After specifying the parameters for analysis, click on **OK** button from **CURVATURE MAP ANALYSIS** dialog box to exit this analysis; refer to Figure-43.

Figure-43. Curvature map analysis of model

Section Analysis

The **Section Analysis** tool is used to define the view of component from inner side. The procedure to use this tool is discussed next.

- Click on the **Section Analysis** tool of **Inspect** drop-down from **Toolbar**; refer to Figure-44. The **SECTION ANALYSIS** dialog box will be displayed; refer to Figure-45.

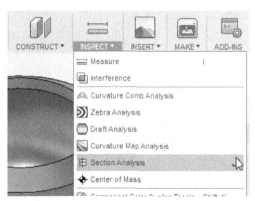

Figure-44. Section Analysis tool

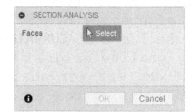

Figure-45. SECTION ANALY-
SIS dialog box

- The **Select** button of **Faces** section is active by default. Click on the face to select.
- On selection of face, the dialog box will be updated along with the preview of model; refer to Figure-46.

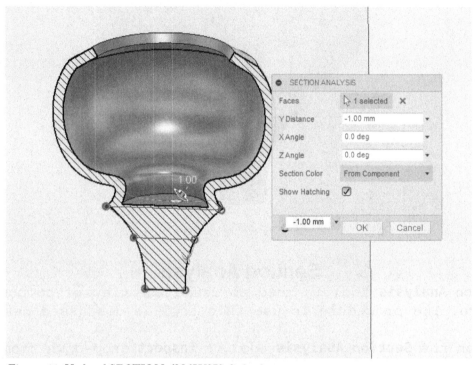

Figure-46. Updated SECTION ANALYSIS dialog box

- Click in the **Y Distance** edit box and specify the location of the section plane along y. You can also set the value by moving the drag handle.
- Click in the **X Angle** edit box and specify the angle of plane about x. You can also set the value by moving the drag handle.
- Click in the **Z Angle** edit box and specify the angle of plane about y. You can also set the value by moving the drag handle.
- Select **From Component** option from **Section Color** section if you want to use the same color as applied to the model.
- Select **Custom** option from **Section Color** section if you want to use the custom color for the section.
- Select the **Show Hatching** check box if you want to show hatching on the section.
- After specifying the parameters for analysis, click on **OK** button from **SECTION ANALYSIS** dialog box to complete the analysis.

Center of Mass

The **Center of Mass** tool is a center point of body use for representing the mean position of matter in a body. The procedure to use this tool is discussed next.

- Click on the **Center of Mass** tool of **Inspect** drop-down from **Toolbar**; refer to Figure-47. The **CENTER OF MASS** dialog box will be displayed; refer to Figure-48.

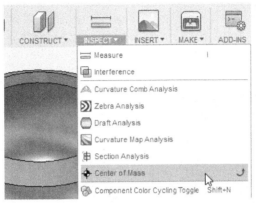

Figure-47. Center of Mass tool

*Figure-48. CENTER OF MASS
dialog box*

- You need to select the body and click on **OK** button from **CENTER OF MASS** dialog box. The center of mass of the body will be displayed; refer to Figure-49.

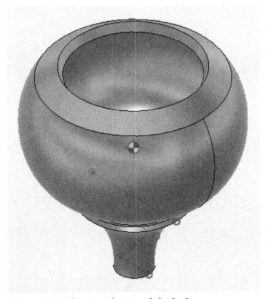

Figure-49. Center of mass of the body

FOR STUDENTS NOTES

FOR STUDENTS NOTES

Chapter 9

Surface Modeling

Topics Covered

The major topics covered in this chapter are:

- *Patch Tool*
- *Surface Trim Tool*
- *Stitch Tool*
- *Reverse Normal*
- *Practical*
- *Practice*

INTRODUCTION

Surface Designing or surfacing is a technique used to create complex shapes. The most general application of surfacing can be seen in the object that are made with good aerodynamics like car body, aeroplanes, ships, and so on. Sometimes, surface modeling is also used to add complex geometry to the models. The models created by surfacing are called surface models and they do not exist in real world.

In Autodesk Fusion 360, there is a separate environment for creating surface designs. We have all the surfacing tools available in the **Patch** workspace; refer to Figure-1.

Figure-1. PATCH Modeling environment

Note that we have already used many tools of **PATCH** Modeling environment in **MODEL** Modeling environment. So, now we will use remaining tools.

Extrude

The **Extrude** tool can also be used to create the extruded surfaces. Using this tool, we can create flat surface. The procedure to use this tool is discussed next.

- Click on the **Extrude** tool of **Create** drop-down from **Toolbar** in **PATCH** Workspace; refer to Figure-2. The **EXTRUDE** dialog box will be displaced; refer to Figure-3.

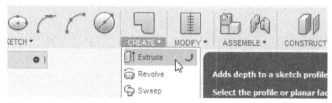

Figure-2. Extrude tool

Figure-3. EXTRUDE dialog box

- The **Select** button of **Profile** section is active by default. Click on the profile to select. The updated **EXTRUDE** dialog box will be displayed; refer to Figure-4.

Note that you are able to select open as well as closed sketch in profile selection option to extrude.

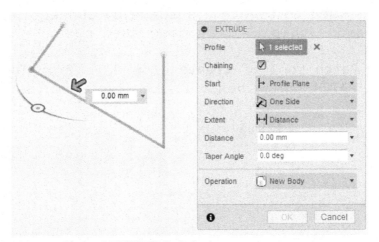

Figure-4. Updated EXTRUDE dialog box

- Select the **Chaining** check box, if you want to select the profile of sketch in chain.
- The other options of **Extrude** dialog box were discussed in **MODEL** workspace.
- After specifying the parameters, preview of extrude will be displayed; refer to Figure-5.

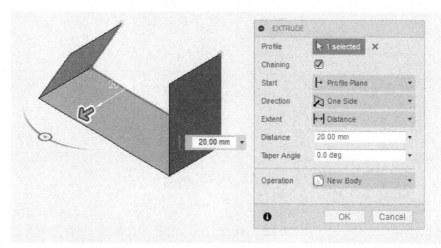

Figure-5. Preview of extrude

- If the preview of extrude is as required, then click on **OK** button **EXTRUDE** dialog box. The extruded surface will be created.

Patch

The **Patch** tool is used to create a surface by connecting edges/curves. The procedure to use this tool is discussed next.

- Click on the **Patch** tool of **Create** drop-down from **Toolbar**; refer to Figure-6. The **PATCH** dialog box will be displayed; refer to Figure-7.

Figure-6. Patch tool

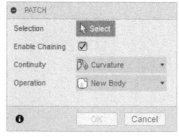

Figure-7. PATCH dialog box

- The **Select** button of **Selection** section is active by default. Click on the edges/curves forming close loop.
- Select the **Enable Chaining** check box if you want to select all the adjacent edges when clicking on one edge.
- Select **Connected** from **Continuity** drop-down if you want to create a surface with G0 edges (connected at an angle).
- Select **Tangent** option from **Continuity** drop-down if you want to create a surface with G1 edges (tangential).
- Select **Curvature** option from **Continuity** drop-down, if you want to creates a surface with G2 edges (blended with continuous curvature).
- After specifying the parameters, the preview of patch will be displayed; refer to Figure-8.

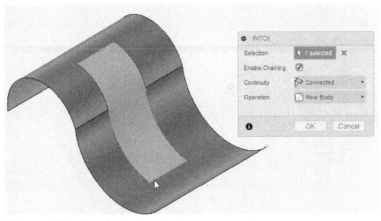

Figure-8. Preview for patch

- If preview of the patch is as desired then click on **OK** button from **PATCH** dialog box to complete the process.

Other Solid Modeling Tools For Surfacing

The other tools in the **Create** drop-down can also be used in the same way as discussed in **Chapter 4: Advanced 3D Modeling** of the book. You can now create surfaces by using **Revolve, Sweep, Loft, Offset, Pattern, Mirror, Thicken** and **Boundary Fill** tools by following the procedure discussed earlier.

EDITING TOOLS

In the above section, we have learned about different surface creation tools used in surface modeling. In this section, we will discuss the surface editing tools.

Trim

The **Trim** tool is used to trim the surface of created model. The procedure to use this tool is discussed next.

- Click on the **Trim** tool of **Modify** drop-down from **Toolbar**; refer to Figure-9. The **TRIM** dialog box will be displayed; refer to Figure-10.

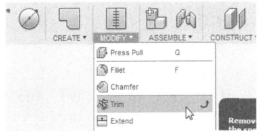

Figure-9. Trim tool

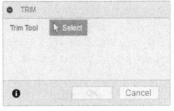

Figure-10. TRIM dialog box

- The **Select** button of **Trim Tool** option is active by default. You need to select an existing sketch to trim the surface.
- On selecting the sketch, preview of trim will be displayed; refer to Figure-11.

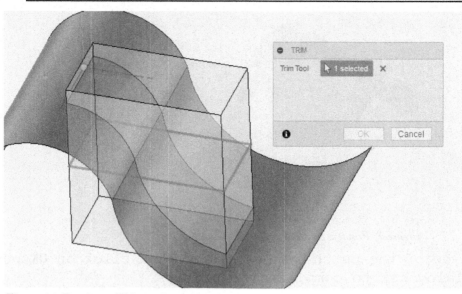

Figure-11. Preview of Trim tool

- Now, you need to select the area for trim. Click on the internal area of sketch to trim; refer to Figure-12.

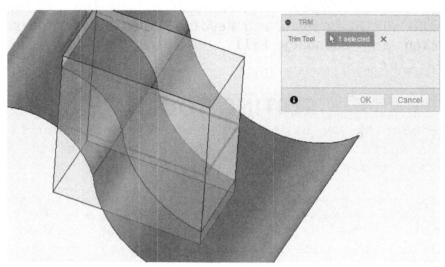

Figure-12. Selection of area for trim

- Click on the **OK** button from **TRIM** dialog box to trim the selected area.

Extend

The **Extend** tool is used to extend the surface edge. The procedure to use this tool is discussed next.

- Click on the **Extend** tool of **Modify** drop-down from **Toolbar**; refer to Figure-13. The **EXTEND** dialog box will be displayed; refer to Figure-14.

Figure-13. Extend tool

Figure-14. EXTEND dialog box

• The **Select** button of **Edges** section is active by default. You need to select the edge to extend. The updated **EXTEND** dialog box will be displayed; refer to Figure-15.

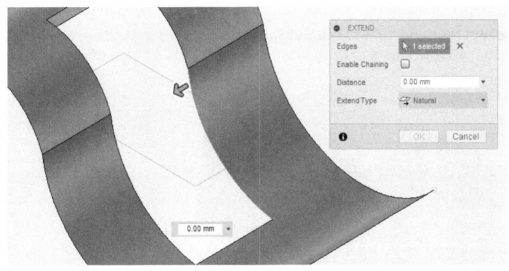

Figure-15. Selection of edge to extend

• Select the **Enable Chaining** dialog box, if you want to select all tangent or adjacent faces on selection of single face.
• Click in the **Distance** edit box and specify the distance for extend. You can also set the distance by moving the manipulator.
• Select **Natural** option from **Extend Type** drop-down, if you want to extend the surface naturally.
• Select **Perpendicular** option from **Extend Type** drop-down, if you want to extend the selected edge perpendicularly.
• Select **Tangent** option from **Extend Type** drop-down, if you want to extend the selected edge tangentially.
• After specifying the parameters, click on **OK** button from **EXTEND** dialog box.

Stitch

The **Stitch** tool is used to combine the surface into one body. The procedure to use this tool is discussed next.

• Click on the **Stitch** tool of **Modify** drop-down from **Toolbar**; refer to Figure-16. The **STITCH** dialog box will be displayed; refer to Figure-17.

Figure-16. Stitch tool

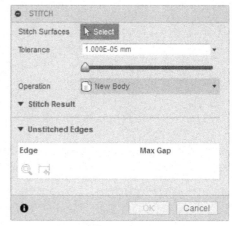

Figure-17. STITCH dialog box

- The **Select** button of **Stitch Surfaces** section is active by default. You need to select the surfaces of model to stitch; refer to Figure-18.

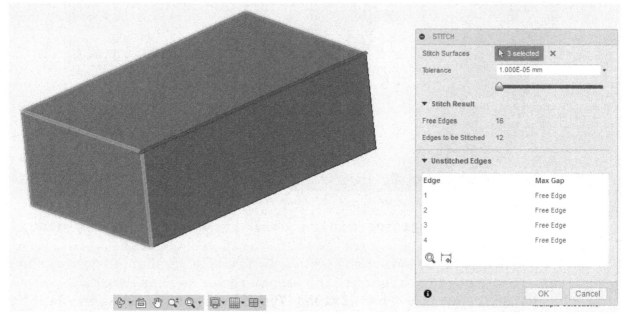

Figure-18. Selection of face for Stitch

- The Green line on the model shows the stitched surface and Red line shows the unstitched surface.
- You need to increase the tolerance from **Tolerance** edit box or **Tolerance** slider to stitch the remaining surfaces which have larger gap between consecutive edges.
- You can also check the result of stitch and unstitch edge from **Stitch Result** section.
- After specifying the parameters, click on the **OK** button from **STITCH** dialog box to complete the process.

Unstitch

The **Unsitch** tool is used to unstitch the earlier stitched surface from model. The procedure to use this tool is discussed next.

- Click on the **Unstitch** tool of **Modify** drop-down from **Toolbar**; refer to Figure-19. The **UNSTITCH** dialog box will be displayed; refer to Figure-20.

Figure-20. UNSTITCH dialog box

Figure-19. Unstitch tool

- The **Select** button of **Faces/ Bodies** section is active by default. You need to select the surface or body to unstitch.
- Select the **Chain Selection** check box to select the adjacent surface of the selected surface.
- After selecting the surfaces, click on **OK** button from **UNSTITCH** dialog box to unstitch the selected surface.

Reverse Normal

The **Reverse Normal** tool is used to flip the normal direction of selected face. When modifying the body or model, some of the faces of model have an unexpected normal direction. To reverse the direction, this tool is used. The procedure to use this tool is discussed next.

- Click on the **Reverse Normal** tool of **Modify** drop-down from **Toolbar**; refer to Figure-21. The **REVERSE NORMAL** dialog box will be displayed; refer to Figure-22.

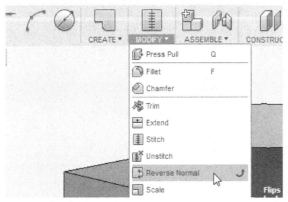

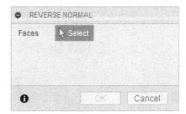

Figure-22. REVERSE NOR-MAL dialog box

Figure-21. Reverse Normal tool

- The **Select** button of **Faces** section is active by default. You need to select the face to flip; refer to Figure-23.

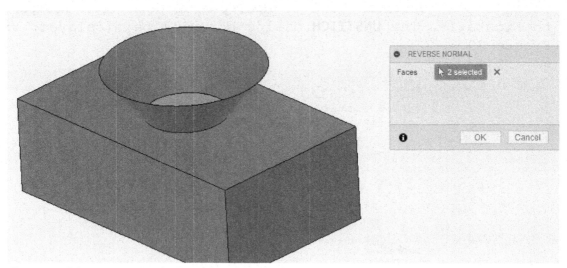

Figure-23. Selection of surface for flip

- After selection of surfaces, click on the **OK** button from **REVERSE NORMAL** dialog box. The selected face will be flipped; refer to Figure-24.

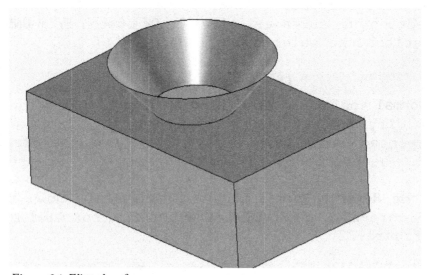

Figure-24. Flipped surface

The other tools of **PATCH** workspace were discussed earlier in this book.

PRACTICAL 1

Create the model of aircraft as shown in Figure-25.

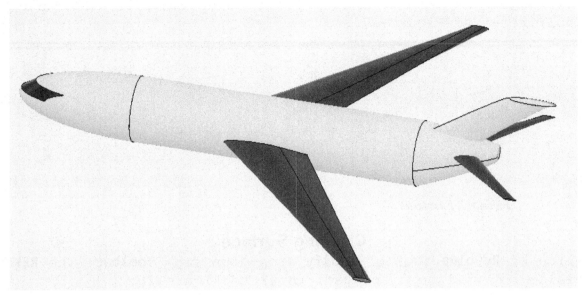

Figure-25. Practical 1 model

Creating first sketch

- Click on the **PATCH** workspace from **Workspace** drop-down; refer to Figure-26. The PATCH workspace will be displayed.

Figure-26. workspace drop-down

- Now create the sketch on Top plane using **Ellipse**, **Line**, and **Trim** tool; refer to Figure-27.

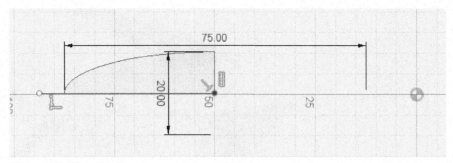

Figure-27. First sketch

Creating Surface

- Click on **Revolve** tool of **Modify** drop-down from **Toolbar**. The **REVOLVE** dialog box will be displayed; refer to Figure-28.

Figure-28. Revolve dialog box

- You need to select the sketch and axis of sketch to apply the revolve feature; refer to Figure-29.

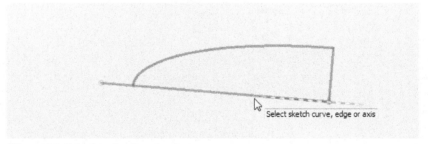

Figure-29. Selecting sketch and axis

- You need to revolve the sketch to 360 degree to create the face of aircraft.
- Click on **Extrude** tool of **Create** drop-down from **Toolbar**. The **EXTRUDE** dialog box will be displayed.
- Select the sketch as displayed to extrude; refer to Figure-30.

Figure-30. Selecting sketch to extrude

- Click in the **Distance** edit box and enter the value as **90**.
- After specifying the parameters, click on **OK** button from **EXTRUDE** dialog box to complete the extrude process.
- Click on **Offset Plane** tool of **Construct** drop-down from **Toolbar**. The **OFFSET PLANE** dialog box will be displayed.
- Select the YZ plane as shown in Figure-31.

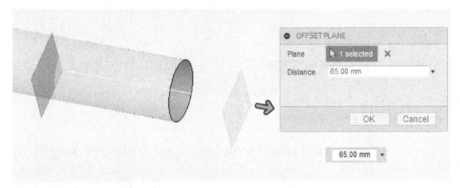

Figure-31. Creating offset plane

- Click in the **Distance** edit box of **OFFSET PLANE** dialog box and enter the value as **65**.
- After specifying the parameters, click on **OK** button from **OFFSET PLANE** dialog box to create the plane.
- Click on the **Circle** tool from **Sketch** drop-down and select the recently created plane as reference for creating sketch.
- Create the circle as shown in Figure-32.

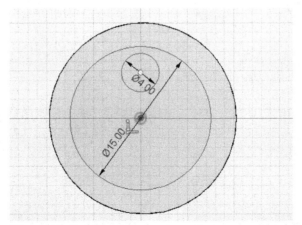

Figure-32. Creating circle

- After creating the sketch, click on **Stop Sketch** button from **Toolbar**.
- Click on **Loft** tool of **Create** drop-down from **Toolbar**. The **LOFT** dialog box will be displayed.
- You need to select the two sketch to create loft feature; refer to Figure-33.

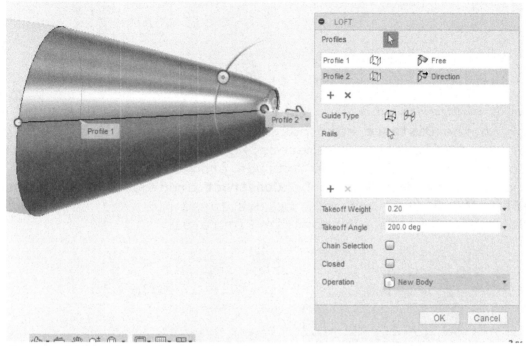

Figure-33. Selecting sketch for loft feature

- Enter the parameters of **Takeoff Weight** and **Takeoff Angle** as displayed in above figure and click on **OK** button from **LOFT** dialog box to complete the process.
- Create an offset plane as discussed earlier in this section. The parameters of offset plane is displayed in Figure-34.

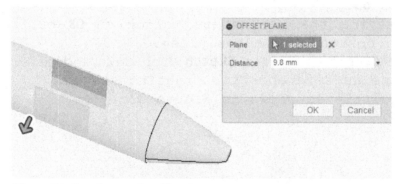

Figure-34. Creating another offset plane

- Click on the **Line** tool and select the recently created sketch. Create the sketch with the help of **Ellipse, Line,** and **Mirror** tool as shown in Figure-35.

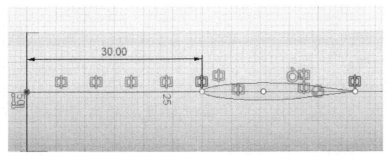

Figure-35. Creating sketch for wings

- After creating the sketch, click on **Stop Sketch** button from **Toolbar**.
- Now again, create the offset plane as shown in Figure-36.

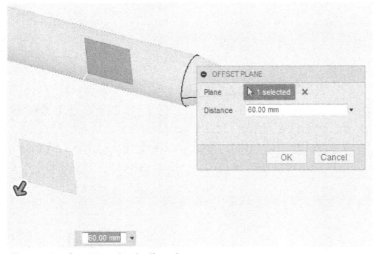

Figure-36. Creating third offset plane

- Click on the **Ellipse** tool from **Sketch** drop-down and select the recently created plane.
- Create the sketch as shown in Figure-37.

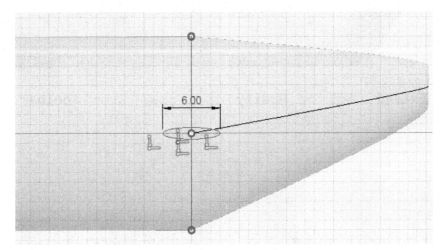

Figure-37. Creating sketch

- Click on the **Stop Sketch** drop-down from **Toolbar**.
- Click on the **Loft** tool from **Create** drop-down. The **LOFT** dialog box will be displayed.
- Select the recently created sketches and create the loft as shown in Figure-38.

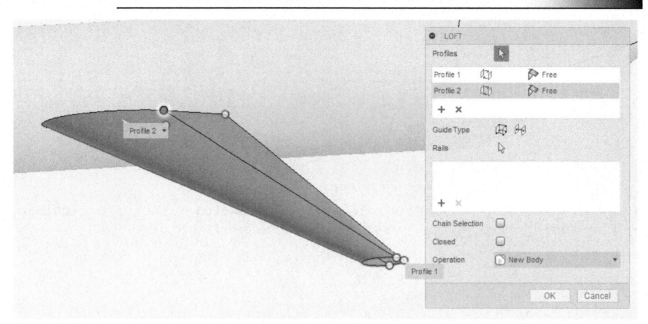

Figure-38. Creating wing

- After selection of profiles for loft, click on **OK** button from **LOFT** dialog box to complete the loft process.
- The wing of aircraft is not attached to the aircraft body; refer to Figure-39. To join the wing, you need to apply the **Extend** tool.

Figure-39. Wing of aircraft

- Click on **Extend** tool of **Modify** drop-down from **Toolbar**. The **EXTEND** dialog box will be displayed; refer to Figure-40.

Figure-40. EXTEND dialog box

- The **Select** button of **Edges** option is active by default. You need to select the open edge of wing to join this to aircraft.
- Click in the **Distance** edit box of **EXTEND** dialog box and enter the value as **0.50**.

- After specifying the parameters, click on the **OK** button from **Extend** dialog box to complete the process; refer to Figure-41.

Figure-41. Joining wing to aircraft

- Click on **Mirror** tool of **Create** drop-down from **Toolbar**. The **MIRROR** dialog box will be displayed; refer to Figure-42.

Figure-42. MIRROR DIALOG box

- Click on **Pattern Type** drop-down from **MIRROR** dialog box and select **Faces** option.
- Click on **Select** button of **Objects section** and select the recently created wing.
- Click on **Select** button of **Mirror Plane** option and select the **XY** plane; refer to Figure-43.

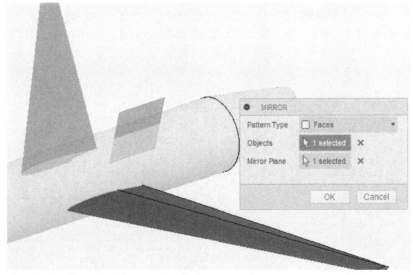

Figure-43. Creating Mirror copy of wing

- Click on **OK** button from **MIRROR** dialog box to complete the process.
- Click on the **Offset Plane** tool of **Create** drop-down from **Toolbar**. The **OFFSET PLANE** dialog box will be displayed.
- Create an offset plane as shown in Figure-44.

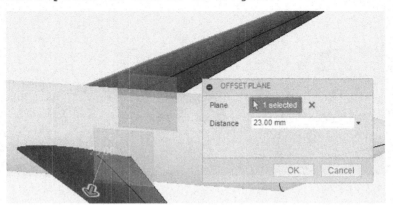

Figure-44. Creating fourth offset plane

- Click on **Ellipse** tool from **Sketch** drop-down and select the XY plane for sketch.
- Create a sketch as shown in Figure-45.

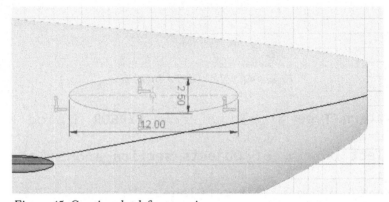

Figure-45. Creating sketch for rear wing

- After creating sketch, click on **Stop Sketch** button from **Toolbar**.
- Again, click on **Ellipse** tool and select the recently created plane .
- Create the sketch on the selected plane as shown in Figure-46.

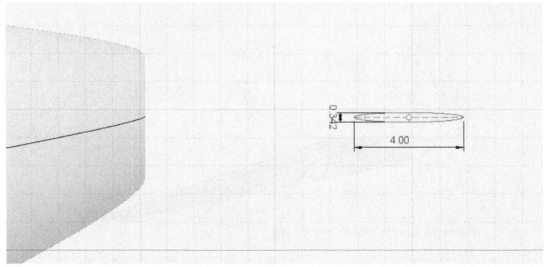

Figure-46. Creting sketch for rear wing

- Click on **Stop Sketch** button from **Toolbar** to exit the sketch.
- Click on **Loft** tool of **Create** drop-down from **Toolbar**. The **Loft** dialog box will be displayed.
- You need to select two recently created sketch in **Profiles** section of **LOFT** dialog box; refer to Figure-47.

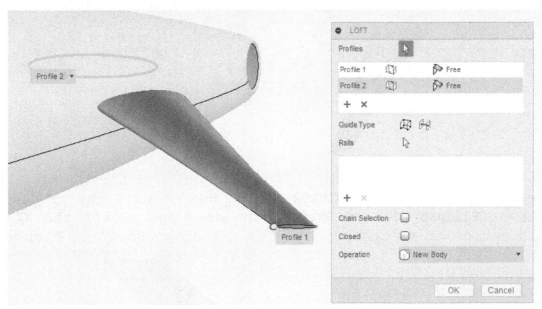

Figure-47. Applying loft tool for rear wing

- After specifying the parameters, click on **OK** button from **LOFT** dialog box to complete the process.
- To fill the open ends of wings, we need to apply the **Patch** tool.
- Click on the **Patch** tool of **Create** drop-down from **Toolbar**. The **PATCH** dialog box will be displayed; refer to Figure-48.

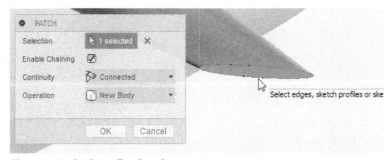

Figure-48. Applying Patch tool

- The **Select** button of **Selection** option is active by default. Click on the area displayed in the above figure.
- Click on the **Continuity** drop-down and select **Connected** option.
- After specifying the parameters, click on the **OK** button from **PATCH** dialog box to complete the process.
- To create other rear wing, click on **Mirror** tool of **Create** drop-down from **Toolbar**. The **MIRROR** dialog box will be displayed.
- Select the recently created wing in Objects section of MIRROR dialog box.
- Click on **Select** button of **Mirror** Plane option of **MIRROR** dialog box and select the **XY** plane as reference; refer to Figure-49.

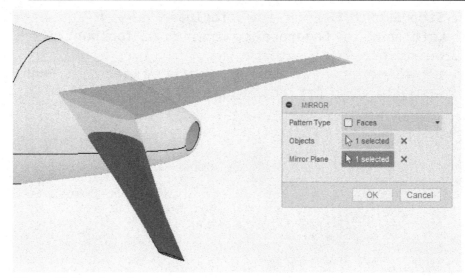

Figure-49. Creating other rear wing

- Click on **OK** button from **MIRROR** dialog box to exit the process.
- Click on **Ellipse** tool of **Sketch** drop-down and select the **XZ** plane for creation of sketch. Create the sketch as shown in Figure-50.

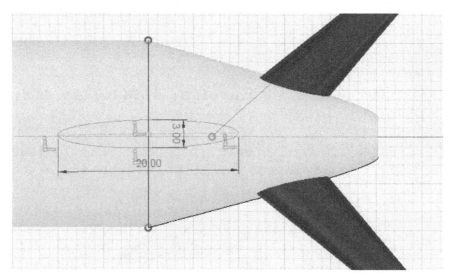

Figure-50. Creating sketch for third wing

- After creating the sketch, click on **Stop Sketch** button from **Toolbar**.
- Click on **Offset Plane** tool from **Construct** drop-down and create a plane with XZ plane as reference; refer to Figure-51.

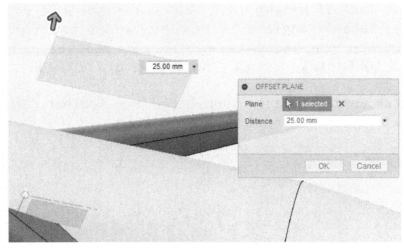

Figure-51. Creating offset plane for third rear wing

- Click on the **Ellipse** tool of **Sketch** drop-down and select the recently created plane for sketch.
- Create the sketch as shown in Figure-52.

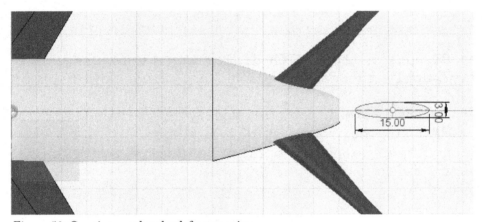

Figure-52. Creating another sketch for rear wing

- Click on **Stop Sketch** button from **Toolbar** to exit the sketch.
- Click on **Loft** tool from **Create** drop-down and select the recently created sketch under **Profiles** section; refer to Figure-53.

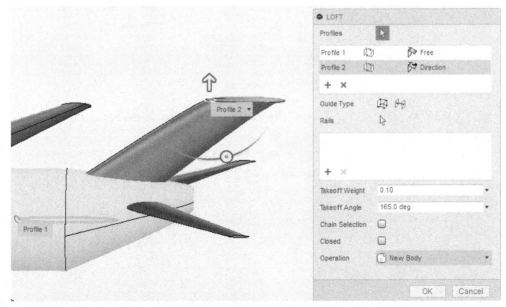

Figure-53. Applying Loft feature

- Click in the **Takeoff Weight** edit box and enter the value as **0.10**.
- Click in the **Takeoff Angle** edit box and enter the value as **165**.
- After specifying the parameters for loft feature, click on the **OK** button from **LOFT** dialog box. The third arm of air craft will be created.
- Click on **Patch** tool of **Create** drop-down from **Toolbar**. The **PATCH** dialog box will be displayed; refer to Figure-54.

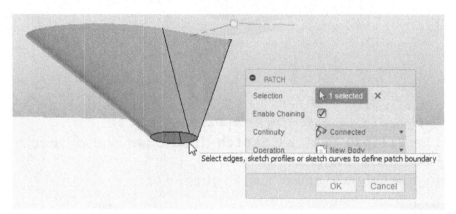

Figure-54. Applying patch on wing

- Click on **OK** button from **PATCH** dialog box to complete the process.
- Similarly, apply the patch tool to all wings and back of aircraft body.
- Click on **Reverse Normal** tool of **Modify** drop-down from **Toolbar**. The **REVERSE NORMAL** dialog box will be displayed.
- The **Select** button of **Faces** option is active by default. Click on the face to select; refer to Figure-55.

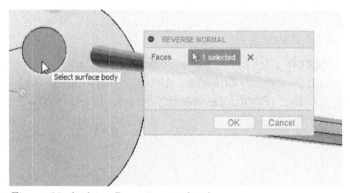

Figure-55. Applying Reverse normal tool

- Click on the **OK** button from **Reverse Normal** dialog box to complete the process.
- Click on **Center Rectangle** tool of **Sketch** drop-down and select the YZ plane for sketch; refer to Figure-56.

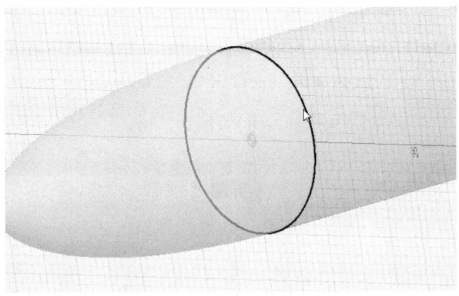

Figure-56. Selecting plane for sketch

- Create a rectangle on the selected plane as shown in Figure-57.

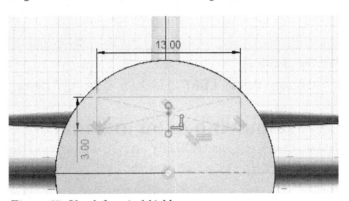

Figure-57. Sketch for windshield

- Click on **Stop Sketch** button from **Toolbar** to exit the sketch.
- Click on **Trim** tool of **Modify** drop-down from **Toolbar**. The **TRIM** dialog box will be displayed.
- The **Select** button of **Trim Tool** option is active by default. Select the recently created rectangular sketch and click on the inner area of trim preview; refer to Figure-58.

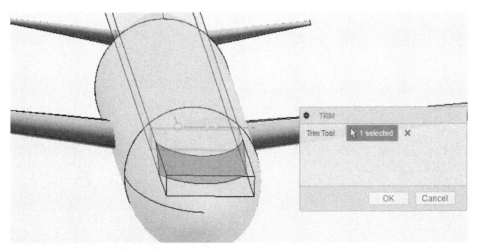

Figure-58. Selection of area for trim

- After selecting, click on the **OK** button of **TRIM** dialog box to exit the **Trim** tool.
- Click on **Loft** tool from **Create** drop-down. The **LOFT** dialog box will be displayed.
- Select the area in **Profiles** section as displayed in Figure-59.

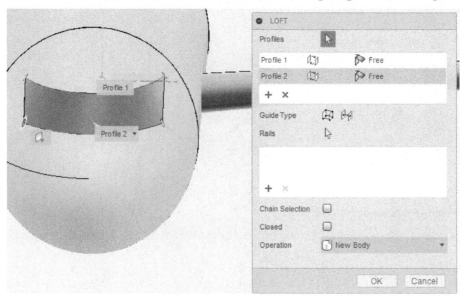

Figure-59. Creating windshield

- Click on the **OK** button from **LOFT** dialog box to complete the process.

Removing extra material

- Click on the **Trim** tool from **MODIFY** drop-down. The **TRIM** dialog box will be displayed; refer to Figure-6.

- The **Select** button of **Trim** Tool is active by default. Click on the extra material to remove; refer to Figure-60.

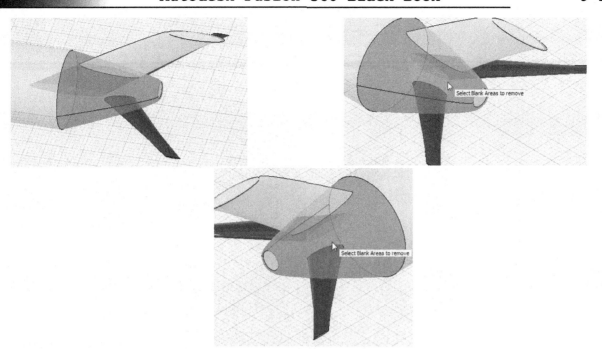

Figure-60. Removing extra material

- After selection of material to remove, click on the **OK** button from **TRIM** dialog box.
- After applying all the tools from above section. The aircraft will be created completely; refer to Figure-61.

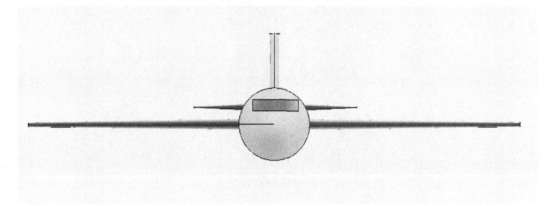

Figure-61. Front view of aircraft

PRACTICAL 2

Create the model of Jug as shown in Figure-62 and Figure-63.

Figure-62. Model of Jug

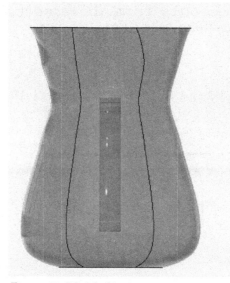

Figure-63. Model of jug1

Starting Autodesk Fusion 360

• Double click on the **Autodesk Fusion 360** software icon from desktop of your PC. The software window will be displayed along with Model Workspace.
• Click on the **Workspace** drop-down from **Toolbar** and select the **PATCH** workspace. The **PATCH** workspace will be displayed.

Creating Sketch

The first step to create jug is to create sketches. The procedure is discussed next.

- Click on the **Circle** tool of **Create** drop-down from **Toolbar** and select the **Top** plane as reference for sketch.
- Create the sketch as shown in Figure-64.

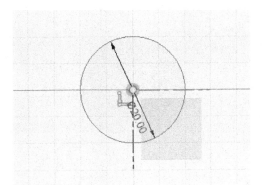

Figure-64. Creating base sketch

- Click on **Offset Plane** tool of **Construct** drop-down from **Toolbar**. The **OFFSET PLANE** dialog box will be displayed.
- The **Select** button of **Plane** option is active by default. Select the **Top** plane as reference.
- Click in the **Distance** edit box and enter the value as **2**; refer to Figure-65.

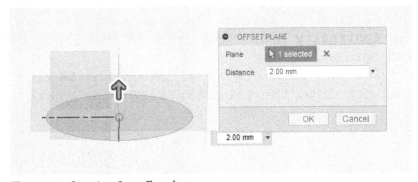

Figure-65. Creating first offset plane

- After specifying the parameters, click on the **OK** button from **OFFSET PLANE** dialog box to complete the process of creating plane.
- Click on the **Center Diameter Circle** tool of **Circle** cascading menu from **Sketch** drop-down and select the recently created offset plane as reference to create sketch.
- Create the sketch as shown in Figure-66.

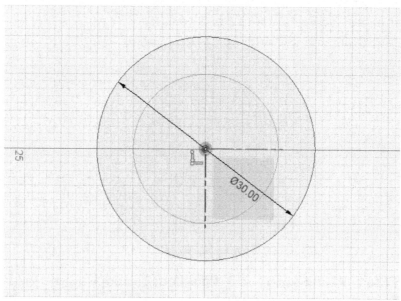

Figure-66. Creating second sketch

- Click on **Stop Sketch** button from **Toolbar**.
- Click on the **Patch** button from **Create** drop-down. The **PATCH** dialog box will be displayed.
- The **Select** button of **Selection** option from **PATCH** dialog box is active by default. You need to select the first sketch to patch.
- Click on the **Continuity** drop-down from **PATCH** dialog box and select the **Connected** option.
- After specifying the parameters, click on the **OK** button from **PATCH** dialog box to complete the process; refer to Figure-67.

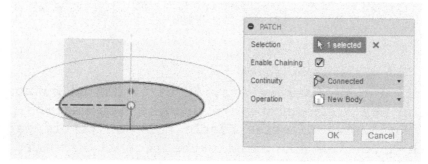

Figure-67. Creating Patch of first sketch

- Click on the **Offset Plane** tool of **Construct** drop-down from **Toolbar**. The **OFFSET PLANE** dialog box will be displayed.
- The **Select** button of **Plane** option is active by default. Select the first offset plane as reference.
- Click in the **Distance** edit box and enter the value as **22**; refer to Figure-68.

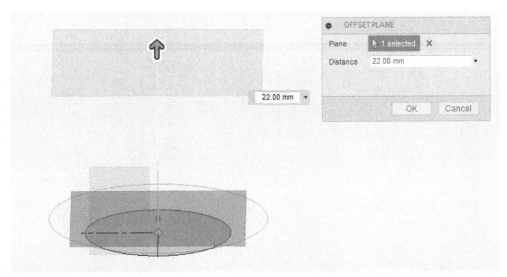

Figure-68. Creating third offset plane

- After specifying the parameter, click on the **OK** button from **OFFSET PLANE** dialog box to create the plane.
- Click on the **Center Diameter Circle** tool of **Circle** cascading menu from **Sketch** drop-down and select the recently created offset plane as reference to create sketch.
- Create the sketch as shown in Figure-69.

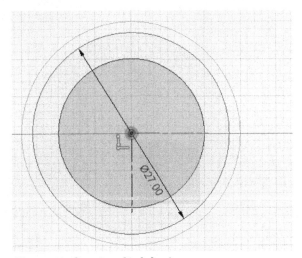

Figure-69. Creating third sketch

- Click on **Stop Sketch** button from **Toolbar** to exit the sketch.
- Click on the **Offset Plane** tool of **Construct** drop-down from **Toolbar**. The **OFFSET PLANE** dialog box will be displayed.
- The **Select** button of **Plane** option is active by default. Select the second offset plane as reference.
- Click in the **Distance** edit box and enter the value as **7**; refer to Figure-70. You can also set the distance by moving the drag handle.

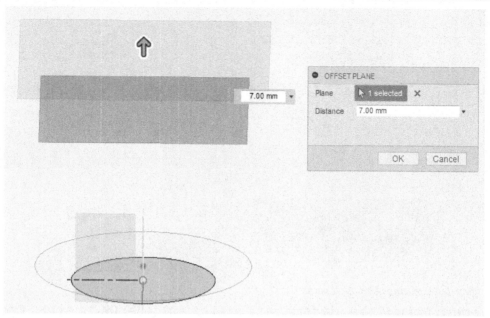

Figure-70. Creating third offset plane

- After specifying the parameter, click on the **OK** button from **OFFSET PLANE** dialog box to create the plane.
- Click on the **Center Diameter Circle** tool of **Circle** cascading menu from **Sketch** drop-down and select the recently created offset plane as reference to create sketch.
- Create the sketch as shown in Figure-71.

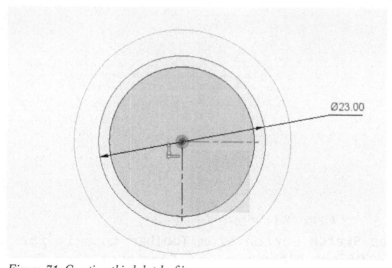

Figure-71. Creating third sketch of jug

- Click on **Stop Sketch** button from **Toolbar** to exit the sketch.
- Click on the **Offset Plane** tool of **Construct** drop-down from **Toolbar**. The **OFFSET PLANE** dialog box will be displayed.
- The **Select** button of **Plane** option is active by default. Select the third offset plane as reference.
- Click in the **Distance** edit box and enter the value as **8**; refer to Figure-72. You can also set the distance by moving the drag handle.

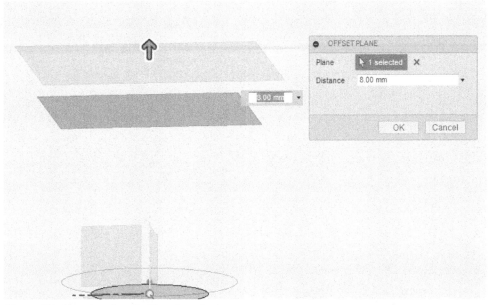

Figure-72. Creating fourth offset plane for jug

- After specifying the parameter, click on the **OK** button from **OFFSET PLANE** dialog box to create the plane.
- Click on the **Center Diameter Circle** tool of **Circle** cascading menu from **Sketch** drop-down and select the recently created offset plane as reference to create sketch.
- Create the sketch as shown in Figure-73.

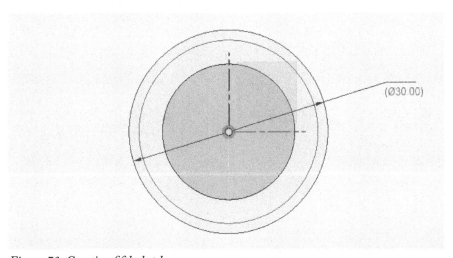

Figure-73. Creating fifth sketch

- Click on **Stop Sketch** button from **Toolbar** to exit the sketch.
- Click on the **Offset Plane** tool of **Construct** drop-down from **Toolbar**. The **OFFSET PLANE** dialog box will be displayed.
- The **Select** button of **Plane** option is active by default. Select the third offset plane as reference.
- Click in the **Distance** edit box and enter the value as **8**; refer to Figure-74. You can also set the distance by moving the drag handle.

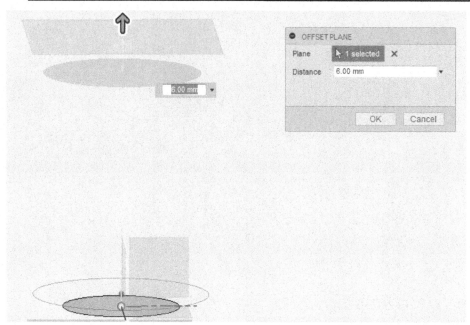

Figure-74. Creating fifth offset plane

- After specifying the parameter, click on the **OK** button from **OFFSET PLANE** dialog box to create the plane.
- Click on the **Circle** tool of **Sketch** drop-down from **Toolbar** and select the recently created plane as reference.
- Create the sketch as shown in Figure-75 with the help of **Circle**, **Trim**, and **Arc** tool.

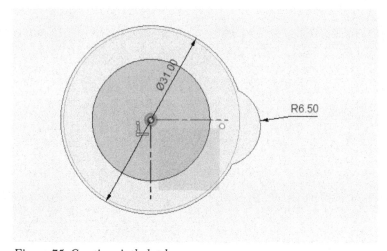

Figure-75. Creating sixth sketch

Creating Loft feature

Till, now we have created the sketch for jug. Now, we will apply the loft feature to the jug. The procedure is discussed next.

- Click on the **Loft** tool of **Create** drop-down from **Toolbar**. The **LOFT** dialog box will be displayed.
- You need to select all the sketch from button to top in **Profiles** section of **LOFT** dialog box; refer to Figure-76.

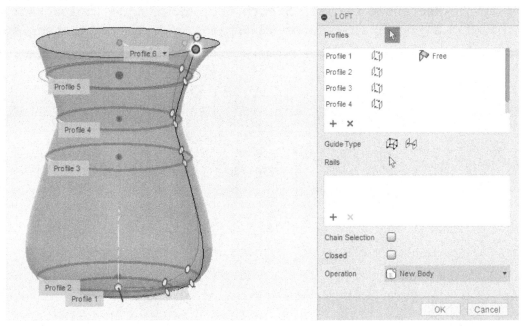

Figure-76. Creating loft feature of jug

- After selecting the profiles, click on the **OK** button from **LOFT** dialog box to complete the loft feature.

Here loft feature is completed. Now, we will create the handle of jug.

- Click on the **Offset Plane** tool of **Construct** drop-down from **Toolbar**. The **OFFSET PLANE** dialog box will be displayed.
- The **Select** button of **Plane** option is active by default. Select the YZ plane as reference.
- Click in the **Distance** edit box and enter the value as **-10** ; refer to Figure-77.

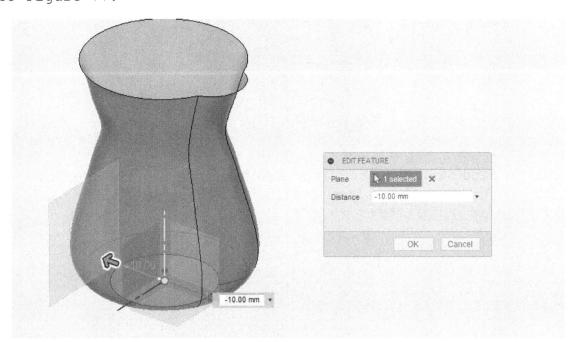

Figure-77. Creating sixth offset plane

- After specifying the parameter, click on the **OK** button from **OFFSET PLANE** dialog box to create the plane.

- Click on the **Rectangle** tool of **Sketch** drop-down from **Toolbar** and create the sketch as shown in Figure-78 with the help of **Rectangle** tool.

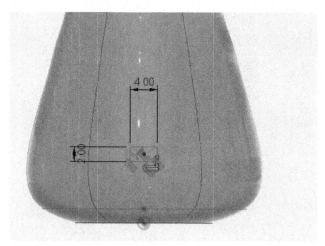

Figure-78. Creting seventh sketch

- Click on **Stop Sketch** button from **Toolbar** to exit the sketch.
- Click on the **Arc** tool from Sketch drop-down and select the **XY** plane.
- Create the sketch as shown in Figure-79.

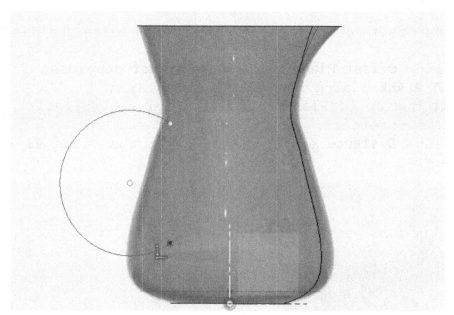

Figure-79. Creating eighth sketch

- Click on **Stop Sketch** button from **Toolbar** to exit the sketch.
- Click on the **Sweep** tool from **Create** drop-down. The **SWEEP** dialog box will be displayed; refer to Figure-80.

Figure-80. Sweep

- The **Select** button of **Profile** option is active by default. You need to select the recently created rectangular sketch.
- Click on the **Select** button of **Path** option and select the recently created arc; refer to Figure-81.

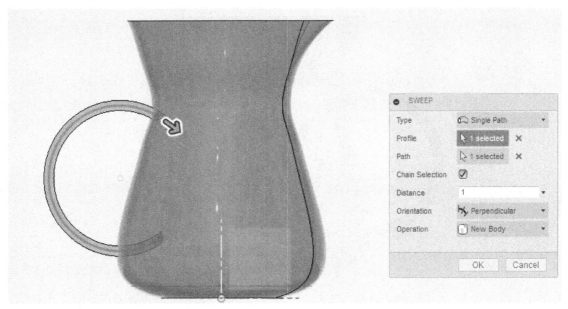

Figure-81. Creating sweep feature for jug

- Click on the **Orientation** drop-down of **SWEEP** dialog box and select the Perpendicular option.
- After specifying the parameters, click on the **OK** button from **SWEEP** dialog box to complete the feature.

Till here, we have created the model of jug but we need to apply the trim tool on the jug to remove the unwanted parts.

- Click on the **Trim** tool of **Modify** drop-down from **Toolbar**. The **TRIM** dialog box will be displayed.
- The **Select** button of **Trim Tool** is active by default. Click on the interior of jug and select the unwanted parts; refer to Figure-82.

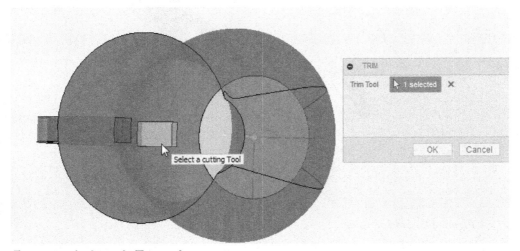

Figure-82. Applying the Trim tool

- After selecting, click on the **OK** button from **Trim** dialog box to trim the material.
- The model of jug is created successfully; refer to Figure-83.

Figure-83. Actual model of jug

- You can apply the physical material or appearance as required to the jug.
- You can also create the model of jug with the help of **Revolve** tool in place of **Loft** tool.

PRACTICE 1

Create the model of mouse as shown in Figure-84 and Figure-85.

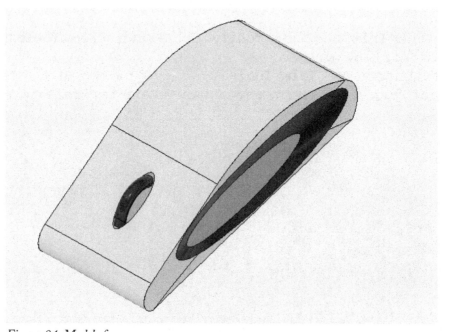

Figure-84. Model of mouse

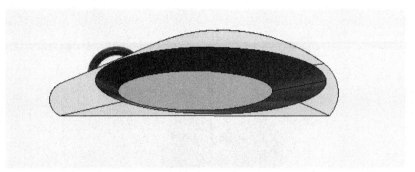

Figure-85. Right view of mouse

PRACTICE 2

Create the model of water bottle as shown in Figure-86 and Figure-87. The drawing view of water bottle is shown in Figure-88.

Figure-86. View of water bottle

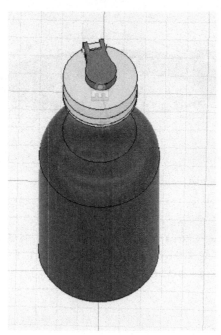

Figure-87. Top view of water bottle

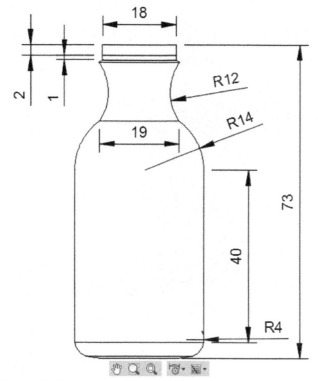

Figure-88. Dimension view of water bottle

For Students Notes

FOR STUDENTS NOTES

Chapter 10

Rendering and Animation

Topics Covered

The major topics covered in this chapter are:

- *Environment Library Tab*
- *Texture Map Control*
- *In-Canvas Render Setting*
- *Render*
- *Animation Workspace*
- *Transform Components*
- *Manual Explode*

INTRODUCTION

When a designer is working on a 3D Model, the model he creates is generally a mathematical representation of points and surfaces in three-dimensions. The conversion of a scene of 3D model through mathematical approximation to a finalized 2D image is known as Rendering. During this process, the entire scene's textural and lighting information are combined to determine the color value of each pixel in the finalized image.

INTRODUCTION TO RENDER WORKSPACE

The **Render Workspace** is used to create the photo-realistic image from a scene of model. There are various tools used for rendering process which are discussed next.

Scene Settings

The **Scene Settings** tool is used to control the lighting, background color, ground effects, and camera rendering. The procedure to use this tool is discussed next.

- Click on the **Scene Settings** tool of **Setup** panel from **Toolbar**; refer to Figure-1. The **SCENE SETTINGS** dialog box will be displayed; refer to Figure-2.

Figure-1. Scene Settings tool

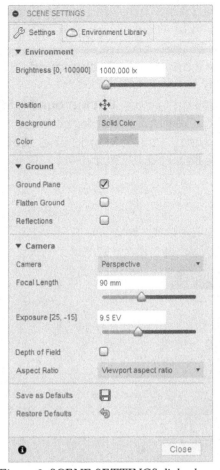

Figure-2. SCENE SETTINGS dialog box

Environment section

- Click on the **Brightness** edit box and specify the value of brightness for model. You can also move the **Brightness** slider to adjust the brightness for the model.
- If you want to control the position and rotation of lights then, click on the **Position** button of **Environment** section from **Settings** tab. The drag handle along with some tools will be displayed on the canvas screen; refer to Figure-3.

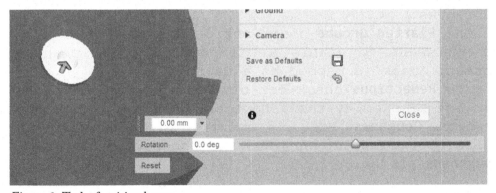

Figure-3. Tools of position button

- Click in the floating window and enter the position of lights. You can also set the position by moving the drag handle from model.
- Click in the **Rotation** edit box and enter the value of rotation of

lights. You can also adjust the value by moving the **Rotation** slider.

- If you want to reset all the changed settings then, click on the **Reset** button from rotation tools.
- Select **Environment** option from **Background** drop-down if you want to use environment image.
- Select **Solid Color** option from **Background** drop-down if you want to set a particular color to the background.
- To select a solid color, click on the **Color** button. A color box will be displayed; refer to Figure-4.

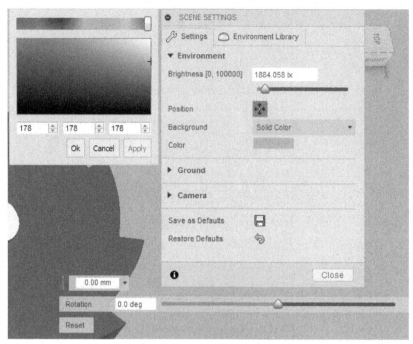

Figure-4. Selecting color for background

- Select the required color from color box and click on the **Apply** button to set the selected color as background color.

Ground section

- Select the **Ground Plane** check box of **Ground** section from **Settings** tab if you want to display the shadow and reflection of model on the ground canvas.
- Select the **Flatten Ground** check box of **Ground** section from **Settings** tab if you want to enable a "textured" ground plane where the environment image is mapped as a texture.
- Select the **Reflections** check box of **Ground** section from **Settings** tab if you want to display the reflection of model on the ground.
- Click in the **Roughness** edit box and enter the value of sharpness of reflection. You can also set the sharpness of reflection by moving the **Roughness** slider.

Camera section

- Select **Orthographic** option of **Camera** drop-down from **Camera** section if you want to set the orthographic view of model.
- Select the **Perspective** option of **Camera** drop-down from **Camera** section if you want to set the perspective view of model.

- Select the **Perspective with Ortho Faces** option from **Camera** drop-down if you want to set the perspective view of model with ortho faces.
- Click in the **Exposure** edit box of **Camera** section from **Settings** tab and enter the value of camera exposure. You can also set the camera exposure by moving the **Exposure** slider.
- Select the **Depth of Field** check box if you want to apply depth of field effect. Depth of field is used to focus on specific area of the model. This effect is only enabled when **Ray Tracing** is enabled. On selecting the **Depth of Field** check box, the dialog box will be updated; refer to Figure-5.

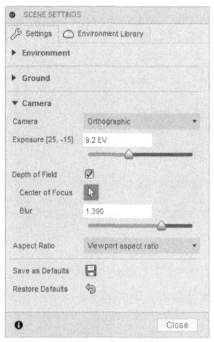

Figure-5. Updated dialog box on selecting depth of field check box

- The **Center of Focus** option is selected by default. You need to select the center of model for focusing.
- Click in the **Blur** edit box and enter the value of blur the model except the selected center of focus. You can also set the value of blur by moving the **Blur** slider.
- Click on the **Aspect Ratio** drop-down and select the required aspect ratio of canvas screen.
- Click on **Save as Default** button to save the setting as default setting.
- Click on the **Restore Defaults** button to restore the changed settings to default.

Environment Library tab

Click on the **Environment Library** tab from **SCENE SETTINGS** dialog box. The tab will be displayed; refer to Figure-6.

Figure-6. Environment library tab

- The **Current Environment** section shows the applied environment on the current model.
- If you want to change the environment of the model then drag a specific environment from **Library** section and drop it to the current model or in the **Current Environment** section. You can download the environments in library by using the **Download** button next to them in the list.
- If you want to apply the custom environment for model then click on the **Attach Custom Environment** button of **Environment Library** tab from **SCENE SETTINGS** dialog box. The **Open** dialog box will be displayed; refer to Figure-7.

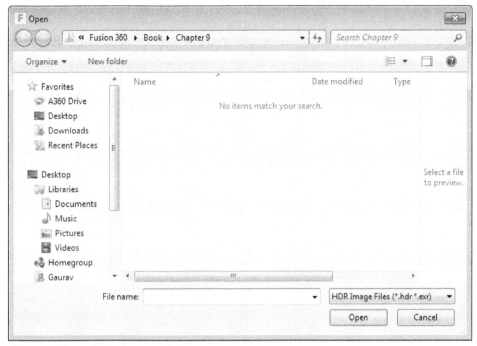

Figure-7. Open dialog box attaching cutom environment

- Select the required file and click on **Open** button. The environment will be applied to model.
- After selecting the required environment from **Environment Library** tab, click on **Close** button to exit the **Scene Settings** tool.

Note- There is no **Timeline** in the **Render** Workspace. The applied tool in the **RENDER** workspace will not be recorded.

Texture Map Control

The **Texture Map Control** tool is used to set the orientation of the texture applied to the face or model. The procedure to use this tool is discussed next.

- Click on the **Texture Map Control** tool of **Setup** panel from **Toolbar**; refer to Figure-8. The **TEXTURE MAP CONTROLS** dialog box will be displayed; refer to Figure-9.

Figure-8. Texture Map Controls tool

Figure-9. TEXTURE MAP
CONTROLS dialog box

- The **Select** button of **Selection** option is selected by default. You
  need to select the texture of model. The updated **TEXTURE MAP CONTROLS**
  dialog box will be displayed refer to Figure-10.

Figure-10. Updated TEXTURE MAP
CONTROLS dialog box

Automatic

- Select the **Automatic** option from **Projection Type** drop-down if you
  want to adjust the texture on the model using model topology.

Planer

- Select the **Planer** option from **Projection Type** drop-down if you want
  to map the image or texture on the model as projection.
- On selecting the **Planer** option, you need to select the axis for
  adjusting the texture. On selecting the texture the dialog box will
  be updated; refer to Figure-11.

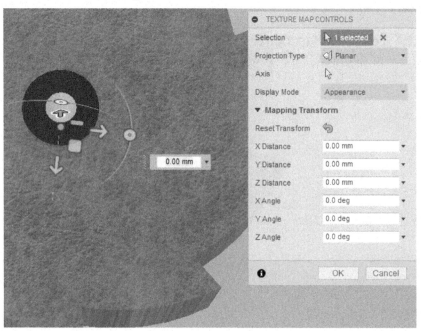

Figure-11. Updated dialog box after selecting axis

- Adjust the image or texture by moving the manipulator or by entering the values in specific edit box from **TEXTURE MAP CONTROLS** dialog box.
- After adjusting the texture, click on the **OK** button from **TEXTURE MAP CONTROLS** dialog box.

Box

- Click on the **Box** option from **Projection Type** drop-down if you want to map the image on model into box like objects.
- On selecting the **Box** option, the manipulator and updated dialog box will be displayed; refer to Figure-12.

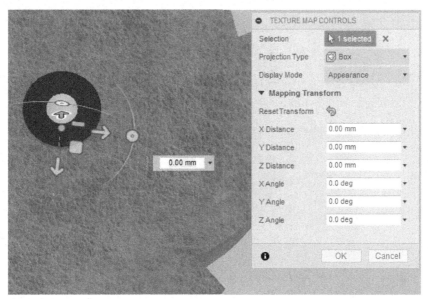

Figure-12. Dialog box on selecting Box-option

- Click in the specific edit box and enter the desired value. You can also set the value by moving the arrow of manipulator.
- If you want to reset the values of **Mapping Transform** or manipulator then click on the **Reset Transform** button.

Spherical

- Select the **Spherical** option from **Projection Type** drop-down if you want to map texture into a spherical object.
- On selecting the **Spherical** option, the manipulator along with updated dialog box will be displayed.
- The options were discussed in last topic.

Cylindrical

- Select the **Cylindrical** option from **Projection Type** drop-down if you want to map texture into a cylindrical   object.
- On selecting the **Cylindrical** option, the manipulator along with updated dialog box will be displayed.
- The options were discussed in last topic.
- After specifying the parameters, click on the **OK** button from **TEXTURE MAP CONTROLS** dialog box to exit the tool.

In-Canvas Render

The **In-Canvas Render** button is used to start and stop the In-canvas rendering process. The procedure to use this tool is discussed next.

- Click on the **In-Canvas Render** button from **IN-CANVAS RENDER** drop-down; refer to Figure-13. The Render quality slider will be displayed to display at the bottom right area of canvas screen; refer to Figure-14.

Figure-13. In-Canvas Render tool

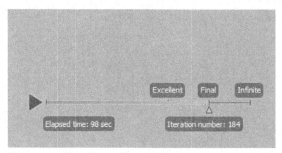

Figure-14. Render quality slider

- The current scene of the model will be changed into the photo-realistic 2D image.
- Clicking again on the **In-Canvas Render** button will stops the rendering process.

In-Canvas Render Setting

The **In-canvas Render Setting** tool is used to change the setting of rendering process. The procedure to use this tool is discussed next.

- Click on the **In-Canvas Render Settings** tool from **In-Canvas Render** panel; refer to Figure-15. The **IN-CANVAS RENDER SETTINGS** dialog box will be displayed; refer to Figure-16.

Figure-15. In-canvas Render Settings tool

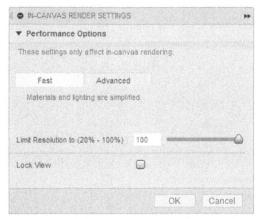

Figure-16. IN-CANVAS RENDER SET-
TINGS dialog box

- Set the quality of rendering by adjusting the **Limit Resolution** to slider. The higher quality rendering takes more time to calculate rendering process and lower quality rendering takes less time to calculate the rendering process.
- Select the **Lock View** check box to lock the current view of model otherwise clear the check box.
- After specifying the parameters, click on the **OK** button from **IN-CANVAS RENDER SETTINGS** dialog box.

Capture Image

The **Capture Image** tool is used to capture the current canvas as an image. The procedure to use this tool is discussed next.

- Click on the **Capture Image** tool of **IN-CANVAS RENDER** panel; refer to Figure-17. The **Image Options** dialog box will be displayed; refer to Figure-18.

Figure-17. Capture Image tool

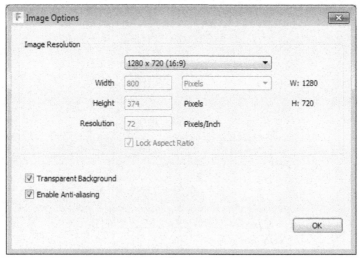

Figure-18. Image Options dialog box

- Click on the **Image Resolution** drop-down and select the required resolution of image.
- Select the **Transparent Background** check box if you want to set the transparent background of model.
- Select the **Enable Anti-Aliasing** check box if you want to activate the anti-aliasing property.
- After specifying the parameters, click on the **OK** button from **Image Options** dialog box. The **Save As** dialog box will be displayed; refer to Figure-19.

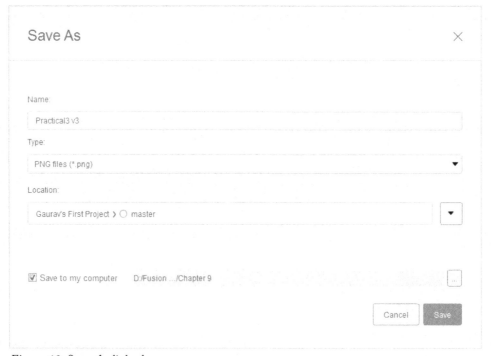

Figure-19. Save As dialog box

- Click in the **Name** edit box of **Save As** dialog box and enter the specific name for image.
- Click on the **Type** drop-down and select the required format of image file.

- Click on the **Location** drop-down button and select the location of saving image.
- Select the **Save to my computer** check box if you want to save the file to computer.
- After specifying the parameters for image, click on the **Save** button from **Save As** dialog box. The image will be saved to the specified location.

Render

The **Render** tool is used to create a high quality rendered image of the current scene. The procedure to use this tool is discussed next.

- Click on the **Render** tool from **Toolbar**; refer to Figure-20. The **RENDER SETTINGS** dialog box will be displayed; refer to Figure-21.

Figure-20. Render tool

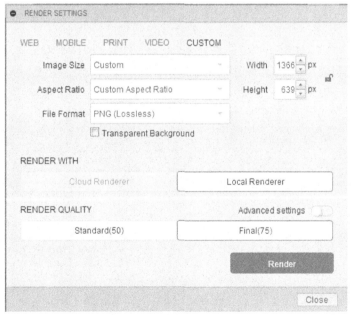

Figure-21. RENDER SETTINGS dialog box

WEB

- Click on the **WEB** tab of **RENDER SETTINGS** dialog box. The various options under **WEB** tab will be displayed; refer to Figure-22.

Figure-22. WEB tab

- Three highlighted options are the size of image in which your rendered file will save. Click on the required size.
- Click on the **Cloud Renderer** option from **RENDER WITH** section, if you want to render the file from Autodesk cloud. On selecting the **Cloud Renderer** button the dialog box will be updated; refer to Figure-23.

Figure-23. Updated WEB dialog box

- With **Local Renderer** option selected, click on the **Advanced settings** toggle button from **WEB** tab to manually set the render quality value. The **Render Quality** slider will be displayed; refer to Figure-24.

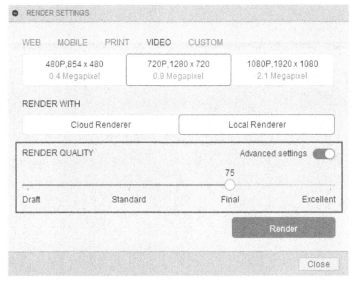

Figure-24. RENDER QUALITY slider for Web tab

- Set the value of quality from slider as required. If you want to automatically set the quality of render then again click on the **Advanced Settings** toggle button. The **Standard(50)** and **Final(75)** option will be displayed.
- Click on the required option from **RENDER QUALITY** section and click on the **Render** button from **WEB** tab. The rendering process will start.
- The completed rendered file will be saved in **RENDERING GALLERY**; refer to Figure-25.

Figure-25. RENDERING GALLERY

- To view the enlarged size of rendered image, click on the particular image. The image will be displayed in **RENDERING GALLERY** dialog box; refer to Figure-26.

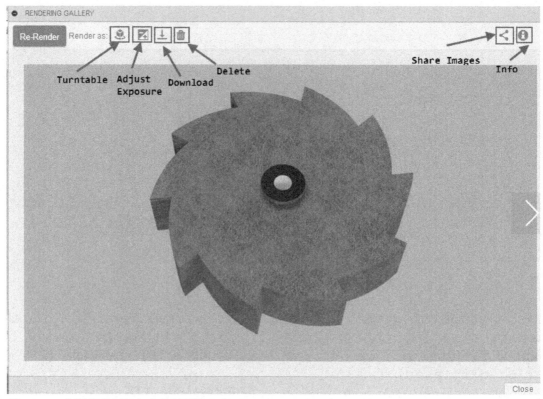

Figure-26. RENDERING GALLERY dialog box

- If you want to create a video of the rotating model then click on the **Turntable** button from **RENDERING GALLERY** dialog box. The **Renders Settings - Turntable** box will be displayed; refer to Figure-27.

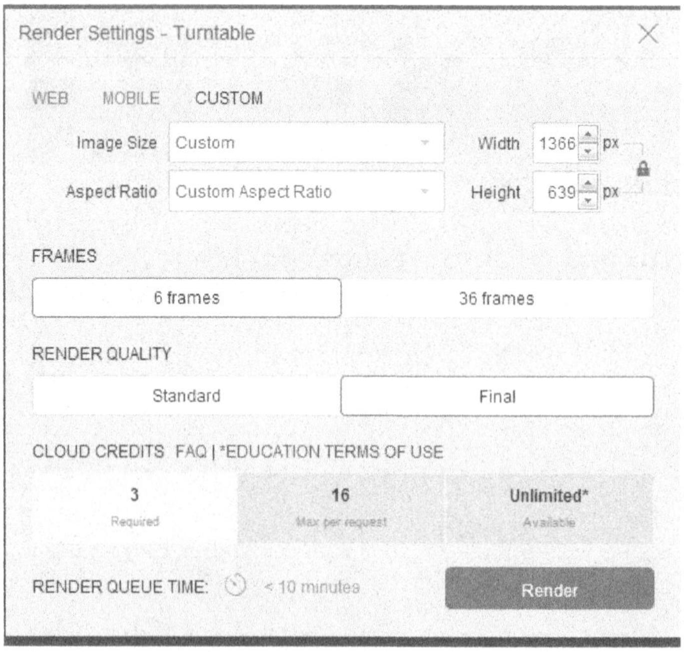

Figure-27. Render Settings-Turntable dialog box

- Specify the desired parameters and click on the **Render** button. The file will be re-created and displayed in **RENDERING GALLERY.**

- If you want to adjust the exposure of model then click on the **Adjust Exposure** button. The **Adjust Exposure** box will be displayed; refer to Figure-28.

Figure-28. Adjust Exposure box

- The value of exposure can be adjusted between -3.00 to 3.00. Adjust the value as required and click on the **Apply** button.
- If you want to download the current file then click on the **Download** button from **RENDERING GALLERY** dialog box. The **Save As** dialog box will be displayed.
- Select the location for saving file and click on the **Save** button from **Save As** dialog box to save the file.
- If you want to delete the current file then click on the **Delete** button from **RENDERING GALLERY** dialog box. The **Delete Image** dialog box will be displayed; refer to Figure-29.

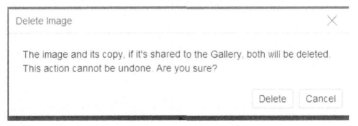

Figure-29. Delete Image box

- If you are sure about deleting the particular file then click on the **Delete** button from **Delete Image** dialog box otherwise click on **Cancel** button.
- If you want to share the image to Fusion 360 Gallery then click on the **Share Images** button from **RENDERING GALLERY** dialog box.
- If you want to see the information of the current file then click on the **Info** button from **RENDERING GALLERY** dialog box.
- After rendering the file, click on the **Close** button from **RENDERING GALLERY** dialog box to exit the rendering process.

ANIMATION

The **Animation** workspace is used to animate the joints of assembly. This workspace is also used to create exploded view of assembly. You can create the exploded view automatically as well as manually in this workspace. To activate the workspace, click on the **ANIMATION** option from the **Workspace** drop-down. The tools to perform animation will be displayed.

New Storyboard

A storyboard is automatically generated for each model. In the process of creating animation, additional storyboard can be added at any time as well. Each new storyboard has a default name "Storyboardx" where x is a number. You can edit this name by double-clicking on it. The procedure to create new storyboard is discussed next.

- Click on the **New Storyboard** button of **Toolbar** from **Animation Workspace**; refer to Figure-30. The **NEW STORYBOARD** dialog box will be displayed; refer to Figure-31.

Figure-30. New Storyboard tool

Figure-31. NEW STORY-BOARD dialog box

- Select the **Clean** option from **Storyboard Type** drop-down if you want to delete all previous actions of animation and want to create a new storyboard.
- Select the **Start from end of Previous** option from **Storyboard Type** drop-down if you want to use position of components at the end of previous animation as the starting position for the new animation.
- The created storyboard will be added at the bottom of **ANIMATION TIMELINE**; refer to Figure-32.

Figure-32. ANIMATION TIMELINE

Transform Components

The **Transform Components** tool is used to move and rotate the component. Note that the movement and rotation performed in this workspace will be recorded in animation. The procedure to use this tool is discussed next.

- Click on the **Transform Components** tool of **TRANSFORM** panel; refer to Figure-33. The **TRANSFORM COMPONENTS** dialog box will be displayed; refer to Figure-34.

Figure-33. Transform Components tool

Figure-34. TRANSFORM COMPONENTS dialog box

- The **Select** button of **Components** option is selected by default. Click on the component that you want to move or rotate. On selection of component, a manipulator will be displayed on the selected component along with updated **TRANSFORM COMPONENTS** dialog box; refer to Figure-35.

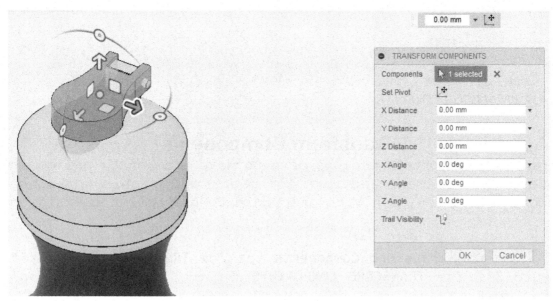

Figure-35. Manipulator on selected component

- The actions like moving or rotating the components are registered as animation in storyboard.
- Click on the specific edit box and enter the desired value. You can also set the parameters by moving the manipulator.
- After specifying the parameters, click on the **OK** button from **TRANSFORM COMPONENTS** dialog box to add the action in animation timeline.
- Similarly move or rotate the other components of model as required.
- To play the animation, you need to click on the **Play** button from **ANIMATION TIMELINE.**

Restore Home

The **Restore Home** tool is used to restore the model position as it is at the end of animation.
- Click on the **Restore Home** button from **TRANSFORM** drop-down; refer to Figure-36. The position of model will be restored as it is in **Modal Workspace.**

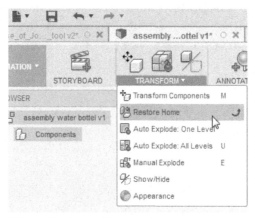

Figure-36. Restore Home tool

Auto Explode: One Level

The **Auto Explode: One Level** tool is used to automatically separate all selected components down to one level in hierarchy. The procedure to use this tool is discussed next.

- Select the parts of assembly to be seperated by holding the **CTRL** key. You can also select all components using window selection; refer to Figure-37.

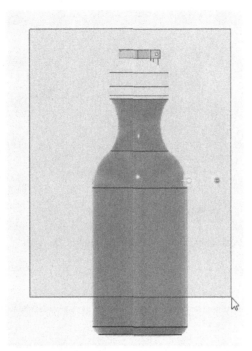

Figure-37. Window selection on component

- Click on the **Auto Explode: One Level** tool from **TRANSFORM** drop-down; refer to Figure-38. The components will be separated on the canvas screen; refer to Figure-39.

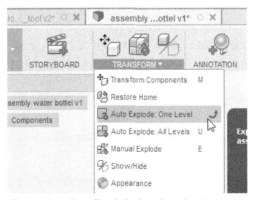

Figure-38. Auto Explode One Level tool

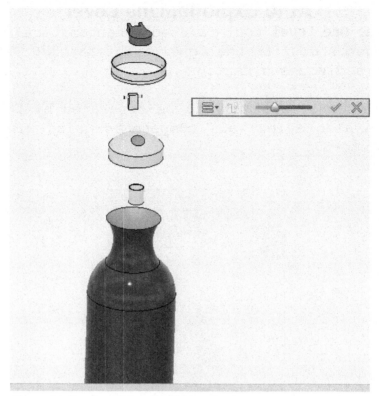

Figure-39. Preview of Auto Explode one level

- By moving the highlighted slider you can set the distance between component.
- After adjusting the space between components, click on the **OK** button from slider.

Auto Explode: All Levels

The **Auto Explode: All Levels** tool is used to To automatically separate all selected components down to all levels in component hierarchy. The procedure to use this tool is discussed next.

- Select the parts of components to separate by holding the **CTRL** key. You can also select the components using window selection.
- Click on the **Auto Explode: All Levels** tool from **TRANSFORM** drop-down; refer to Figure-40. The components of selected model will be separated.

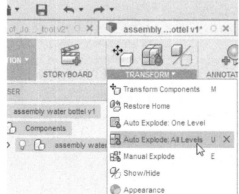

Figure-40. Auto Explode All Levels tool

- Click on the screen to view the separated component properly; refer to Figure-41.

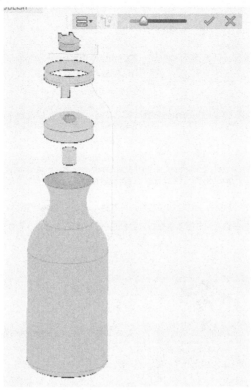

Figure-41. Preview of Explode

- Move the slider to adjust the distance between separated components of models.
- After adjusting the components, click on the ✅ **OK** button. The exploded view of assembly will be displayed; refer to Figure-42.

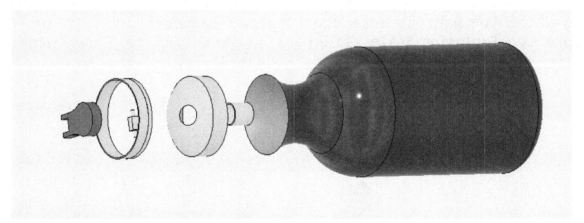

Figure-42. Exploded view of water bottle assembly

Manual Explode

The **Manual Explode** tool is used to manually separate the selected component. The procedure to use this tool is discussed next.

- Click on the **Manual Explode** tool of **Transform** drop-down from **Toolbar**; refer to Figure-43.

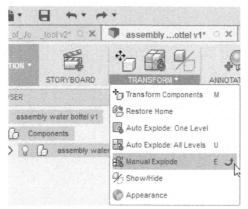

Figure-43. Manual Explode tool

- You need to select the components to be moved during explosion. Click on the component to separate; refer to Figure-44.

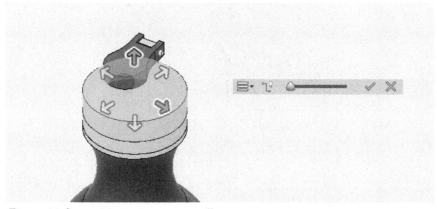

Figure-44. Seperating component manually

- On selecting a component, number of arrows will be displayed upon the selected component. You need to click on the arrow of direction in which you want to move the selected component.
- Move the **Explosion Scale** slider to separate the selected component from remaining model; refer to Figure-45.

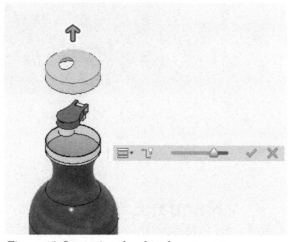

Figure-45. Separating the selected component

Show/Hide

The **Show/Hide** tool is used to show or hide the selected component. The procedure to use this tool is discussed next.

- Click on the desired component from model to show or hide.
- Now, click on the **Show/Hide** tool from **TRANSFORM** drop-down; refer to Figure-46. The selected component will be shown or hide based on its condition.

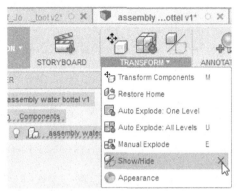

Figure-46. Show Hide tool

- To view the hidden component, click on light bulb from **Browser**; refer to Figure-47.

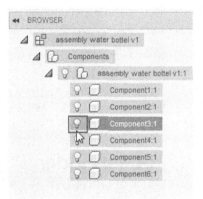

Figure-47. Viewing the hidden component

Create Callout

The **Create Callout** tool is used to add annotation to the animation. The procedure to use this tool is discussed next.

- Click on the **Create Callout** tool from **ANNOTATION** panel; refer to Figure-48. The callout symbol will be attached to the cursor.

Figure-48. Create Callout Tool

- Click on the component to add the callout. A text box will be displayed; refer to Figure-49.

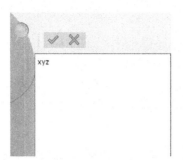

Figure-49. Callout Text box

- Enter the desired text in the **Text** box and click on **OK** button. The annotation callout will be added at the selected component.

View

The **View** tool is used to start and stop the recording of the actions or commands occurring on the model or component. When **View** tool is activated then it captures the actions like move for animation and when the **View** tool is deactivated it does not record the process; refer to Figure-50.

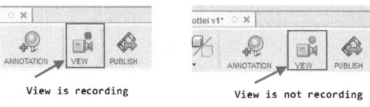

Figure-50. View tool

Publish Video

The **Publish Video** tool is used to publish the video. The procedure to use this tool is discussed next.

- Click on the **Publish Video** tool from **Toolbar**; refer to Figure-51. The **Video Options** dialog box will be displayed; refer to Figure-52

Figure-51. PUBLISH tool

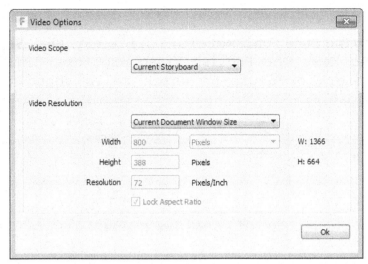

Figure-52. Video Options dialog box

- Select **Current Storyboard** option from **Video Scope** section if you want to create a video of current storyboard animation.
- Select **All Storyboard** option from **Video Scope** section if you want to create a video of all storyboard.
- Click on the **Current Document Window Size** drop-down from **Video Resolution** section or select the required video resolution.
- After specifying the parameters, click on the **OK** button from **Video Options** dialog box. The **Save As** dialog box will be displayed; refer to Figure-53.

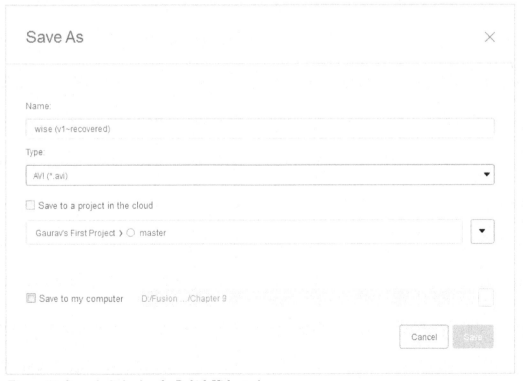

Figure-53. Save As dialog box for Pubish Video tool

- Click in the **Name** edit box and specify the name of file.
- Select the **Save to a project in the cloud** check box if you want to save the file online or in you Autodesk account.

- Select the **Save to my computer** check box if you want to save the file in your computer.
- After specifying the parameters, click on the **Save** button from **Save As** dialog box. The **Publish Video** progress dialog box will be displayed; refer to Figure-54 which tells about the progress of publishing video.

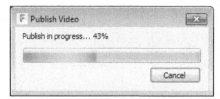

Figure-54. Publish Video progress dialog box

The video will be published online and saved to the computer at preferred location.

FOR STUDENT NOTES

For Student Notes

For Student Notes

Chapter 11

Drawing

Topics Covered

The major topics covered in this chapter are:

- *Drawing View*
- *Projected view*
- *Section View*
- *Detail View*
- *Centerline*
- *Datum Identifier*
- *Spline Ballon*
- *Output DWG, CSV, and PDF*

INTRODUCTION

Drawing is the engineering representation of a model on the paper. For manufacturing a model in the real world, we need some means by which we can tell the manufacturer what to manufacture. For this purpose, we create drawings from the models. These drawings have information like dimensions, material, tolerances, objective, precautions and so on.

STARTING DRAWING WORKSPACE

In Autodesk Fusion 360, we create drawings by using the Drawing environment. The procedure to open a file in **DRAWING Workspace** is discussed next.

- Open the desired model from **Data Panel** or saved directory.
- Click on **From Design** option from **DRAWING** cascading menu in **Workspace** drop-down. The **MODEL Workspace** will displayed along with model and **CREATE DRAWING** dialog box; refer to Figure-1

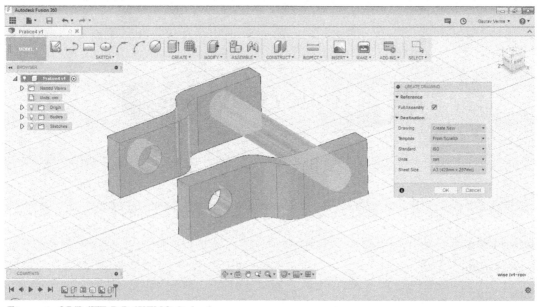

Figure-1. CREATE DRAWING dialog box

- Select the **Full Assembly** check box from **CREATE DRAWING** dialog box if you want to create drawing using whole assembly. Clear the check box if you want to select the component of a model.
- Click on the **Standard** drop-down from **Create Drawing** dialog box and select the standard for the drawing. **ASME** option uses the third angle projection and **ISO** option uses first angle projection.
- Click on the **Units** drop-down from **CREATE DRAWING** dialog box and select the unit for drawing.
- Click on the **Sheet Size** drop-down from **CREATE DRAWING** dialog box and select the size of sheet for drawing.

- After specifying the parameters, click on the **OK** button from **CREATE DRAWING** dialog box. The **DRAWING Workspace** will be displayed along with **DRAWING VIEW** dialog box; refer to Figure-2.
- The model will be attached to the cursor. Click on the drawing sheet to place the base view.

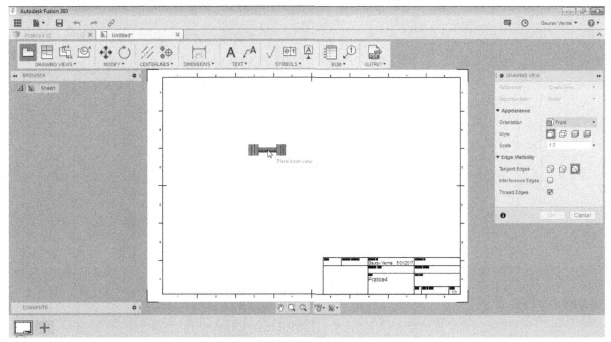

Figure-2. DRAWING Workspace

DRAWING VIEW dialog box

- Click on the **Orientation** drop-down and select the base view of model. Here **Front** view is default option. When you change the view from **Front** to **Top**, the view of model will automatically changed.
- Select the style of model from the **Style** section.
- Click in the **Scale** drop-down and select the required scaling factor for model.
- Select the required option from **Tangent Edges** options to apply the edges with the selected view.
- Select the **Interference Edges** check box if you want to choose the interference edges preference within the selected view.
- Select the **Thread Edges** check box if you want to choose the thread edges preferences within the selected view.
- After specifying the parameters in **DRAWING VIEW** dialog box, click on **OK** button. The Top view of model will be displayed; refer to Figure-3.

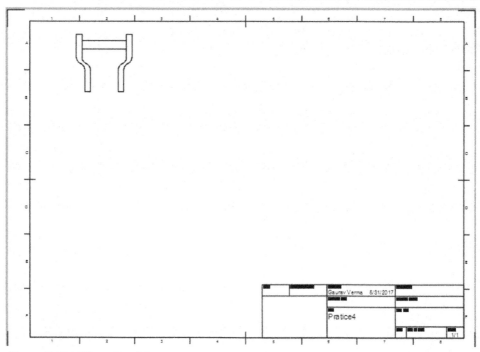

Figure-3. Top View of model

CREATING BASE VIEW

The **Base View** tool is used to create the drawing views that are derived directly from a 3D model. You can create various views of your model. They are projected views of 3D models, viewed from a specific direction. The procedure to use this tool is discussed next.

* Click on the **Base View** tool from **DRAWING VIEWS** panel; refer to Figure-4. The **DRAWING VIEW** dialog box will be displayed.

Figure-4. Base View tool

The options of **DRAWING VIEW** dialog box have been discussed earlier.

CREATING PROJECTED VIEW

The **Projected View** tool is used to create drawing views generated from an existing base or drawing view. The procedure to use this tool is discussed next.

* Click on the **Projected View** tool from **DRAWINGS VIEWS** panel; refer to Figure-5. The tool will be activated.

Figure-5. Projected View tool

- Click on the pre-existing base view from drawing sheet to select the view as parent view; refer to Figure-6.

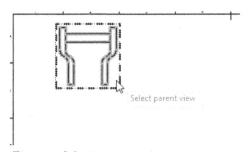

Figure-6. Selecting parent view

- Click to select the parent view. The projected view will be attached to the cursor; refer to Figure-7.

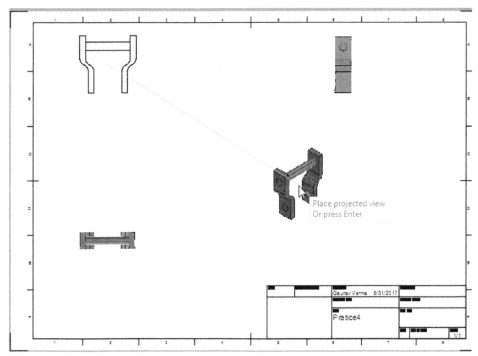

Figure-7. Placing projected view

- Place the projected view of model as desired and press **ENTER** key. The projected view(s) are created on drawing sheet; refer to Figure-8.

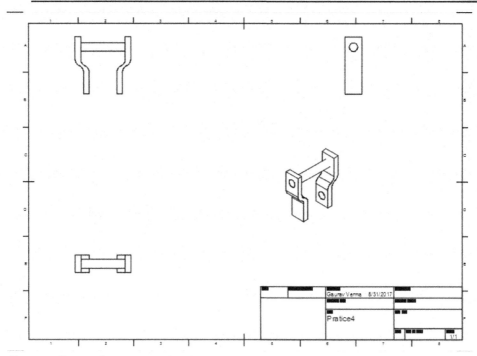

Figure-8. Projected views created

CREATING SECTION VIEW

The **Section View** tool is used to create full, half, offset, or aligned section views. The procedure to use this tool is discussed next.

*   Click on the **Section View** tool from **DRAWING VIEWS** panel; refer to Figure-9. The tool will be activated.

Figure-9. Section View tool

*   Click on the projected or base view from drawing sheet to select as parent view. The **DRAWING VIEW** dialog box will be displayed.
*   You need to divide the selected parent view into two parts to create half section half; refer to Figure-10.

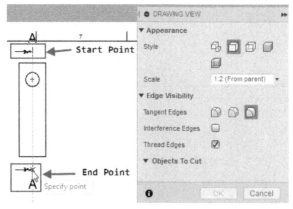

Figure-10. Creating section view

- Click on the parent view to specify the start point and end point. You can create different shape of section line to create desired type of section. On specifying the start and end point, click on **ENTER** key. The section view will be created and attached to the cursor; refer to Figure-11.

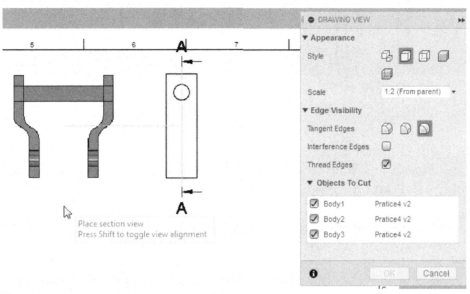

Figure-11. Section view attached to the cursor

- Place the section view at desired location and click on **OK** button from **DRAWING VIEW** dialog box. The section view will be created on drawing sheet; refer to Figure-12.

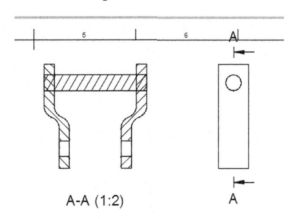

Figure-12. Created section view

- Note that the hatching is created automatically in the section view. To edit hatching double-click on it.
- If you want to edit the section annotation then double-click on it. The **TEXT** dialog box will be displayed; refer to Figure-13.

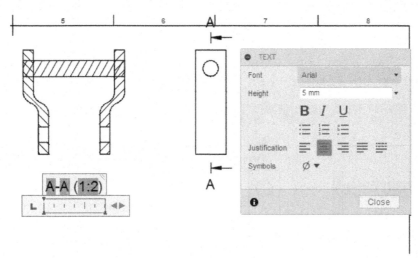

Figure-13. TEXT dialog box for annotation

- Click on the **Font** drop-down and select the required font from the list.
- Click in the **Height** edit box and specify the height of text and select the writing style of text from **TEXT** dialog box.
- Click on the desired text alignment from **Justification** section.
- If you want to write any special symbol then click on the required symbol from **Symbols** section.
- After specifying the parameters for the text, click in the text box from drawing sheet and enter the required text.
- Click on the **Close** button from **TEXT** dialog box to complete the process. The text will be displayed below the section view.

CREATING DETAIL VIEW

The **Detail View** tool is used to create detail view from any projected or base view. The procedure to use this tool is discussed next.

- Click on the **Detail View** tool of **DRAWING VIEWS** panel; refer to Figure-14. The tool will be activated.

Figure-14. Detail View tool

- You need to select the parent view from drawing sheet; refer to Figure-15. The **DRAWING VIEW** dialog box will be displayed.

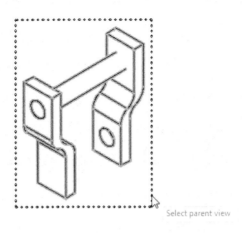

Figure-15. Selecting parent view for detail view

- You need to select the center point on parent view to specify the boundary; refer to Figure-16. Create the boundary on parent view as required; refer to Figure-17.

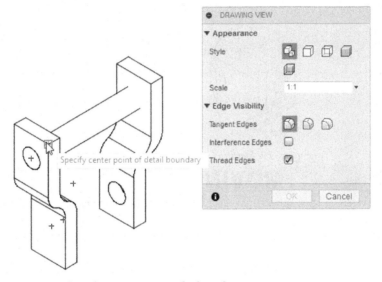

Figure-16. Specifying center point for boundary

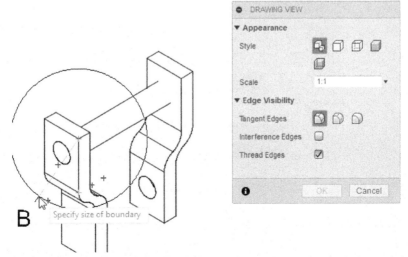

Figure-17. Creating boundary

- On creating the circular boundary, the detail view of model will be attached to the cursor; refer to Figure-18. Place it as desired.

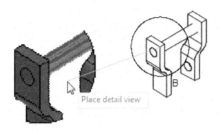

Figure-18. Placing detail view

- After placing the view, set the parameters of detail view from **DRAWING VIEW** dialog box and click on **OK** button. The detailed view will be created; refer to Figure-19.

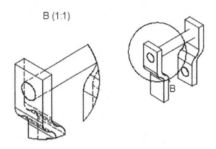

Figure-19. Detailed view

MOVING VIEWS

The **Move** tool is used to move a selected view from one place to another. The procedure to use this tool is discussed next.

- Click on the **Move** tool from **MODIFY** drop-down; refer to Figure-20. The **MOVE** dialog box will be displayed; refer to Figure-21.

Figure-20. Move tool

Figure-21. MOVE dialog box

- The **Select** button of **Selection** option in **MOVE** dialog box is selected by default. Click on the view to move from one place to other. The updated **MOVE** dialog box will be displayed; refer to Figure-22.
- Click on **Transform** button from **MOVE** dialog box and select a base point from the selected view. The selected view will get attached to the cursor; refer to Figure-23.

Figure-22. Updated Move dialog box

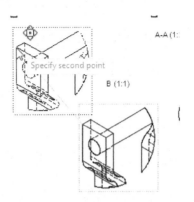

Figure-23. Specifying second point

- Click on the sheet to place the view. The view will be moved from previous position to new position.
- Click on the **OK** button from **MOVE** dialog box to complete the process.

ROTATING VIEWS

The **Rotate** tool is used to rotate the selected view as desired. The procedure to use this tool is discussed next.

- Click on the **Rotate** tool from **MODIFY** panel; refer to Figure-24. The **ROTATE** dialog box will be displayed; refer to Figure-25.

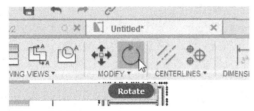

Figure-24. Rotate tool

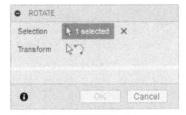

Figure-25. ROTATE dialog box

- The **Select** button of **Selection** option from **ROTATE** dialog box is selected by default. Click on the view to be rotated.
- Click on **Transform** button from **ROTATE** dialog box and select the base point from selected view. The view will attached to the cursor; refer to Figure-26.

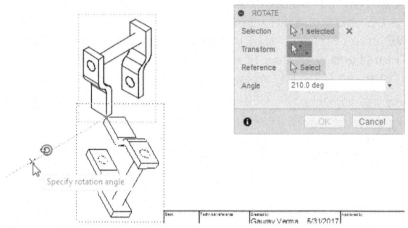

Figure-26. Rotating view

- You can adjust the angle of rotation of view by moving the cursor around parent view and place as required. If you want to specify the value of rotation then click in the **Angle** edit box of **ROTATE** dialog box and specify the value.
- You can also set the reference for rotation by clicking on **Select** button of **Reference** section and selecting the reference point from parent view.
- After specifying the parameters, click on **OK** button from **ROTATE** dialog box to complete the process.

CREATING CENTERLINE

The **Centerline** tool is used to create centre line. The procedure to use this tool is discussed next.

- Click on the **Centerline** tool from **CENTERLINES** panel; refer to Figure-27. The tool will be activated.
- You need to select parallel edges from a view to create a centerline. Click on the edge to select; refer to Figure-28. You will be asked to select the second edge.

Figure-27. Centerline tool

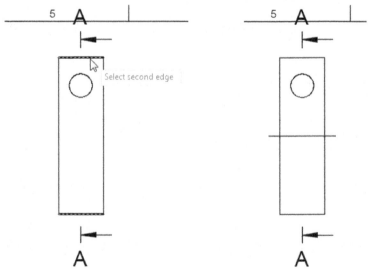

Figure-28. Creating centerline

- On selecting second edge, the centerline will be created.

CREATING CENTER MARK

The **Center Mark** tool is used to indicate the round edges of the selected object. The procedure to use this too is discussed next.

- Click on the **Center Mark** tool from **CENTERLINES** panel; refer to Figure-29. The tool will be activated.
- You need to select a round edge from any view.
- On selecting the edge, the center mark will be created at the selected edge; refer to Figure-30.

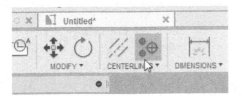

Figure-29. Center Mark tool

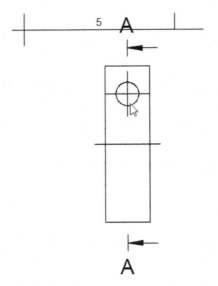

Figure-30. Creating center mark

EDGE EXTENSION

The **Edge Extension** tool is used to create an extension of associative edge for two intersecting (unequal) edges within an existing drawing view. The procedure to us this tool is discussed next.

- Click on the **Edge Extension** tool from **GEOMETRY** panel; refer to Figure-31. The tool will be activated.

Figure-31. Edge Extension tool

- You need to select the two non-parallel edges to create the extension. Click on the edge to select; refer to Figure-32.

- On selecting the second edge, the extended edge will be displayed; refer to Figure-33.

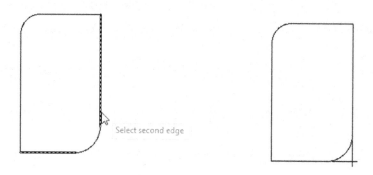

Figure-32. Selecting second edge   *Figure-33. Created edge extension*

DIMENSIONING

The **Dimension** tool is used to apply dimension to model. The procedure to use this tool is discussed next.

- Click on the **Dimension** tool from **DIMENSIONS** panel; refer to Figure-34. The tool will be activated.

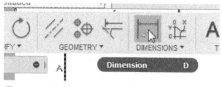

Figure-34. Dimension tool

- When you hover the cursor over on any geometry, the measurement of that dimension will be displayed; refer to Figure-35.

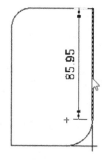

Figure-35. Checking dimension

- If you want to mention that dimension then click on the line. The dimension will attached to the cursor.
- Place the dimension at a specific distance as required.

Ordinate Dimensioning

The **Ordinate Dimension** tool is used to apply ordinate dimensioning. . The procedure to use this tool is discussed next.

- Click on the **Ordinate Dimension** tool from **DIMENSIONS** panel; refer to Figure-36. The tool will be activated and you will be asked to specify origin for ordinate dimensioning.
- Click at the desired location to place origin of ordinate dimensioning. The ordinate dimension will be attached to the cursor; refer to Figure-37.

Figure-36. Ordinate Dimension tool

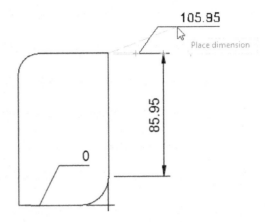

Figure-37. Calculating ordinate dimension

- Click at the desired location to place dimension text. Similarly, you can dimension horizontally/vertically.

CREATING TEXT

The **Text** tool is used to insert text into the active drawing. The procedure to use this tool is discussed next.

- Click on the **Text** tool from **TEXT** panel; refer to Figure-38. The **Text** tool is activated and you will be asked to draw a text box.
- Click to specify the first corner.
- After specifying the first corner, click on the screen to specify the second corner; refer to Figure-39. The **TEXT** dialog box will be displayed; refer to Figure-40.

Figure-38. Text tool

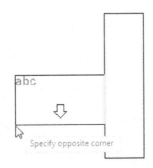

Figure-39. Specifying second corner

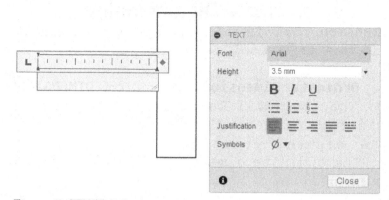

Figure-40. TEXT dialog box

- Click on the **Font** drop-down and select the required font for text.
- Click in the **Height** edit box and specify the height of text.
- Set the other parameters as required and type the desired text in the Text box.
- Click on **Close** button. The text will be added in drawing sheet.

CREATING LEADERS

The **Leader** tool is similar to **Text** tool but these notes are intended to associate with a specific component. The procedure to use this tool is discussed next.

- Click on the **Leader** tool from **TEXT** panel; refer to Figure-41. The tool will be activated.

- Click on the drawing to select the start point. An arrow is attached to the cursor.
- Click on the screen to place text box. The text box will be displayed along with **TEXT** dialog box; refer to Figure-42.

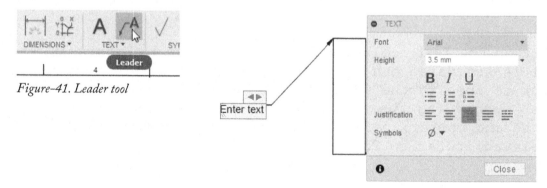

Figure-41. Leader tool

Figure-42. Leader text box

- Set the desired parameter from **TEXT** dialog box and enter the text in text box.
- After specifying the parameters, click on **Close** button from **TEXT** dialog box.

APPLYING SURFACE TEXTURE SYMBOL

The **Surface Texture** tool is used to apply surface finish to objects and existing document. The procedure to use this tool is discussed next.

- Click on the **Surface Texture** tool from **Symbols** panel; refer to Figure-43. The tool will be activated.

Figure-43. Surface Texture tool

- Click on the model in a drawing view to select object. An arrow will be displayed on the selected part.
- Click to select the second point or press **ENTER** key. The **SURFACE TEXTURE** dialog box will be displayed; refer to Figure-44.

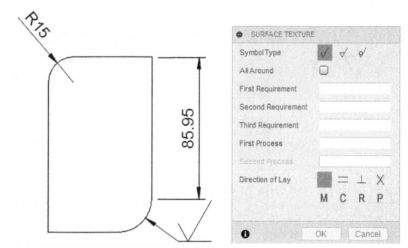

Figure-44. SURFACE TEXTURE dialog box

- Select the desired surface finish symbol from **Symbol Type** option.
- Select the **All Around** check box from **SURFACE TEXTURE** dialog box if you want to create a circle on the surface finish symbol.
- Click in the **First Requirement** edit box of **SURFACE TEXTURE** dialog box and enter the first parameter of surface finish symbol. Similarly, specify the other parameters in the edit boxes of the dialog box.
- Select the lay direction button as desired from **Direction of Lay** section.
- After specifying the parameters, click on the **OK** button from **SURFACE TEXTURE** dialog box; refer to Figure-45.

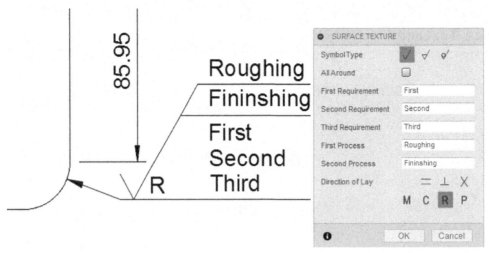

Figure-45. Applying Surface Texture

Surface Texture Symbols or Surface Roughness Symbols

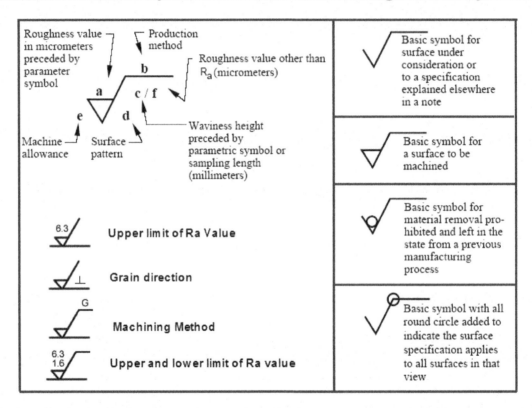

CREATING FEATURE CONTROL FRAME

The **Feature Control Frame** contains data for controlled specifications. It shows the characteristics symbol, datum reference, and tolerance value. It is a rectangular frame which is divided into two or more sections. The feature control frame is also known as GD&T box in laymen's language. The method to insert Feature Control Frame in drawing is same as discussed for Surface Finish symbol. In GD&T, a feature control frame is required to describe the conditions and tolerances of a geometric control on a part's feature. The feature control frame consists of four pieces of information:

1. GD&T symbol or control symbol
2. Tolerance zone type and dimensions
3. Tolerance zone modifiers: features of size, projections...
4. Datum references (if required by the GD&T symbol)

This information provides everything you need to know about geometry of part like what geometrical tolerance needs to be on the part and how to measure or determine if the part is in specification; refer to Figure-46. The common elements of feature control frame are discussed next.

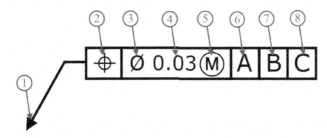

Figure-46. Feature control frame

1. **Leader Arrow** – This arrow points to the feature that the geometric control is placed on. If the arrow points to a surface than the surface is controlled by the GD&T. If it points to a diametric dimension, then the axis is controlled by GD&T. The arrow is optional but helps clarify the feature being controlled.
2. **Geometric Symbol** – This is where your geometric control is specified.
3. **Diameter Symbol (if required)** – If the geometric control is a diametrical tolerance then the diameter symbol (Ø) will be in front of the tolerance value.
4. **Tolerance Value** – If the tolerance is a diameter you will see the Ø symbol next to the dimension signifying a diametric tolerance zone. The tolerance of the GD&T is in same unit of measure that the drawing is written in.
5. **Feature of Size or Tolerance Modifiers (if required)** – This is where you call out max material condition or a projected tolerance in the feature control frame.
6. **Primary Datum (if required)** – If a datum is required, this is the main datum used for the GD&T control. The letter corresponds to a feature somewhere on the part which will be marked with the same letter. This is the datum that must be constrained first when measuring the part. Note: The order of the datum is important for measurement of the part. The primary datum is usually held in three places to fix 3 degrees of freedom
7. **Secondary Datum (if required)** – If a secondary datum is required, it will be to the right of the primary datum. This letter corresponds to a feature somewhere on the part which will be marked with the same letter. During measurement, this is the datum fixated after the primary datum.

8. **Tertiary Datum (if required)** – If a third datum is required, it will be to the right of the secondary datum. This letter corresponds to a feature somewhere on the part which will be marked with the same letter. During measurement, this is the datum fixated last.

Reading Feature Control Frame

The feature control frame forms a kind of sentence when you read it. Below is how you would read the frame in order to describe the feature.

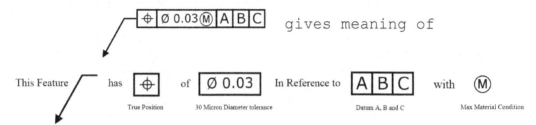

Meaning of various geometric symbols are given in Figure-47.

SYMBOL	CHARACTERISTICS	CATEGORY
—	Straightness	Form
▱	Flatness	
○	Circulatity	
⌀	Cylindricity	
⌒	Profile of a Line	Profile
⌓	Profile of Surface	
∠	Angularity	Orientation
⊥	Perpendicularity	
//	Parallelism	
⊕	Position	Location
◎	Concentricity	
≡	Symmetry	
↗	Circular Runout	Runout
↗↗	Total Runout	

Figure-47. Geometric Symbols

Figure-48 and Figure-49 shows the use of geometric tolerances in real-world.

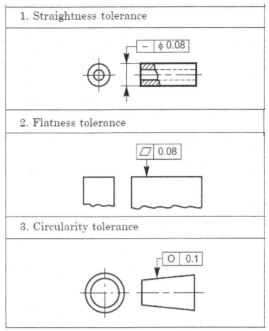

Figure-48. Use of geometric tolerance 1

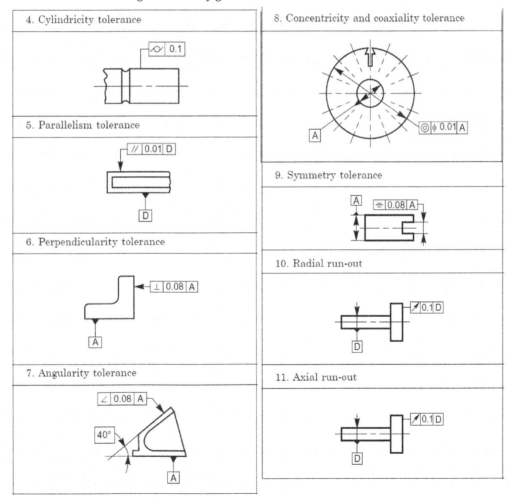

Figure-49. Use of geometric tolerance 2

Note that in applying most of the Geometrical tolerances, you need to define a datum plane like in Perpendicularity, Parallelism and so on. There are a few dimensioning symbols also used in geometric dimensioning and tolerances, which are given in Figure-50.

Symbol	Meaning	Symbol	Meaning
Ⓛ	LMC – Least Material Condition	◄─⊕	Dimension Origin
Ⓜ	MMC – Maximum Material Condition	⎵	Counterbore
Ⓣ	Tangent Plane	∨	Countersink
Ⓟ	Projected Tolerance Zone	⊽	Depth
Ⓕ	Free State	⌀	All Around
⌀	Diameter	◄─►	Between
R	Radius	✕	Target Point
SR	Spherical Radius	▷	Conical Taper
SⓄ	Spherical Diameter	▱	Slope
CR	Controlled Radius	☐	Square
ⓈⓉ	Statistical Tolerance		
77	Basic Dimension		
(77)	Reference Dimension		
5X	Places		

Figure-50. Dimensioning symbols

The procedure to use this tool is discussed next.

- Click on the **Feature Control Frame** tool from **SYMBOLS** panel; refer to Figure-51. The tool will be activated.

Figure-51. Feature Control Frame tool

- Click on the drawing to select the object. An arrow will be attached to the cursor; refer to Figure-52.

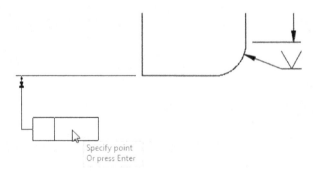

Figure-52. Placing the rectangular box

- Click on the screen to place the rectangular box and press **ENTER** key. The **FEATURE CONTROL FRAME** dialog box will be displayed; refer to Figure-53.

- Click in the **Top Note** edit box and enter the note. If you want to add any symbol then click on the drop-down button. A list of symbols will be displayed; refer to Figure-54.

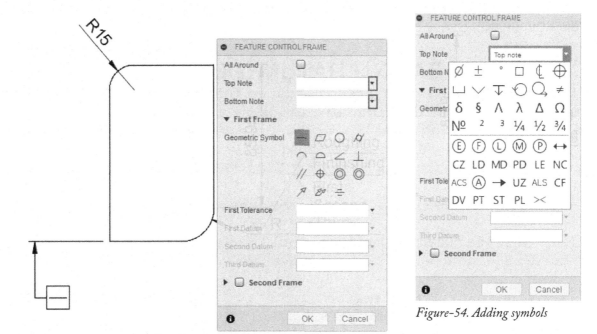

Figure-53. FEATURE CONTROL FRAME dialog box

Figure-54. Adding symbols

- Click on the required symbol. The selected symbol will be displayed on screen.
- Select the required symbol from **Geometric Symbol** section. The selected symbol will be displayed in the feature frame rectangular box.
- Click in the **First Tolerance** edit box and enter the value of tolerance. You can also add the required symbol; refer to Figure-55.

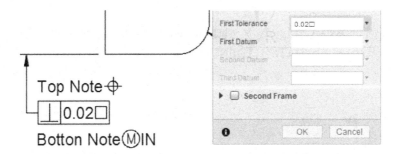

Figure-55. Added tolerance

- Click on the **First Datum** edit box and enter the value of datum as required; refer to Figure-56.

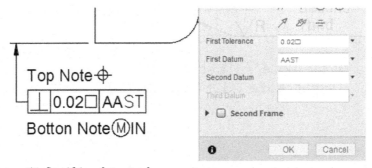

Figure-56. Specifying datum value

- Similarly enter the remaining parameters in respective edit box.
- Select the **Second Frame** check box of **FEATURE CONTROL FRAME** dialog box if you want to add second frame for the drawing.
- After specifying the parameters, click on the **OK** button from **FEATURE CONTROL FRAME** dialog box.

CREATING DATUM IDENTIFIER

The **Datum Identifier** tool identifies a datum feature for a feature control frame symbol. Datum identifier are circular frames divided in two parts by a horizontal line. The lower half represents the datum feature, and the upper half is for additional information, such as dimensions of the datum target area; refer to Figure-57.

The procedure to use this tool is discussed next.

- Click on the **Datum Identifier** tool from **SYMBOLS** panel; refer to Figure-58. The tool will be activated.

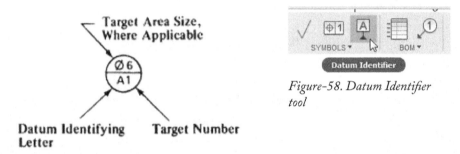

Figure-57. Datum identifier symbol

Figure-58. Datum Identifier tool

- Click on the drawing to select object. An arrow will be attached to the cursor.
- Specify the point and press **ENTER** key. The datum ID symbol will be displayed along with **DATUM IDENTIFIER** dialog box; refer to Figure-59

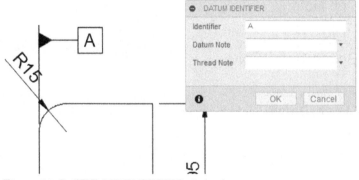

Figure-59. DATUM IDENTIFIER dialog box

- Click in the **Datum Note** edit box and enter the desired note.
- Click in the **Thread Note** edit box and enter the desired text. You can also add the required symbol from **Symbol** drop-down.

- After specifying the parameters, click on the **OK** button from **DATUM IDENTIFIER** dialog box; refer to Figure-60.

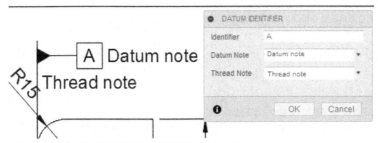

Figure-60. DATUM IDENTIFIER dialog box

GENERATING PARTS LIST

The **Parts List** tool is used to generate a parts list of the model linked with drawing. The procedure to use this tool is discussed next.

- Click on the **Parts List** option from **BOM** drop-down. The **PARTS LIST** dialog box will be displayed along with attached parts list; refer to Figure-61.

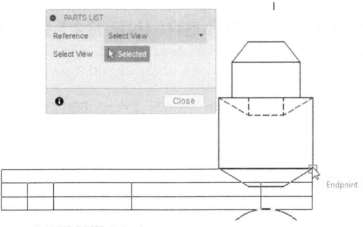

Figure-61. PARTS LIST dialog box

- Select the file name option from **Reference** drop-down if you want to create a part list of particular file.
- Select **Select View** option from **Reference** drop-down if you want to select a view manually which should be used to create part list.
- On selecting the desired option from **PARTS LIST** dialog box, place the parts list as desired refer to Figure-62.

Parts List				
Item	Qty	Part Number	Description	Material
1	1	Practice3 v1		Steel

Figure-62. Parts list of file

GENERATING BALLOON

The **Balloon** tool is used to generate a balloon for component in drawing. The procedure to use this tool is discussed next.

* Click on the **Balloon** tool from **BOM** drop-down; refer to Figure-63. The tool will be activated.
* Click to select edge of component. On selection of point, an arrow will be attached to the cursor.
* Again click at a desired location to place the balloon as required; refer to Figure-64.

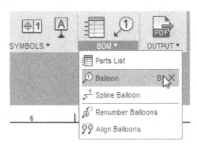

Figure-63. Balloon tool

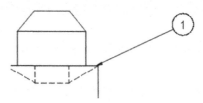

Figure-64. Placing ballon

GENERATING SPLINE BALLOON

The **Spline Balloon** tool is used to create balloons with curved leaders. The procedure to use this tool is discussed next.

* Click on the **Spline Balloon** tool from **BOM** drop-down; refer to Figure-65. The tool will be activated.
* Click on the edge of component to select. An arrow will be attached to the cursor.
* Click at a desired distance and then click again on the screen to place the balloon; refer to Figure-66.

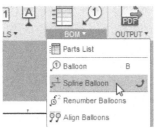
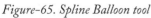

Figure-65. Spline Balloon tool

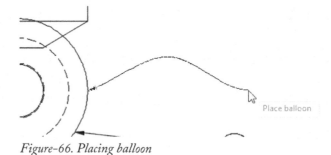

Figure-66. Placing balloon

RENUMBERING BALLOONS

The **Renumber Balloons** tool is used to renumber the existing balloons. The procedure to use this tool is discussed next.

* Click on the **Renumber Balloons** tool from **BOM** drop-down; refer to Figure-67. The **RENUMBER BALLOONS** dialog box will be displayed; refer to Figure-68.

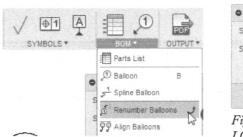

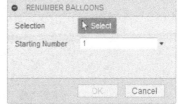

Figure-68. RENUMBER BAL-
LOONS dialog box

Figure-67. Renumber Balloons tool

- Click in the **Starting Number** edit box and enter the desired number.
- Click on **Select** button from **Selection** section and click on the balloons as per their sequences to renumber them.. The balloon numbers will be changed; refer to Figure-69.

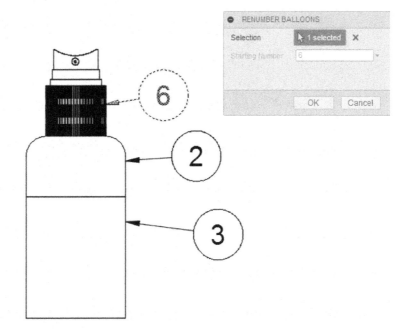

Figure-69. Changed ballon number

- After specifying the parameters, click on **OK** button from **RENUMBER BALLOONS** dialog box to complete the process.

ALIGNING BALLOONS

The **Align Balloons** tool is used to align all the pre-existing balloons. The procedure to use this tool is discussed next.

- Click on the **Align Balloons** tool from **BOM** drop-down; refer to Figure-70. The tool will be activated.

Figure-70. Align Balloons tool

*   Click to select all the balloons to align; refer to Figure-71 and press **ENTER**.

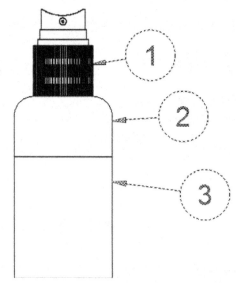

Figure-71. Selection of balloons for allignment

*   Click on the screen to specify the start point. All balloons will be attached to cursor.
*   Click on the screen to align as desired; refer to Figure-72.

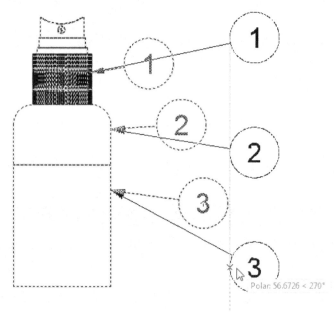

Figure-72. Performing alignement process

GENERATING PDF

The **Output PDF** tool is used to output the drawing in PDF format. The procedure to use this tool is discussed next.

*   Click on the **Output PDF** tool from **OUTPUT** drop-down; refer to Figure-73. The **OUTPUT PDF** dialog box will be displayed; refer to Figure-74

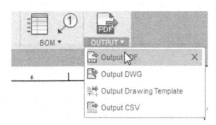

Figure-73. Output PDF tool

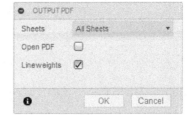

Figure-74. OUTPUT PDF
dialog box

- Click on the **Sheets** drop-down from **OUTPUT PDF** dialog box and select the required sheet for output. You can also use SHIFT key or CTRL key to select multiple sheets.
- Select the **Open PDF** check box if you want to open the output PDF file.
- Select the **Lineweights** check box if you want to make bold lines of drawing otherwise clear it.
- After specifying the parameters, click on the **OK** button from **OUTPUT PDF** dialog box. The **Output PDF** dialog box will be displayed; refer to Figure-75.

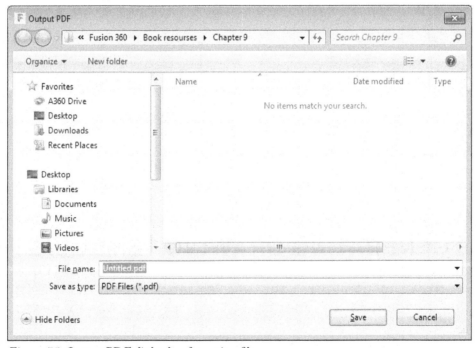

Figure-75. Output PDF dialog box for saving file

- Click in the **File name** edit box and enter the desired text.
- Select the saving location of file and click on the **Save** button. The PDF file will be created and opened if you had selected the **Open PDF** check box from **OUTPUT PDF** dialog box.

GENERATING DWG

The **Output DWG** tool is used to create an output file of drawing in DWG format. The procedure to use this tool is discussed next.

- Click on the **Output DWG** tool from **OUTPUT** drop-down; refer to Figure-76. The **Output DWG** dialog box will be displayed; refer to Figure-77

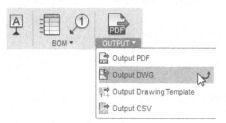

Figure-76. Output DWG tool

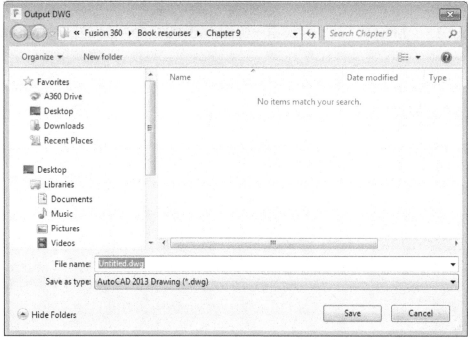

Figure-77. Output DWG dialog box

- Select the desired location to save the file.
- After specifying the parameter, click on the **Save** button from **Output DWG** dialog box to save the file.

GENERATING DRAWING TEMPLATE

The **Output Drawing Template** tool is used to create a drawing template of the current drawing configuration which can be used later in to create drawing. The procedure to use this tool is discussed next.

- Click on the **Output Drawing Template** tool from **OUTPUT** drop-down; refer to Figure-78. The **SAVE** dialog box will be displayed; refer to Figure-79.

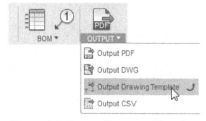

Figure-78. Output Drawing Template tool

Save   ✕

The Save dialog is using the same location as the Data Panel while in offline mode.   ✕

Name:

Untitled

Location:

Gaurav's First Project ❯ ○ master

Cancel   Save

Figure-79. Save Dialoge box

- Click in the **Name** edit box and enter the desired text.
- Select the desired location in **Location** option.
- After specifying the parameters, click on the **Save** button from **Save** dialog box. The template will be saved at the desired location.

GENERATING CSV FILE

The **Output CSV** tool is used to create the output of the current file in CSV format. CSV is a format used by database management software. The procedure to use this tool is discussed next.

- Click on the **Output CSV** tool from **OUTPUT** drop-down; refer to Figure-80. The **Output Parts List or Table** dialog box will be displayed; refer to Figure-81.

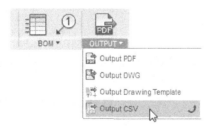

Figure-80. Output CSV tool

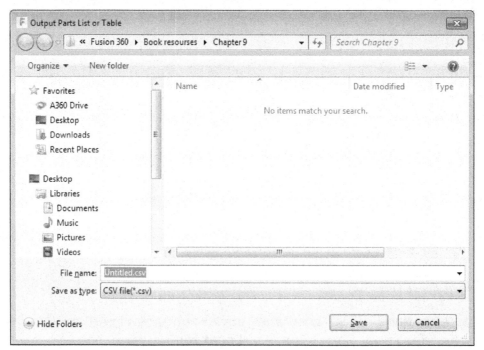

Figure-81. Output Parts List or Table dialog box

- Select the location of file and enter the desired name in **File Name** edit box.
- After specifying the parameters, click on the **Save** button to save the file.

Chapter 12

Sculpting

Topics Covered

The major topics covered in this chapter are:

- *Plane*
- *Cylinder*
- *Quadball*
- *Face*
- *Loft*
- *Edit Form*
- *Merge Edge*
- *Weld Vertices*
- *Fill Hole*
- *Crease and Uncrease*
- *Pull and Flatten*
- *Freeze and UnFreeze*

INTRODUCTION

Sculpting is a workspace that offers tools to push, pull, grab, or pinch to make the component as real life object. In this chapter, we will learn various commands and tool of **SCULPT Workspace** which are used to create or modify the object

OPENING THE SCULPT WORKSPACE

The components or parts created in Sculpt are easily converted in solid bodies. Generally, the **SCULPT Workspace** is not present in **Workspace** drop-down. It is displayed in **MODEL Workspace.** The procedure is discussed next.

- Click on the **MODEL Workspace** from **Workspace** drop-down. The **MODEL Workspace** will be displayed in **Autodesk Fusion 360** window.
- Click on **Create Form** tool of **CREATE** drop-down from **Toolbar**; refer to Figure-1. The **SCULPT Environment** notification dialog box will be displayed; refer to Figure-2.

Figure-1. Create Form tool

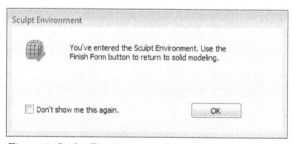

Figure-2. Sculpt Environment dialog box

- Click on **OK** button from **Sculpt Environment** dialog box. The **SCULPT Workspace** will be displayed; refer to Figure-3.

Figure-3. SCULPT Workspace application window

CREATIONS TOOLS

In this section, we will discuss the creation tools used to create sculpt form.

Box

The **Box** tool is used to create a rectangular body on the selected plane or face. The procedure to use this tool is discussed next.

- Click on the **Box** tool of **CREATE** drop-down from **Toolbar**; refer to Figure-4. The **BOX** dialog box will be displayed; refer to Figure-5 .

Figure-4. Box tool

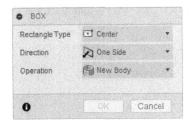

Figure-5. BOX dialog box1

- You will be asked to select the plane or face. Click to select plane or face.

- Select **Center** option from **Rectangular Type** drop-down if you want to create a center rectangle as reference for box.
- Select **2-Point** option from **Rectangular Type** drop-down if you want to create a 2-Point rectangle as a reference for creating box. In our case, we are selecting the **Center** option.
- Specify the desired parameters in **Direction** and **Operation** drop-down as discussed earlier in **MODEL Workspace**.
- Click on the screen to specify the center point of rectangle.
- Enter the desired dimension of length and width in the respective floating window and click on the screen. The updated **BOX** dialog box will be displayed; refer to Figure-6.

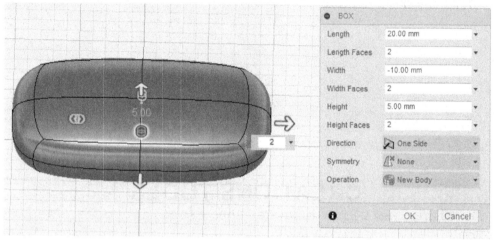

Figure-6. Updated BOX dialog box

- If you want to change the length of box then click on the **Length** edit box and enter the desired value.
- Click in the **Length Faces** edit box of **BOX** dialog box and enter the number of faces in which surface of box will be divided along the length.
- If you want to change the width of plane then click on the **Width** edit box and enter the desired value.
- Click in the **Width Faces** edit box of **BOX** dialog box and enter the number of faces in which surface of box will be divided along the width.
- Click in the **Height** edit box and enter the desired value of height of box.
- Click in the **Height Faces** edit box and enter the number of faces in which height of plane will be divided along the height.
- Select the **Mirror** option from **Symmetric** drop-down if you want to create a symmetric box. On selecting the **Mirror** option the updated dialog box will be displayed; refer to Figure-7.

Figure-7. Updated Box dialog box

• Select the **Length Symmetry** check box if you want to apply mirror symmetry along length of sculpt object; refer to Figure-8.

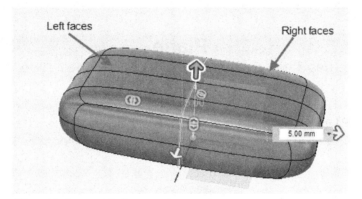

Figure-8. Symmetric faces

• Select the **Width Symmetry** check box if you want to apply mirror symmetry along width of sculpt object; refer to Figure-9. Note that later while editing, if you will move one face then the other symmetric face will also move accordingly.

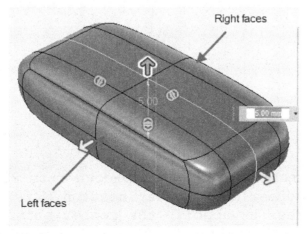

Figure-9. Width Symmetry

- Select the **Height Symmetry** check box if you want to apply mirror symmetry between upper and lower faces of sculpt object; refer to Figure-10.

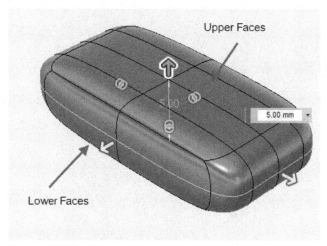

Figure-10. Height Symmetry

- After specifying the parameters, click on **OK** button from **BOX** dialog box to complete the process.

Plane

The **Plane** tool is used to create T-Spline plane. The procedure to use this tool is discussed next.

- Click on the **Plane** tool from **CREATE** drop-down; refer to Figure-11. The **PLANE** dialog box will be displayed; refer to Figure-12.

Figure-11. Plane tool

Figure-12. PLANE dialog box

- The options of **Rectangular Type** drop-down from **PLANE** dialog box are same as discussed earlier in last section.
- Click on the screen to specify the center point.
- Enter the desired dimension of length and width in respective input boxes. The updated **PLANE** dialog box will be displayed; refer to Figure-13.

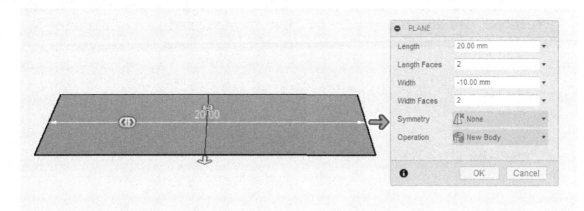

Figure-13. Updated PLANE dialog box

- If you want to change the length of plane then click in the **Length** edit box and enter the desired value.
- Click in the **Length Faces** edit box of **PLANE** dialog box and enter the number of faces in which surface of plane will be divided along the length.
- If you want to change the width of plane then click on the **Width** edit box and enter the desired value.
- Click in the **Width Faces** edit box of **PLANE** dialog box and enter the number of faces in which surface of plane will be divided along the width.
- After specifying the parameters, click on the **OK** button from **PLANE** dialog box to complete the process of creating plane. The plane will be displayed; refer to Figure-14.

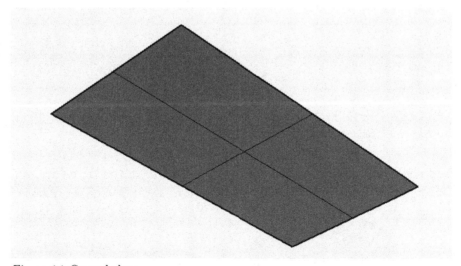

Figure-14. Created plane

Cylinder

The **Cylinder** tool is used to create a cylindrical body by defining diameter and depth. The procedure to use this tool is discussed next.

- Click on the **Cylinder** tool from **CREATE** drop-down; refer to Figure-15. The **CYLINDER** dialog box will be displayed; refer to Figure-16.

Figure-15. Cylinder tool

Figure-16. Cylinder dialog box

- You need to select the base plane or face to create the cylinder. Click on the plane to select.
- Click on the screen to specify the center point and enter the desired radius of cylinder; refer to Figure-17.

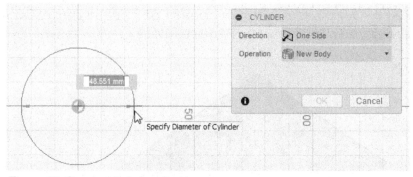

Figure-17. Creating circle for cylinder

- After specifying the diameter, click on the screen. The updated **CYLINDER** dialog box will be displayed; refer to Figure-18.

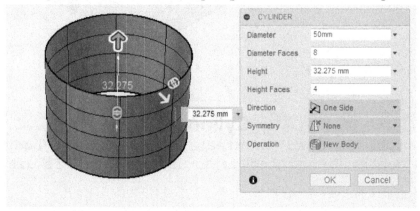

Figure-18. Entering parameters for cylinder

- If you want to change the diameter of cylinder then click on the **Diameter** edit box of **CYLINDER** dialog box and enter the desired diameter.
- Click in the **Diameter Faces** edit box and enter the number of faces in which the round surface of cylinder will be divided.
- Click in the **Height** edit box and enter the desired value of height of cylinder.
- Click in the **Height Faces** edit box and enter the number of faces in which height of cylinder will be divided along the height.
- The options in the **Direction** drop-down have been discussed earlier.
- Select **Circular** option from **Symmetry** drop-down if you want to apply circular symmetry between the faces of sculpt object.
- Click on the **Symmetric Faces** edit box and enter the desired value of circular symmetry on round surface.

When we move or edit one face of circular symmetry, the other faces will move accordingly; refer to Figure-19.

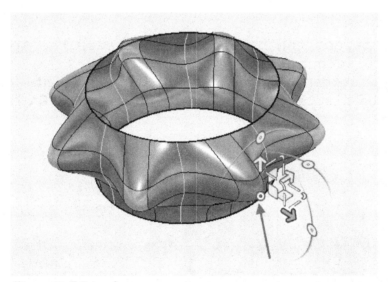

Figure-19. Editing form

- After specifying the parameters, click on the **OK** button from **CYLINDER** dialog box to complete the process.

Sphere

The **Sphere** tool is used to create a T-Spline sphere. The procedure to use this tool is discussed next.

- Click on the **Sphere** tool from **CREATE** drop-down; refer to Figure-20. The **SPHERE** dialog box will be displayed; refer to Figure-21.

Figure-20. Sphere tool

Figure-21. SPHERE dialog box

- You need to select the base plane to create the sphere. Click on the screen to select.
- Click on the screen to specify the center point. The preview of sphere will be displayed along with updated **SPHERE** dialog box; refer to Figure-22.

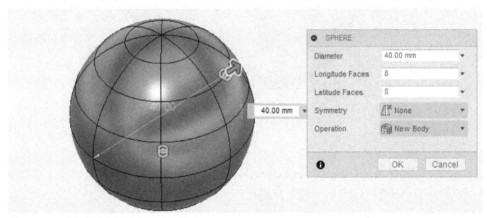

Figure-22. Updated SPHERE dialog box

- Click in the **Diameter** edit box and enter the desired value of sphere diameter.
- Click in the **Longitude Face** edit box and enter the desired number of faces displayed on longitude of sphere.
- Click in the **Latitude Face** edit box and enter the desired number of faces displayed on latitude of sphere.
- The options of **Symmetry** drop-down have been discussed earlier in **Box** and **Cylinder** tools.
- After specifying the parameters, click on the **OK** button from **SPHERE** dialog box to complete the process.

Torus

The **Torus** tool is used to create a T-Spline torus. The procedure to use this tool is discussed next.

- Click on the **Torus** tool from **CREATE** drop-down; refer to Figure-23. The **TORUS** dialog box will be displayed; refer to Figure-24.

Figure-23. The Torus tool

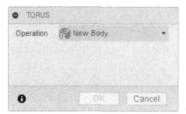

Figure-24. The TORUS dialog box

- You need to select the base plane to create the torus. Click on the screen to select.
- Click on the screen to specify the center point and enter the desired diameter of torus in the input box. The preview of torus will be displayed along with updated **TORUS** dialog box; refer to Figure-25.

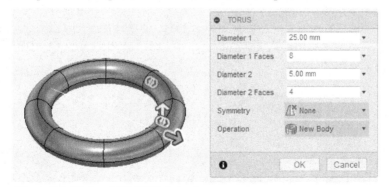

Figure-25. Updated TORUS dialog box

- Click in the **Diameter 1** edit box of **TORUS** dialog box and enter the desired value of diameter for center circle of torus.
- Click in the **Diameter 1 Faces** edit box and enter the number of faces in which round torus will be divided horizontally.
- Click in the **Diameter 2** edit box and enter the diameter of torus tube.
- Click in the **Diameter 2 Faces** edit box and enter the number of faces in which round torus will be divided vertically.

- After specifying the parameters, click on the **OK** button from **TORUS** dialog box.

Quadball

The **Quadball** tool is used to create a T-Spline quadball. The procedure to use this tool is discussed next.

- Click on the **Quadball** tool from **CREATE** drop-down; refer to Figure-26. The **QUADBALL** dialog box will be displayed; refer to Figure-27.

Figure-26. Quadball tool

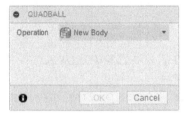

Figure-27. QUADBALL dialog box

- You need to select the base plane to create the quadball. Click on the screen to select.
- Click on the screen to specify the center point. The preview of quadball will be displayed along with updated **QUADBALL** dialog box; refer to Figure-28.

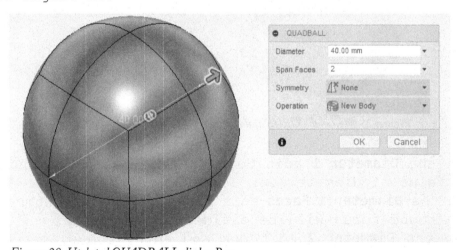

Figure-28. Updated QUADBALL dialog Box

- Click in the **Diameter** edit box and enter the desired value of quadball diameter.

- Click in the **Span Faces** dialog box and enter the number of faces in which quadball will be divided.
- The options of **Symmetry** and **Operation** drop-down were discussed earlier.
- After specifying the parameter, click on **OK** button from **QUADBALL** dialog box. The quadball will be displayed; refer to Figure-29.

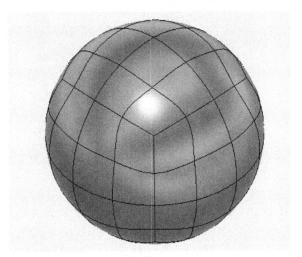

Figure-29. Created quadball with 4 span faces

Pipe

The **Pipe** tool is used to create complex pipe based on selected sketch. The procedure to use this tool is discussed next.

- Click on the **Pipe** tool from **CREATE** drop-down; refer to Figure-30. The **PIPE** dialog box will be displayed; refer to Figure-31.

Figure-30. The Pipe tool

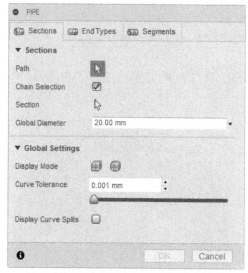

Figure-31. The PIPE dialog box

- The **Path** selection button is active by default. Click on the path to select. You can select sketch lines/curves or edges of the model. You can also use window selection to create pipe.
- Select the **Chain selection** check box if you want to select the nearby geometries while selecting the one.
- On selection of sketch, the preview of pipe will be displayed; refer to Figure-32.

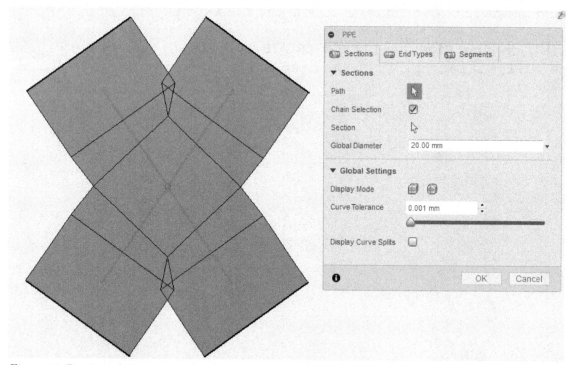

Figure-32. Preview of pipe

- Click on the **Section** button of **Sections** tab if you want to modify the sections of pipe. On selecting the button, the sections of pipe will be displayed; refer to Figure-33.

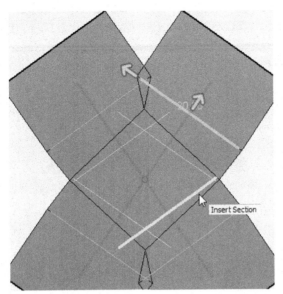

Figure-33. Sections of pipe

- Click on the desired section to change the diameter, angle, and position.
- Click on the **Diameter** edit box and enter the desired value of particular section diameter; refer to Figure-34.

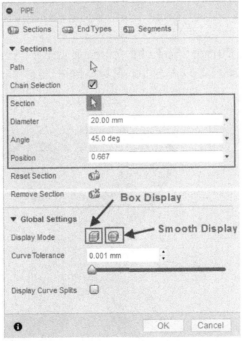

Figure-34. Sections options from PIPE dialog box

- Similarly, specify the value of angle and position in their respective edit box as desired. You can also move the arrow displayed on the selected section to specify the value.
- Click on the **Reset Section** button to reset all the changed value of sections.
- Click on the **Remove Section** button to remove the selected section.

- Click on **Box Display** button of **Display mode** options from **Global Settings** section to display the rectangular T-Spline pipe of the selected sketch.
- Select **Smooth Display** button of **Display mode** option from **Global Settings** section to display a smooth circular pipe; refer to Figure-35.

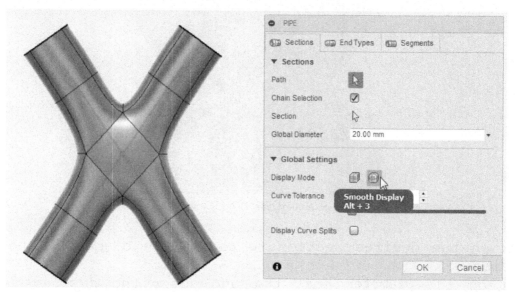

Figure-35. Smooth pipe

- Click in the **Curve Tolerance** edit box and enter the value of pipe tolerance. You can also set the tolerance by adjusting the **Curve Tolerance** slider.
- Select the **Display Curve Splits** check box if you want to see the curve splits of pipe; refer to Figure-36.

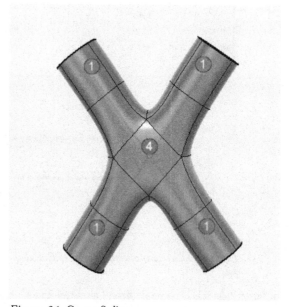

Figure-36. Curve Splits

- The **Handle** selection button of **End Type** section from **End Types** tab is active by default and you are asked to select the end handle of pipe to modify its shape.
- Select the desired end handle from the model. Select **Open** option of **End Type** section from **End Types** tab if you want to keep all the

ends of pipe open. Select **Square** option of **End Type** section from **End Types** tab if you want to close all the ends of pipe in square like structure; refer to Figure-37.

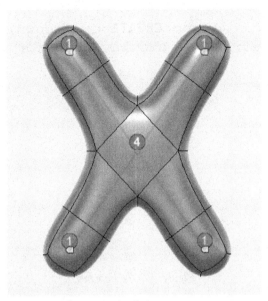

Figure-37. Square ends of pipe

- Select **Spike** option of **End Type** section from **End Types** tab if you want to close all the ends of pipe in spike like structure; refer to Figure-38.

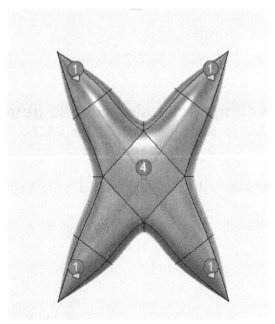

Figure-38. Spike ends of pipe

- The **Segment** selection button of **Segments** section from **Segments** tab is active by default.
- Move the **Density** slider to increase or decrease the density of pipe segment.
- After specifying the parameters, click on the **OK** button from **PIPE** dialog box to complete the process.

Face

The **Face** tool is used to create individual face. The procedure to use this tool is discussed next.

- Click on the **Face** tool from **CREATE** drop-down; refer to Figure-39. The **FACE** dialog box will be displayed; refer to Figure-40.

Figure-39. The Face tool

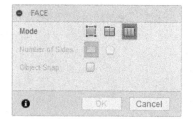

Figure-40. The FACE dialog box

Creating Face using Simple button

- Select **Simple** button of **Mode** option from **FACE** dialog box if you want to create a simple face by selecting four vertices on a specific plane.
- On selecting the **Simple** button, you need to select plane for creating the face. Click to select the plane.
- Now, specify the four corner on the plane as desired; refer to Figure-41.

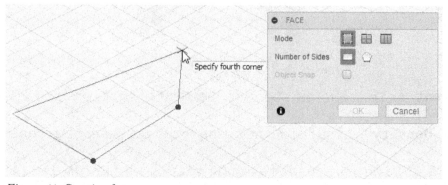

Figure-41. Creating face

- On selecting the fourth corner, the face will be created; refer to Figure-42.

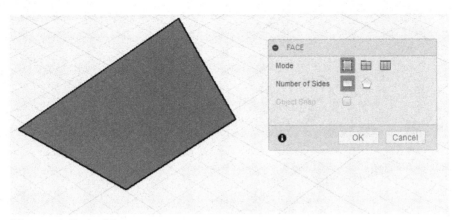

Figure-42. Created face

Creating Face using Edge button

The **Edge** button is used to create a face by selecting a edge. The procedure is discussed next.

- Click on the **Edge** button of **Mode** option from **FACE** dialog box. You need to select the edge for creating face.
- Now, you need to select the plane on which you want to create a face. Click to select the plane.
- Click on the screen to specify third and fourth point.
- On creating fourth point, the face will be created; refer to Figure-43.

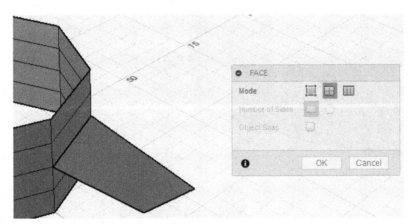

Figure-43. Created face by edge button

Creating Chain of Faces using Edges

The **Chain** button is used to create multiple faces continuously. The procedure is discussed next.

- Click on the **Chain** button of **Mode** option from **FACE** dialog box. You will be asked to select an edge for creating face.
- Select the desired edge. Now, you need to select the plane on which you want to create a face. Click to select the plane.
- Specify third and fourth corner to create face; refer to Figure-44.

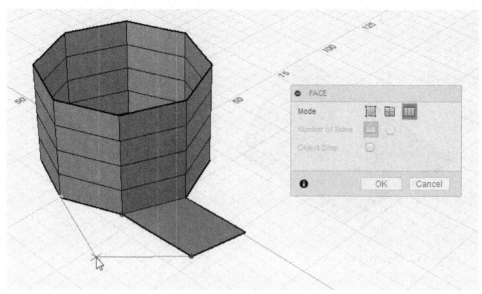

Figure-44. Creating Chain face

- After creating a face, click on the screen to create another face; refer to Figure-45. Similarly, create multiple faces as desired.

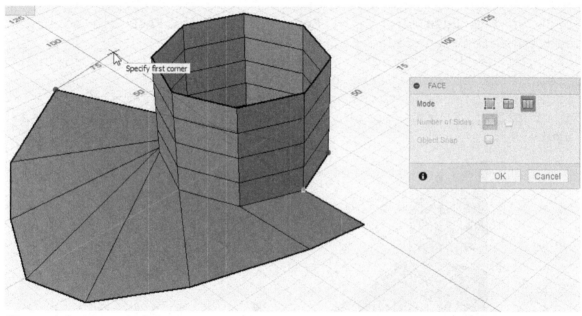

Figure-45. Creating multiple faces

Extrude

The **Extrude** tool is used to extrude the selected sketch up to the desired depth or height. The procedure to use this tool is discussed next.

- Click on the **Extrude** tool from **CREATE** drop-down; refer to Figure-46. The **EXTRUDE** dialog box will be displayed; refer to Figure-47.

Figure-46. Extrude tool

Figure-47. EXTRUDE dialog box

- The **Profile** option of **EXTRUDE** dialog box is active by default. You are asked to select the sketch for extrude.
- Select the **Profile Chain check** box if you want to select the nearby geometries.
- Select **Uniform** button from **Spacing** drop-down if you want to position the faces evenly around the profile.
- Select **Curvature** button from **Spacing** drop-down if you want to position the faces based on the curvature of the profile. The more the curvature area more the faces.
- Click in the **Faces** edit box and enter the desired number of faces in which curvature of extruded sketch will be divided.
- Click in the **Distance** edit box and enter the distance for extrusion.
- Click in the **Angle** edit box and enter the desired angle of extrusion.
- Click in the **Front Faces** edit box and enter the desired number of faces in which surface of extrude will be divided along the height.
- Select the **Maintain Crease Edges** check box from **EXTRUDE** dialog box if you want to keep the crease at merged edges.
- After specifying the parameter, click on the **OK** button from **EXTRUDE** dialog box to complete the extrusion process; refer to Figure-48. Note that you can also select face of any other sculpt body to extrude but some of the options in this dialog box will not be available in that case.

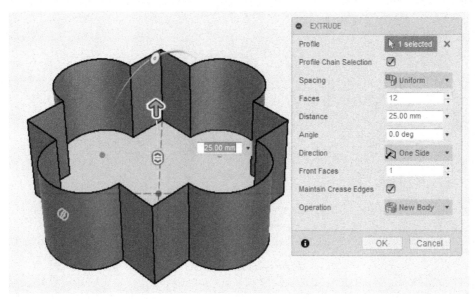

Figure-48. Creating extrude

Revolve

The **Revolve** tool is used to create a revolve feature by sweeping the selected sketch around the selected axis. The procedure to use this tool is discussed next.

- Click on the **Revolve** tool from **CREATE** drop-down; refer to Figure-49. The **REVOLVE** dialog box will be displayed; refer to Figure-50.

Figure-49. Revolve tool

Figure-50. The REVOLVE dialog box

- The **Profile** option of **REVOLVE** dialog box is active by default. Click on the sketch to apply revolve feature.
- Click on **Select** button of **Axis** option and select the axis of rotation of sketch; refer to Figure-51.

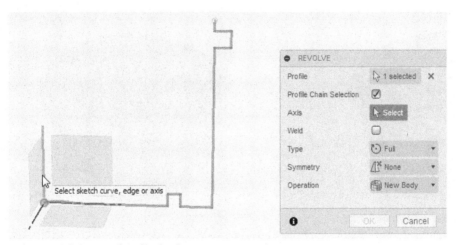

Figure-51. Selecting Axis for sketch

- On selecting the axis, the preview of revolve will be displayed along with updated **REVOLVE** dialog box; refer to Figure-52.

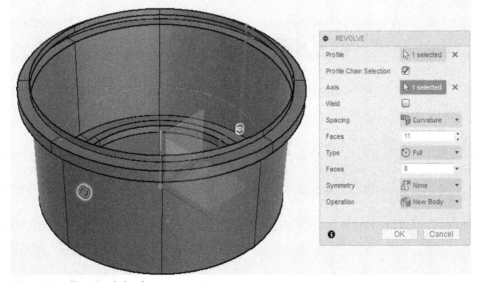

Figure-52. Revolved sketch

- Click in the **Faces** edit box below **Spacing** option and enter the desired number of faces in which curvature of revolved sketch will be divided along flat faces; refer to Figure-53.

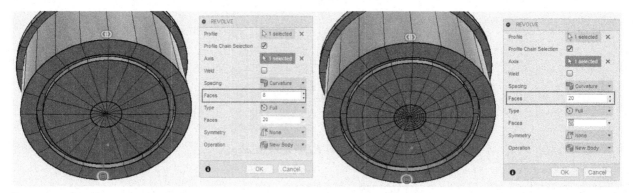

Figure-53. Entering the value of Faces

- Click in the **Faces** edit box below **Type** option and enter the desired number of faces in which curvature of revolved sketch will be divided along round faces; refer to Figure-54.

Figure-54. Entering teh value of faces1

- After specifying the parameters, click on the **OK** button from **REVOLVE** dialog box to complete the process.

Similarly, you can use other tools of **CREATE** drop-down as discussed in **MODEL Workspace**.

MODIFYING TOOLS

Till now, we have learned about various tools to create the sculpt object. In this section we will learn various tools for modifying the sculpt object.

Edit Form

The **Edit Form** tool is used to move, scale, or rotate the selected geometry. The procedure to use this tool is discussed next.

- Click on the **Edit Form** tool of **MODIFY** drop-down; refer to Figure-55. The **EDIT FORM** dialog box will be displayed; refer to Figure-56.

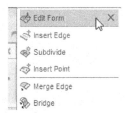

Figure-55. Edit Form tool

Figure-56. EDIT FORM dialog box

- The **Select** button of **T-Spline Entity** is active by default. Click on any face of the sculpt object to edit.
- Click on **Multi** button from **Transform Mode** option to edit the face with the help of 3D Manipulator; refer to Figure-57.

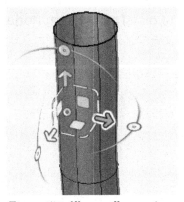

Figure-57. All controller manipulator

- Click on **Translation** button from **Transform Mode** option to edit the face with the help of translation manipulator; refer to Figure-58.

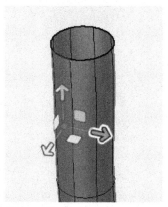

Figure-58. Translation manipulator

- Click on **Rotation** button from **Transform Mode** option to edit the face with the help of rotation manipulator; refer to Figure-59.

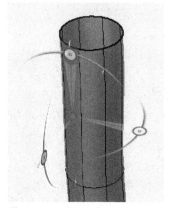

Figure-59. Rotational manipulator

- Click on **Scale** button from **Transform Mode** option to edit the face with the help of scale manipulator; refer to Figure-60.

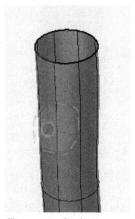

Figure-60. Scale manipulator

- Select the **World Space** button from **Coordinate Space** option to orient the manipulator relative to the origin of model.
- Select the **View Space** button from **Coordinate Space** option to orient the manipulator relative to the current view of model.

- Select the **Local** button from **Coordinate Space** option to orient the manipulator relative to the selected object.
- Select **Vertex** button from **Selection Fibre** option to select the vertex of object for editing purpose. On selecting this button, only vertices will available for selection.
- Select **Edge** button from **Selection Fibre** option to select the edge of object. On selecting this button, only edges of object will be available for selection.
- Select **Face** button from **Selection Fibre** option to select the face of object. On selecting this button, only faces of object will be available for selection.
- Select **All** button from **Selection Fibre** option to select the edge, vertex, or face from a object. On selecting this button edge, vertex, or face will available for selection.
- Select **Body** button from **Selection Fibre** option to select the body for editing.
- Select the **Object Snap** check box and specify the desired value in the **Offset** edit box to specify the limit within which the selected vertex can move.

Soft Modification

Select the **Soft Modification** check box if you want to control the influence of changes on the surrounding area of object. On selecting, the updated **EDIT FORM** dialog box will be displayed; refer to Figure-61.

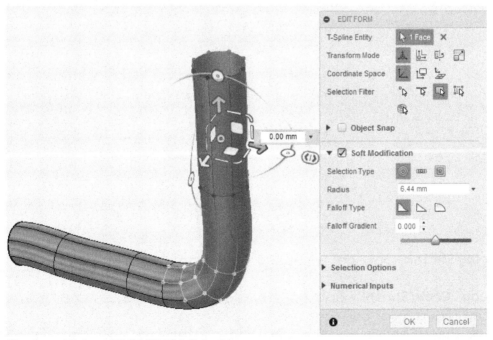

Figure-61. Updated EDIT FORM dialog box

- When selected, the vertices of body is visually represented with red and white vertices highlighting that can be adjusted with the gradient slider.
- Select **Circular** button from **Selection Type** option to specify the radius for controlling the influence of round region. With the help of vertices displayed on object, you can control the amount of

change they undergo and the shape of the affected region.

- Select **Rectangle** button from **Selection Type** option to specify the number of width and length faces.
- Select **Grow** button from **Selection Type** option to specify the number of faces far from selected object.
- Select the desired shape of the affected region from **Falloff Type** option.
- Click in the **Falloff Gradient** check box and enter the value of amount of influence applied in the affected region. The value of gradient will be in between -1 to 1. You can also adjust the value of gradient by moving the **Falloff Gradient** slider.

Selection Options

- Click on the **Selection Options** node from **EDIT FORM** dialog box. The options of **Selection Options** node will be displayed; refer to Figure-62.

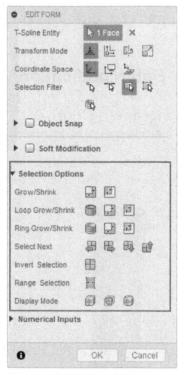

Figure-62. Selection Options

- The **Grow/Shrink** option is used contract and expand the selected region.
- The **Loop Grow/Shrink** option is used to expand and contract the selected loop for the current location.
- The **Ring Grow/Shrink** option is used to expand and contract the ring after selecting the adjacent edge of the selected edge.
- The **Select Next** button is used to move the selected vertex, edge, or face to the next adjacent object. The movement to adjacent faces is based upon the current camera position.
- The **Invert Selection** button is used for reverse the selection. It means all the selected object will be deselected and deselected will be selected.

- The **Range Selection** button is used to select all the faces between two selected faces.
- The **Display Mode** option is used to select **Box display, Control Frame display,** and **Smooth Display** mode for the object.

Numerical Input

- Click on the **Numerical Inputs** node from **EDIT FORM** dialog box to enter the numerical value of various parameters, refer to Figure-63.

Figure-63. Numerical Inputs

- After specifying the various parameters, click on **OK** button from **EDIT FORM** dialog box to complete the process.

Insert Edge

The **Insert Edge** tool is used to insert an edge at a specified distance from the selected edge. The procedure to use this tool is discussed next.

- Click on the **Insert Edge** tool from **MODIFY** drop-down; refer to Figure-64. The **INSERT EDGE** dialog box will be displayed; refer to Figure-65.

*Figure-64. Insert
Edge tool*

*Figure-65. INSERT EDGE
dialog box*

- The **Select** button of **T-Spline Edge** option is active by default. Click on the edge from model to select.
- Select **Simple** option from **Insertion Mode** drop-down if you do not want to move any point but may be it change the surface shape.
- Select **Exact** option from **Insertion Mode** drop-down to maintain the exact surface shape by adding points.
- Select **Single** option from **Insertion Side** drop-down to insert the edge in only one side to the selected edge.
- Select **Both** option from **Insertion Side** drop-down to insert the edge on both side to the selected edge.
- Click in the **Insert Location** edit box and enter the value of location for inserting edge. The value of **Insert Location** lie between 0 to 1.
- Select the **Object Snap** check box to move the new vertices to the closest point. These vertices can snap to solid, surface, and mesh bodies.
- Click in the **Offset** edit box and enter the desired value.
- After specifying the parameters, click on the **OK** button from **INSERT EDGE** dialog box to complete the process; refer to Figure-66.

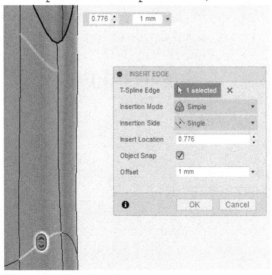

Figure-66. Inserting edge

Subdivide

The **Subdivide** tool is used to divide the selected face in four or more faces. The procedure to use this tool is discussed next.

- Click on the **Subdivide** tool from **MODIFY** drop-down; refer to Figure-67. The **SUBDIVIDE** dialog box will be displayed; refer to Figure-68.

Figure-67. Subdivide tool

Figure-68. SUBDIVIDE dialog box

- The **Select** button of **T-Spline Face** option is active by default. Click on the face from model to select. Selected face will be divided into four face; refer to Figure-69.

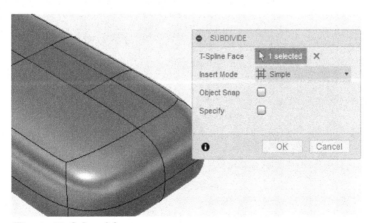

Figure-69. Selected face

- Select **Simple** option from **Insert Mode** drop-down to add the minimum number of control points to the subdivide face.
- Select **Exact** option from **Insert Mode** drop-down if you do not want to change the surface body. To maintain the shape extra control points are added.
- Select the **Specify** check box to enter the number of faces along length and width of the selected face.
- After specifying the parameters, click on the **OK** button from **SUBDIVIDE** dialog box to complete the process.

Insert Point

The **Insert Point** tool is used to insert control points at the selected locations. The procedure to use this tool is discussed next.

• Click on the **Insert Point** tool from **MODIFY** drop-down; refer to Figure-70. The **INSERT POINT** dialog box will be displayed; refer to Figure-71.

Figure-70. Insert Point tool

Figure-71. INSERT POINT dialog box

• The **Insertion Point** button is active by default. You need to click on the edge of the model to create point; refer to Figure-72.

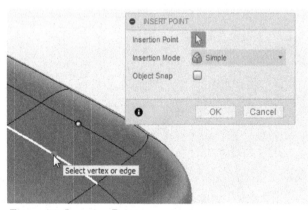

Figure-72. Inserting Point

• The options of **Insertion Mode** and **Object Snap** options were discussed earlier in this book.
• After specifying the parameters, click on the **OK** button from **INSERT POINT** dialog box to complete the process; refer to Figure-73.

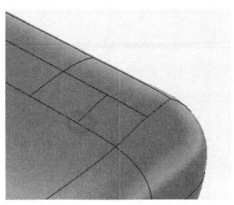

Figure-73. Created point

Merge Edge

The **Merge Edge** tool is used to connect two bodies by joining their edge. The procedure to use this tool is discussed next.

- Click on the **Merge Edge** tool from **MODIFY** drop-down; refer to Figure-74. The **MERGE EDGE** dialog box will be displayed; refer to Figure-75.

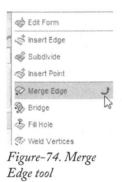

*Figure-74. Merge
Edge tool*

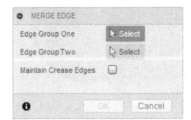

*Figure-75. MERGE EDGE
dialog box*

- The **Select** button of **Edge Group One** option is active by default. You need to the select the edges or boundaries of a body.
- The **Select** button of **Edge Group Two** option is active by default. Click to select the boundary of other group; refer to Figure-76.
- Double click on the edge of a face to select the adjacent edges also.

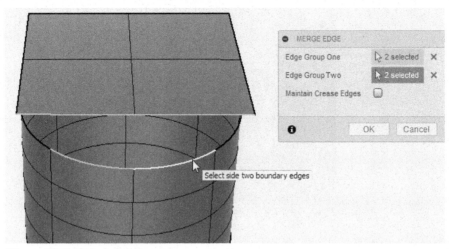

Figure-76. Selecting edge to merge

- Select the **Maintain Crease Edges** check box from **MERGE EDGE** dialog box to keep the crease at the merged edges.
- After specifying the parameters, click on the **OK** button from **MERGE EDGE** dialog box to complete the process; refer to Figure-77.

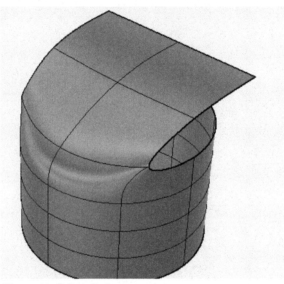

Figure-77. Merged edges

Bridge

The **Bridge** tool is used to connect two bodies by adding their intermediate faces. The procedure to use this tool is discussed next.

- Click on the **Bridge** tool from **MODIFY** drop-down; refer to Figure-78. The **BRIDGE** dialog box will be displayed; refer to Figure-79.

Figure-78. Bridge tool

Figure-79. BRIDGE dialog box

- Click on the **Side One** button from **BRIDGE** dialog box and select the desired faces.
- Click on the **Side two** button from **BRIDGE** dialog box and select the faces from the model to join.
- Click on the **Follow Curve** button from **BRIDGE** dialog box and select a curve for the bridge to follow.
- Select the **Preview** check box to display a mesh preview of the bridge from the selected faces; refer to Figure-80.

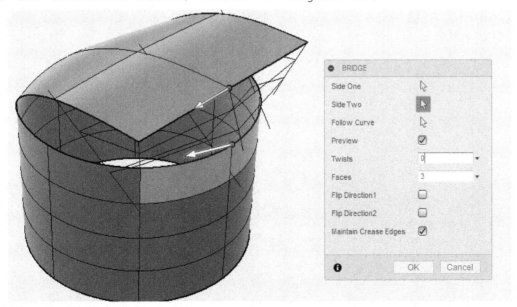

Figure-80. Preview of bridge

- Click in the **Twists** edit box and specify the number of rotation of bridge between two selected different faces (side one and side two).
- Click in the **Faces** edit box and specify the number of faces created between two selected sides.
- Select the **Flip Direction 1** check box to flip the bridge direction of side one.
- Select the **Flip Direction 2** check box to flip the bridge direction of side two.
- After specifying the parameter, click on the **OK** button from **BRIDGE** dialog box to complete the process; refer to Figure-81.

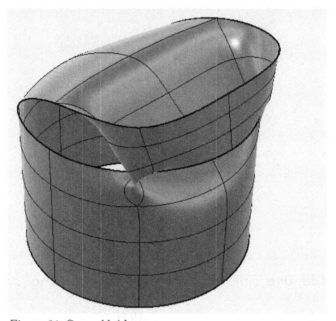

Figure-81. Created bridge

Fill Hole

The **Fill Hole** tool is used to close the opening of a T-Spline model. The procedure to use this tool is discussed next.

- Click on the **Fill Hole** tool from **Modify** drop-down; refer to Figure-82. The **FILL HOLE** dialog box will be displayed; refer to Figure-83.

Figure-82. Fill Hole tool

Figure-83. FILL HOLE dialog box

- The **Select** button of **T-Spline Edge** option is active by default. Click on the edge of hole to select. A preview will be displayed; refer to Figure-84.

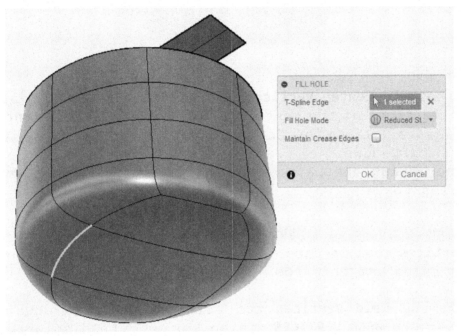
Figure-84. Preview of fill hole

- Select **Reduced Star** option from **Fill Hole Mode** drop-down to fill the hole by adding minimum number of star points of face.
- Select **Fill Star** option from **Fill Hole Mode** drop-down to fill the hole using single face. Due to this option the star points will created at each vertex; refer to Figure-85.

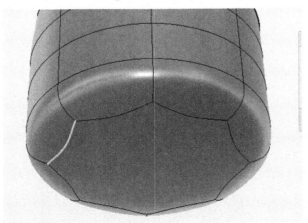
Figure-85. Fill Star option

- Select **Collapse** option from **Fill Hole Mode** drop-down to fill the hole by collapsing all the vertices of selected edge to the center point of hole face; refer to Figure-86.

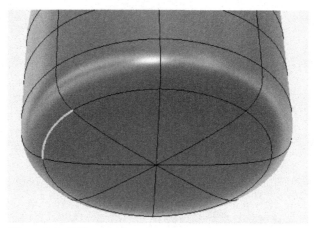

Figure-86. Collapse option

- Select the **Weld Center Vertices** option from **FILL HOLE** dialog box to weld the vertices at the center of hole face. The **Weld Center Vertices** option is available when Collapse option is selected in the Fill Hole Mode drop-down.
- After specifying the parameters, click on the **OK** button from **FILL HOLE** dialog box.

Weld Vertices

The **Weld Vertices** tool is used to join two vertices into single vertex. The procedure to use this tool is discussed next.

- Click on the **Weld Vertices** tool from **MODIFY** drop-down; refer to Figure-87. The **WELD VERTICES** dialog box will be displayed; refer to Figure-88.

Figure-87. Weld Vertices tool

Figure-88. WELD VERTICES
dialog box

- The **Select** button of **T-Spline Vertices** is active by default. Click on the vertices to join together; refer to Figure-89.

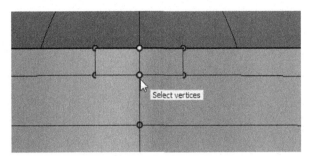

Figure-89. Selecting vertices

- Select **Vertex to Vertex** option from **Weld Mode** drop-down to move the first vertex to the second vertex.
- Select **Vertex to Midpoint** option from **Weld Mode** drop-down to move the two vertices to the midpoint of the selections.
- Select **Vertex to Tolerance** option from **Weld Mode** drop-down to weld all the visible and invisible vertices together with in a specified tolerance.
- Click in the **Weld Tolerance** edit box and enter the value of tolerance as required.
- After specifying the parameters, click on the **OK** button from **WELD VERTICES** dialog box; refer to Figure-90.

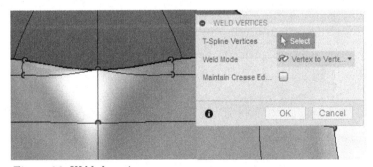

Figure-90. Welded vertices

UnWeld Edges

The **UnWeld Edges** tool is used to detach an edge or loop. The procedure to use this tool is discussed next.

- Click on the **UnWeld Edges** tool from **MODIFY** drop-down; refer to Figure-91. The **UNWELD EDGES** dialog box will be displayed; refer to Figure-92.

*Figure-91. UnWeld
Edges tool*

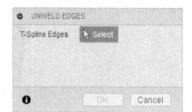

*Figure-92. UNWELD EDGES
dialog box*

- The **Select** button of **T-Spline Edges** option is active by default. Click on the edge or loop to select; refer to Figure-93. Double-Click on the edge to select the loop.

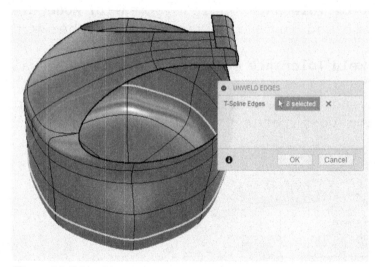

Figure-93. Selected loop

- After selecting the edge or loop, click on the **OK** button from **UNWELDED EDGES** dialog box to complete the process; refer to Figure-94.

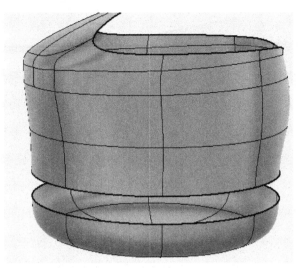

Figure-94. Seperated body

Crease

The **Crease** tool is used to add a sharp crease on the selected T-Spline body. The procedure to use this tool is discussed next.

- Click on the **Crease** tool from **MODIFY** drop-down; refer to Figure-95. The **CREASE** dialog box will be displayed; refer to Figure-96.

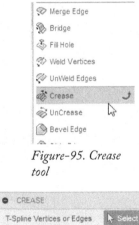

Figure-95. Crease tool

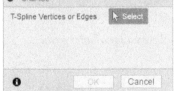

Figure-96. CREASE dialog box

- The **Select** button of **T-Spline Vertices or Edges** option is active by default. Click to select the edge. You can also use window selection to select multiple edges; refer to Figure-97.

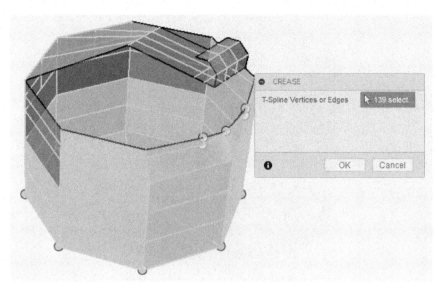

Figure-97. Selection of edges for crease

After selection of edges for crease, click on **OK** button from **CREASE** dialog box. The selected edges will be converted into creased edges; refer to Figure-98.

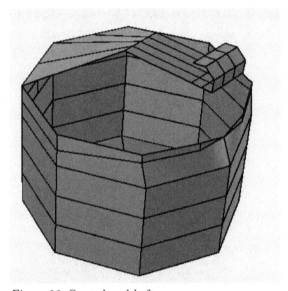

Figure-98. Created model after crease

UnCrease

The **UnCrease** tool is used to remove crease from the selected body, edge, vertex, or body. The procedure to use this tool is discussed next.

- Click on the **UnCrease** tool from **MODIFY** drop-down; refer to Figure-99. The **UNCREASE** dialog box will be displayed; refer to Figure-100.

Figure-99. UnCrease tool

Figure-100. UNCREASE dialog box

- The **Select** button of **T-Spline Vertices or Edges** option is active by default. Click to select the edge. You can also use window selection to select multiple edges; refer to Figure-101.

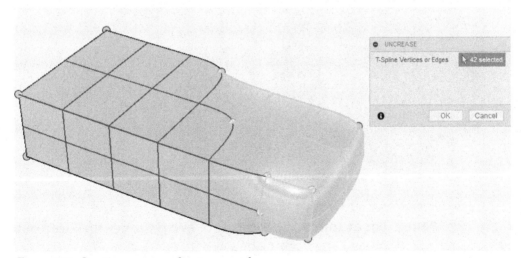

Figure-101. Creating uncrease edges to crease edges

- After selection of edges for uncrease, click on **OK** button from **UNCREASE** dialog box to complete the process.

Bevel Edge

The **Bevel Edge** tool is used to flatten the area of body by selecting a specific edge. The procedure to use this tool is discussed next.

- Click on the **Bevel Edge** tool from **MODIFY** drop-down; refer to Figure-102. The **BEVEL EDGE** dialog box will be displayed; refer to Figure-103.

Figure-102. Bevel Edge tool

Figure-103. BEVEL EDGE tool

- The **Select** button of **T-Spline Edge** option is selected by default. Click on the edge of a body to select; refer to Figure-104.

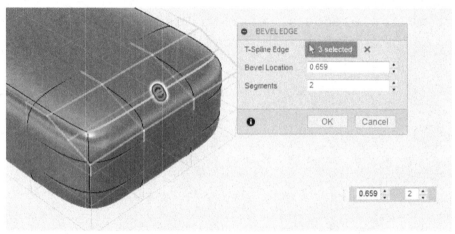

Figure-104. Selection of edges for bevel edges

- Click in the **Bevel Location** edit box and enter the value to position the new edge in decimal percentage.
- Click in the **Segments** edit box and enter the desired number of faces to be inserted in between new edges.
- After specifying the parameters, click on the **OK** button from **BEVEL EDGE** dialog box. The selected edge will be flatten; refer to Figure-105.

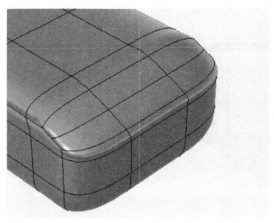

Figure-105. Flatten edges

Slide Edge

The **Slide Edge** tool is used to move two edges closer together or farther apart. The procedure to use this tool is discussed next.

- Click on the **Slide Edge** tool from **MODIFY** drop-down; refer to Figure-106. The **SLIDE EDGE** dialog box will be displayed; refer to Figure-107.

Figure-106. Slide Edge tool

Figure-107. SLIDE EDGE dialog box

- The **Select** button of **T-Spline Edge** option is selected by default. Click on the edge of a body to select; refer to Figure-108.

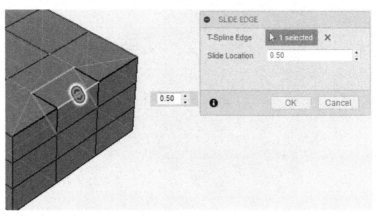

Figure-108. Selecting edge for slide edge

- Click in the **Slide Location** edit box and enter the value to position the new edge in decimal percentage. You can also set the value of **Slide Location** option by moving the manipulator displayed on the selected edge.
- After specifying the parameters, click on the **OK** button from **SLIDE EDGE** dialog box. The slide edge will be created; refer to Figure-109.

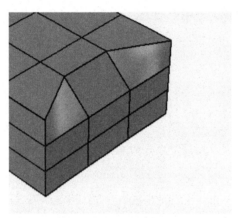

Figure-109. Created slide edge

Pull

The **Pull** tool is used for moving the selected vertices to the target body. The procedure to use this tool is discussed next.

- Click on the **Pull** tool from **MODIFY** drop-down; refer to Figure-110. The **PULL** dialog box will be displayed; refer to Figure-111.

Figure-110. Pull tool

Figure-111. PULL dialog box

- The **Select** button of **T-Spline Vertices** option is active by default. You need to click on the specific vertex to pull the vertex upto the targeted body; refer to Figure-112.

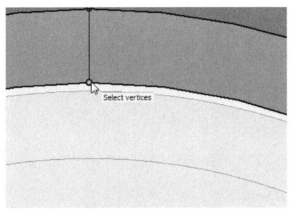 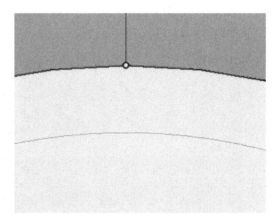

Figure-112. Pulling the vertex

- Select **Auto** option from **Target Select** drop-down to automatically pull the selected vertex.
- Select the **Select Targets** option from **Target Select** drop-down to manually select the target body. The updated **PULL** dialog box will be displayed; refer to Figure-113.

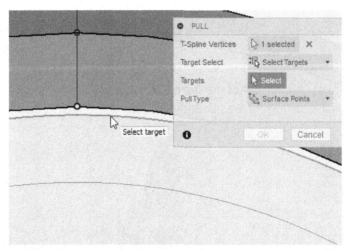

Figure-113. Updated PULL dialog box

- The **Select** button of **Targets** option is active by default. You need to click on the target body to pull the selected vertex.
- Select **Surface Points** option from **Pull Type** drop-down to move the surface points to the target body.
- Select **Control Points** option from **Pull Type** drop-down to move the control points to the target body.

- After specifying the parameters, click on the **OK** button from **PULL** dialog box to complete the process.

Note- The **Pull** tool will be applied when the bodies are near to each other.

Flatten

The **Flatten** tool is used for moving the selected control point to the plane for flatten the selected surface. The procedure to use this tool is discussed next.

- Click on the **Flatten** tool from **MODIFY** drop-down; refer to Figure-114. The **FLATTEN** dialog box will be displayed; refer to Figure-115.

Figure-114. Flatten tool

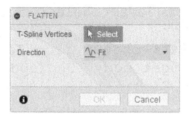

Figure-115. FLATTEN dialog box

- The **Select** button of **T-Spline Vertices** option is active by default. Select the vertices of the face which you want to flatten; refer to Figure-116.

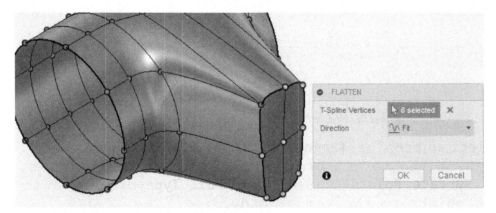

Figure-116. Selecting vertices for flatten surface

- On selecting the all vertices, the preview of flatten surface will displayed; refer to Figure-117.

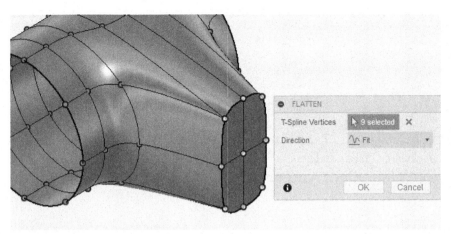

Figure-117. Flatten surface

- Select **Fit** option from **Direction** drop-down to use the best fit plane for flatten the selected surface.
- Choose **Select Plane** option from **Direction** drop-down to manually select the plane to flatten all control points. You are required to select the plane.
- Select **Select Parallel Plane** option from **Direction** drop-down to move the control points to the selected parallel plane. You are required to select the plane.
- After specifying the parameters, click on the **OK** button from **FLATTEN** dialog box.

Match

The **Match** tool is used to align the selected T-Spline edge with a sketch, face, or edge. The procedure to use this tool is discussed next.

- Click on the **Match** tool from **MODIFY** drop-down; refer to Figure-118. The **MATCH** dialog box will be displayed; refer to Figure-119.

Figure-118. Match tool

Figure-119. MATCH dialog box

- The **Select** button of **T-Spline Edges** option is active by default. You need to select the edge. You can also select the loop by double clicking on the edge; refer to Figure-120.

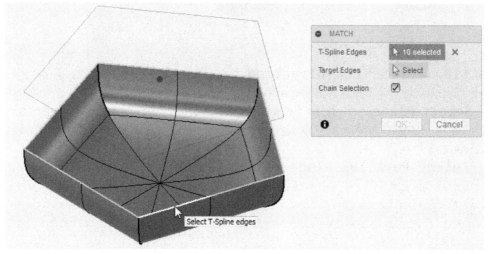

Figure-120. Selected loop for match

- Click on the **Select** button of **Target Edges** option and select the target edge or sketch.
- On selecting, the preview of alignment will be displayed along with updated **MATCH** dialog box; refer to Figure-121.

Figure-121. Updated MATCH dialog box

- Click on the **Flip Direction** button from **MATCH** dialog box to flip the direction of alignment; refer to Figure-122.

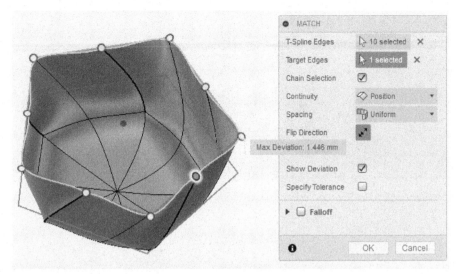

Figure-122. Flipped directipn of object

- Select the **Show Deviation** check box from **Tolerance** node to display the amount and location of maximum deviation.
- Select the **Specific Tolerance** check box from **Tolerance** node to specifies the value of how close the T-Spline edges need to be to the target edge.
- Select the **Falloff** check box from **MATCH** dialog box to determine the value of surface is affected by the match.
- Click in the **Falloff Range** check box and enter the value of surface is affected by match.
- After specifying the parameters, click on the **OK** button from **MATCH** dialog box to complete the process.

Interpolate

The **Interpolate** tool is used to move the T-Spline control points or surface points for improving fitting. The procedure to use this tool is discussed next.

- Click on the **Interpolate** tool from **MODIFY** drop-down; refer to Figure-123. The **INTERPOLATE** dialog box will be displayed; refer to Figure-124.

Figure-123. Interpo-late tool

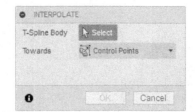

Figure-124. INTERPOLATE
dialog box

- The **Select** button of **T-Spline body** option is active by default. Click on the body to select; refer to Figure-125.

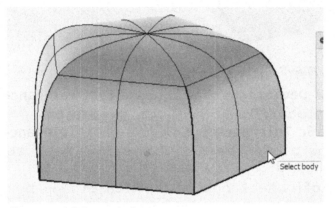

Figure-125. Selecting body

- On selecting the body, the preview will be displayed; refer to Figure-126.

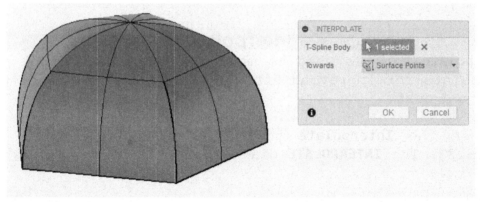

Figure-126. Selected body

- Select **Surface Points** from **Towards** drop-down to move the control points towards the surface.
- Select **Control Points** from **Towards** drop-down of you want to fit the surface through existing control points.
- After specifying the parameters, click on the **OK** button from **INTERPOLATE** dialog box.

Thicken

The **Thicken** tool is used to apply thickness to the sculpt faces. The procedure to use this tool is discussed next.

- Click on the **Thicken** tool from **MODIFY** drop-down; refer to Figure-127. The **THICKEN** dialog box will be displayed; refer to Figure-128.

Figure-127. *Thicken tool1*

Figure-128. *THICKEN dialog box*

- The **Select** button of **T-Spline body** option is active by default. Click on the body to select. The updated **THICKEN** dialog box will be displayed.
- Click in the **Thickness** edit box and enter the desired thickness. You can also move the arrow displaying on the selected model to adjust the thickness; refer to Figure-129.

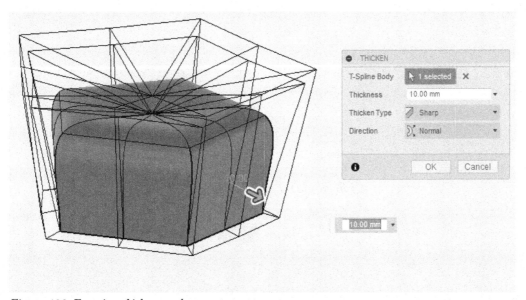

Figure-129. *Entering thickness value*

- Select the **Sharp** button from **Thicken Type** drop-down to connect the surface with the straight face.
- Select the **Soft** button from **Thicken Type** drop-down to connect the surface with the round face.

- Select the **No Edge** button from **Thicken Type** drop-down to not connect the surfaces.
- Select **Normal** button from **Direction** drop-down to create a new surface perpendicular to the selected surface.
- Select **Axis** button from **Direction** drop-down to create a new surface perpendicular to a selected axis.
- After specifying the parameters, click on the **OK** button from **THICKEN** dialog box. The selected body will be thicken; refer to Figure-130.

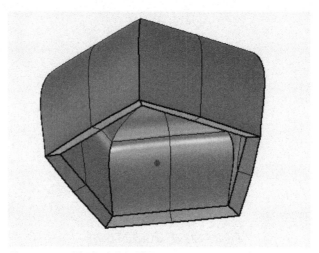

Figure-130. Thickened surface

Freeze

The **Freeze** tool is used to freeze the selected edges or faces to prevent changes. The procedure to use this tool is discussed next.

- Click on the **Freeze** tool of **Freeze** cascading menu from **MODIFY** drop-down; refer to Figure-131. The **FREEZE** dialog box will be displayed; refer to Figure-132.

Figure-131. Freeze tool

Figure-132. FREEZE dialog box

- The **Select** button of **T-Spline Faces/Edges** option is active by default. Click on the edges/ faces to select; refer to Figure-133

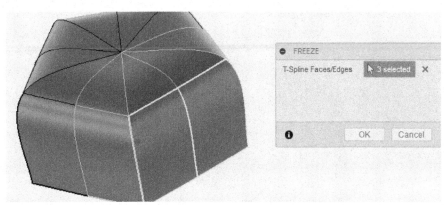

Figure-133. Selection of faces and edges to freeze

- After selection of edges, click on the **OK** button from **FREEZE** dialog box. The selected edges or faces will be frozen.

UnFreeze

The **UnFreeze** tool is used to unfreeze the frozen edges of faces. The procedure to use this tool is discussed next.

- Click on the **UnFreeze** tool of **Freeze** cascading menu from **MODIFY** drop-down; refer to Figure-134. The **UNFREEZE** dialog box will be displayed; refer to Figure-135.

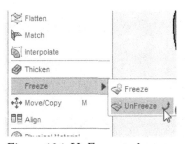

Figure-134. UnFreeze tool

Figure-135. UNFREEZE dialog box

- The **Select** button of **T-Spline Faces/Edges** option is active by default. Click on the edges/faces to unfreeze; refer to Figure-136.

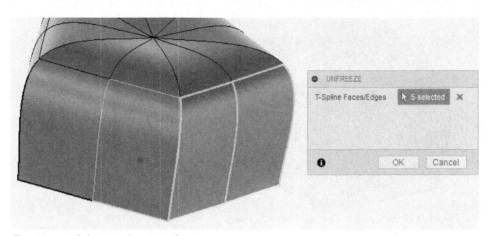

Figure-136. Selecting edges to unfreeze

- After selecting, click on the **OK** button from **UNFREEZE** dialog box. The frozen edges will be unfreeze.

PRACTICAL

In this practical, we will create the model shown in Figure-137.

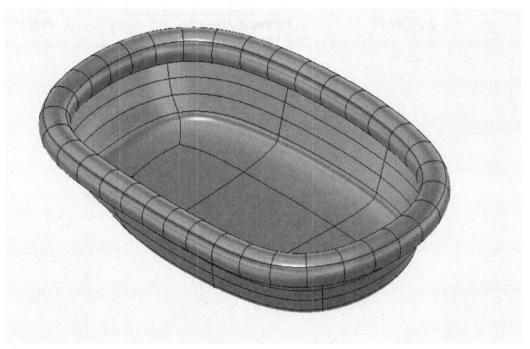

Figure-137. Created practical 1

Creating Sketch

- Open the **Sculpt** workspace as discussed earlier.
- Click on the **Center Rectangle** tool of **Rectangle** cascading menu from **SKETCH** drop-down and create a rectangle as shown in Figure-138.

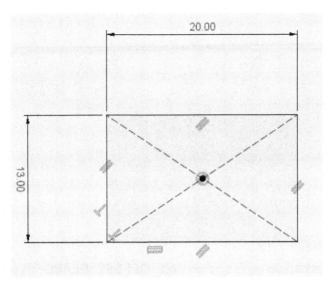

Figure-138. Creating first sketch

- Click on the **Fillet** tool of **SKETCH** drop-down from **Toolbar** and apply the fillet of 3 as shown in Figure-139.

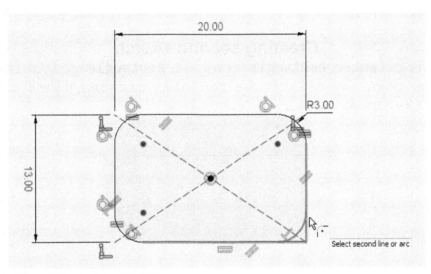

Figure-139. Applying the fillet

- Click on the **Stop Sketch** button from **Toolbar** to exit the sketch.

Creating plane

- Click on the **Offset Plane** tool of **CONSTRUCT** drop-down from **Toolbar**. The **OFFSET PLANE** dialog box will be displayed.
- The **Select** button of **Plane** option is active by default. You need to select the recently created sketch plane as reference; refer to Figure-140.

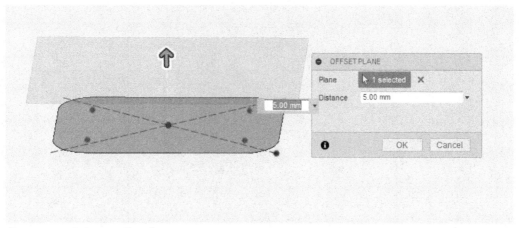

Figure-140. Creating offset plane

- Click in the **Distance** edit box of **OFFSET PLANE** dialog box and enter the value as **5**.
- After specifying the parameters, click on the **OK** button from **OFFSET PLANE** edit box. The plane will be created and displayed above the first sketch.

Creating second sketch

- Click on the **Center Rectangle** tool of **Rectangle** cascading menu from **SKETCH** drop-down and select the recently created plane as reference. Create a sketch as shown in Figure-141.

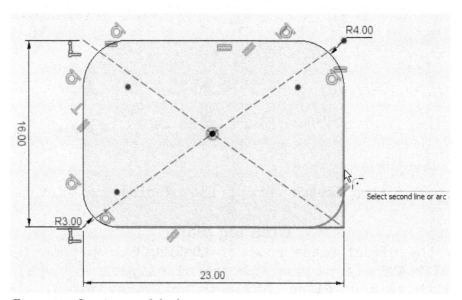

Figure-141. Creating second sketch

- Click on the **Stop Sketch** button from **Toolbar** to exit the sketch.

Creating Loft Feature

- Click on the **Loft** tool of **CREATE** drop-down from **Toolbar**. The **LOFT** dialog box will be displayed.
- Click on the **Chain Selection** check box of **LOFT** dialog box to select the edges in chain.
- Select the first and second created sketch to select in **Profiles** section; refer to Figure-142.

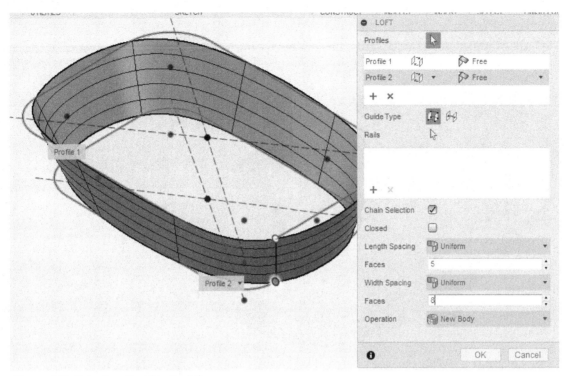

Figure-142. Selection of profiles for loft

- After specifying the parameters displayed on above figure, click on the **OK** button from **LOFT** dialog box.

Applying Fill Hole

- Click on the **Fill Hole** tool of **MODIFY** drop-down from **Toolbar**. The **FILL HOLE** dialog box will be displayed.
- Select the edge from the model as displayed; refer to Figure-143.

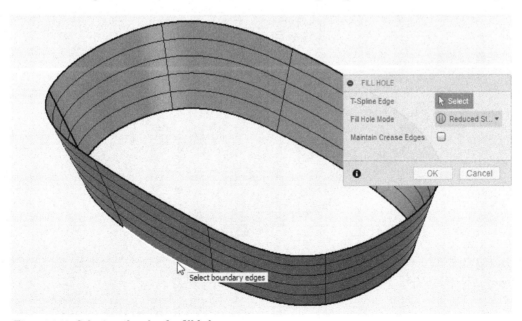

Figure-143. Selecting the edge for fill hole

- The selected hole will be filled.
- Click on the **PIPE** tool of **CREATE** drop-down from **Toolbar**. The **PIPE** dialog box will be displayed.

- Click on the path of model to select as a path for pipe; refer to Figure-144.

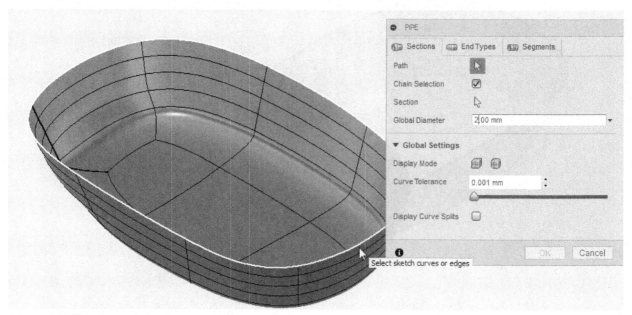

Figure-144. Selecting path for pipe

- Click on the **Smooth Display** button of **Global Settings** section from **End Types** tab of **PIPE** dialog box and enter the parameters as displayed in above figure.
- After specifying the parameters, the model will be displayed as shown in Figure-145.

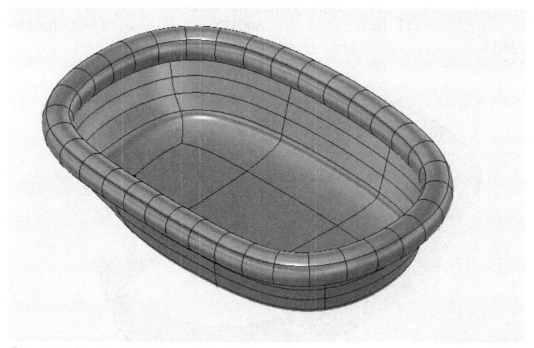

Figure-145. Created practical 1

PRACTICE 1

Create a wooden tool as displayed in Figure-146 and Figure-147. As a primitive structure, use the Sculpt cylinder.

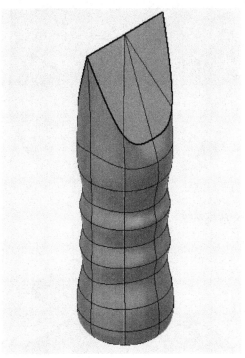

Figure-146. Practice 1

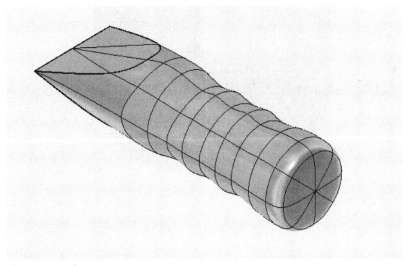

Figure-147. Practice 2

PRACTICE 2

Create the model as shown in Figure-148 and Figure-149 by using the tools of Sculpt Workspace.

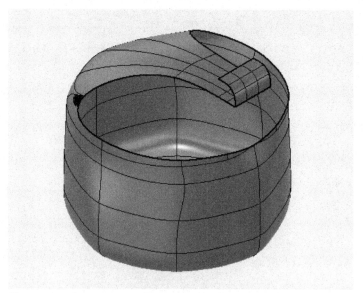

Figure-148. Practice 2

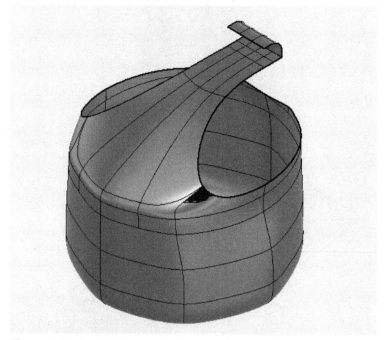

Figure-149. Practice 2

Chapter 13

Sculpting-2

Topics Covered

The major topics covered in this chapter are:

- *Mirror Tools*
- *Duplicate Tools*
- *Utility Tools*
- *Repair Mode*
- *Convert Tools*

INTRODUCTION

In the last chapter, we have learned to create and modify the sculpt object by using various tools. In this chapter we will discuss the symmetric and utilities tool use in **Sculpt Workspace**.

SYMMETRY TOOLS

Symmetry tools are used to create symmetric copies of the selected sculpt features in the Sculpting environment. These tools are discussed next.

Mirror - Internal

The **Mirror - Internal** tool is used to create an internal mirror symmetry in the T-Spline body on selecting an edge, face, and vertex. The procedure to use this tool is discussed next.

- Click on the **Mirror - Internal** tool from **SYMMETRY** drop-down; refer to Figure-1. The **MIRROR - INTERNAL** dialog box will be displayed; refer to Figure-2.

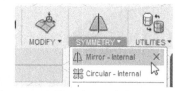

Figure-1. MIRROR-INTERNAL tool

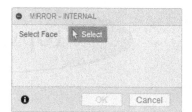

Figure-2. MIRROR-INTERNAL dialog box

- The **Select** button of **Select Face** section is active by default. Click on the face from master side; refer to Figure-3.

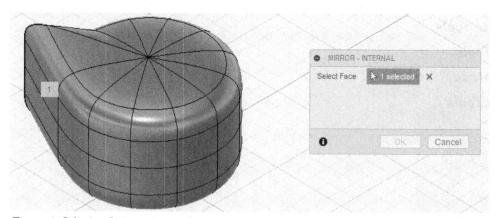

Figure-3. Selecting face on master side

- Click on the face to be made mirror symmetric. The preview of mirror will be displayed; refer to Figure-4.

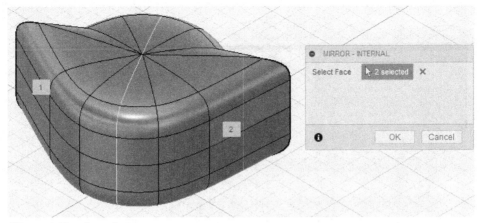

Figure-4. Preview of mirror

- If the preview of mirror symmetry is as required, then click on the **OK** button from **MIRROR - INTERNAL** dialog box.

Circular - Internal

The **Circular - Internal** tool is used to create an internal circular symmetry based on selected face, edge, or vertex. The procedure to use this tool is discussed next.

- Click on the **Circular - Internal** tool from **SYMMETRY** drop-down; refer to Figure-5. The **CIRCULAR - INTERNAL** dialog box will be displayed; refer to Figure-6.

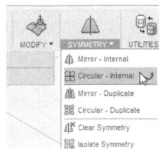

Figure-5. Circular-Internal tool

Figure-6. CIRCULAR-INTERNAL dialog box

- The **Select** button of **Select Face** section is active by default. Click on the face to select for symmetry; refer to Figure-7

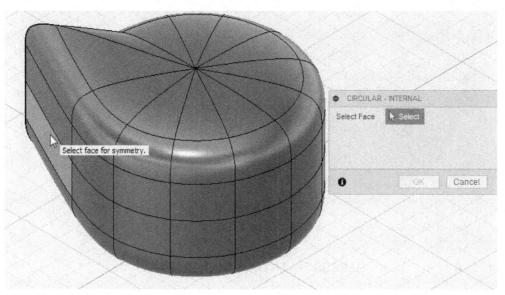

Figure-7. Selecting face for symmetry

- After selecting, the updated **CIRCULAR - INTERNAL** dialog box will be displayed along with the preview; refer to Figure-8.

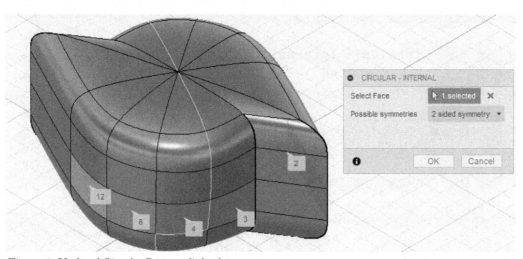

Figure-8. Updated Circular Pattern dialog box

- Select the desired option from **Possible symmetries** drop-down to define number of symmetric sections and click on the **OK** button from **CIRCULAR - INTERNAL** dialog box; refer to Figure-9.

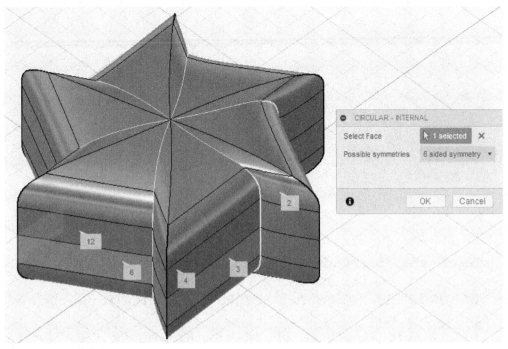

Figure-9. 6-sided mirror symmetry

Mirror - Duplicate

The **Mirror - Duplicate** tool is used to create a new T-Spline body or surface based on the selected plane or face. The procedure to use this tool is discussed next.

- Click on the **Mirror - Duplicate** tool from **SYMMETRY** drop-down; refer to Figure-10. The **MIRROR - DUPLICATE** dialog box will be displayed; refer to Figure-11.

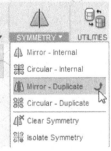

Figure-10. Mirror-Duplicate tool

Figure-11. MIRROR-DUPLI-CATE dialog box

- The **Select** button of **T-Spline Body** section is active by default. Click on the sculpt body whose mirror copy is to be created; refer to Figure-12.

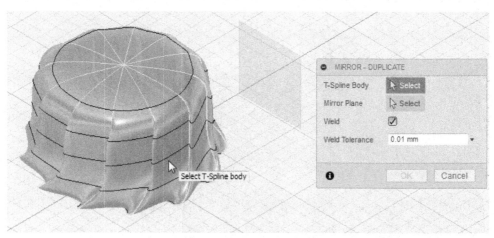

Figure-12. Selection of body for mirror

- Now, select the desired plane to mirror the selected body. The preview of mirror will be displayed.
- Select the **Weld** check box from **MIRROR - DUPLICATE** check box to weld the symmetric edges together.
- Click in the **Weld Tolerance** check box and enter the desired value of tolerance; refer to Figure-13.

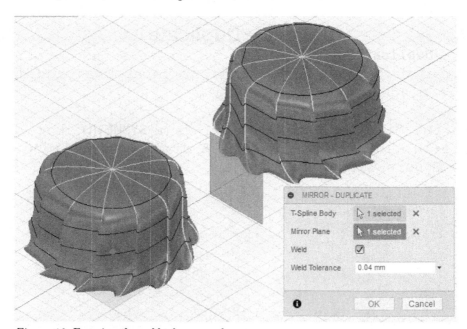

Figure-13. Entering the weld tolerance value

- After specifying the parameters, click on the **OK** button from **MIRROR - DUPLICATE** check box.

Circular - Duplicate

The **Circular - Duplicate** tool is used to create circular symmetric copies of the selected body around an axis. The procedure to use this tool is discussed next.

- Click on the **Circular - Duplicate** tool from **SYMMETRY** drop-down; refer to Figure-14. The **CIRCULAR - DUPLICATE** dialog box will be displayed; refer to Figure-15.

Figure-14. Circular Duplicate
tool

Figure-15. CIRCULAR-DUL-
PICATE dialog box

- The **Select** button of **T-Spline Body** section is active by default. Click on the body from canvas to select.
- Click to select the axis; refer to Figure-16.

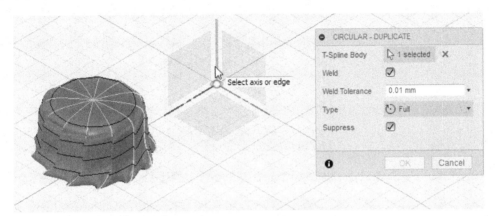

Figure-16. Selecting the axis for rotation

- Select the **Weld** check box if required.
- Click in the **Quantity** edit box and enter the desired number of duplicate copy you want to create.
- After specifying the parameter, click on the **OK** button from **CIRCULAR -DUPLICATE** dialog box; refer to Figure-17.

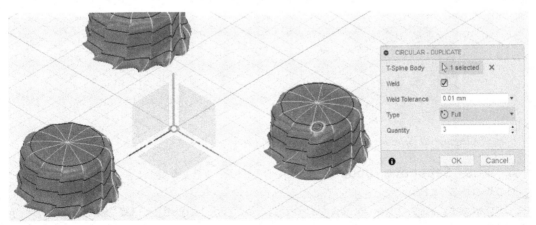

Figure-17. Created circular copy

Clear Symmetry

The **Clear Symmetry** tool is used to delete the created symmetry created earlier on the body. The procedure to use this tool is discussed next.

- Click on the **Clear Symmetry** tool from **SYMMETRY** drop-down; refer to Figure-18. The **CLEAR SYMMETRY** dialog box will be displayed; refer to Figure-19.

Figure-18. Clear Symmetry tool

Figure-19. CLEAR SYMME-TRY dialog box

- Click on the T-Spline body to select; refer to Figure-20.

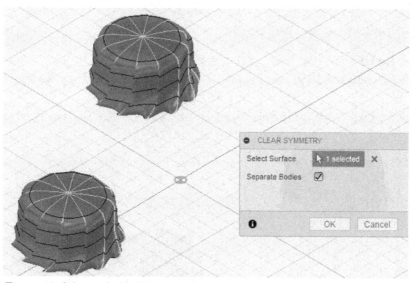

Figure-20. Selecting bodies for clearing symmetry

- Select the **Separate Bodies** check box to create a new bodies for disjointed surface.
- After specifying the parameters, click on the **OK** button from **CLEAR GEOMETRY** check box; refer to Figure-21.

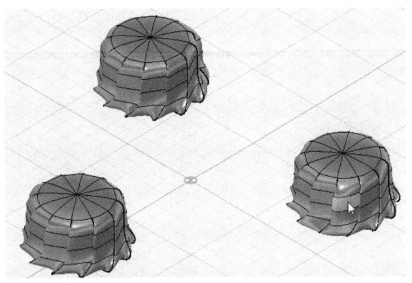

Figure-21. Deleted symmety from bodies

Isolate Symmetry

The **Isolate Symmetry** tool is used to remove the symmetry condition from the selected face, edge, or vertex but the selected geometry will still symmetric to the other duplicate bodies. The procedure to use this tool is discussed next.

- Click on the **Isolate Symmetry** tool from **SYMMETRY** drop-down; refer to Figure-22. The **ISOLATE SYMMETRY** dialog box will be displayed; refer to Figure-23.

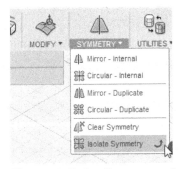

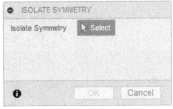

Figure-23. ISOLATE SYMME-TRY dialog box

Figure-22. Isolate Symmetry tool

- The **Select** button of **Isolate Symmetry** section is active by default. Click on the symmetry to select; refer to Figure-24.

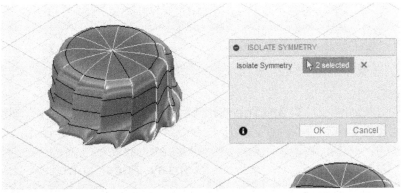

Figure-24. Selecting faces

- After specifying the geometry, click on the **OK** button from **ISOLATE SYMMETRY** dialog box.
- If you edit or modify the isolated face, the symmetry faces will modified accordingly; refer to Figure-25.

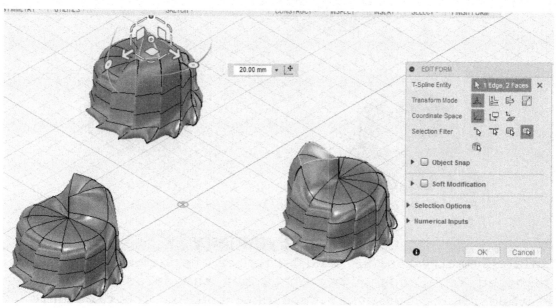

Figure-25. Modifying the selected Face

UTILITIES

Till now, we have discussed various symmetric tool. In this section we will discuss various utility tools used in **Sculpt Workspace**.

Display Mode

The **Display Mode** tool is used to switches the view of selected body to box or smooth display. The procedure to use this tool is discussed next.

- Click on the **Display Mode** tool from **UTILITIES** drop-down; refer to Figure-26. The **DISPLAY MODE** dialog box will be displayed; refer to Figure-27.

Figure-26. Display Mode tool

Figure-27. DISPLAY MODE dialog box

- The **Select** button of **T-Spline Entity** is active by default. You need to select the face, edge, body or vertex. You can also use window selection for selecting the whole geometry.
- Select **Box Display** button from **Display Mode** section to display the

control points of the T-Spline body.

- Select **Control Frame Display** from **Display Mode** section to display the rounded frame body with the control frame around it.
- Select **Smooth** from **Display Mode** section to display the rounded shape of the T-Spline body.
- After selecting the required display, click on the **OK** button from **DISPLAY MODE** dialog box.

Repair Mode

The **Repair Body** tool is used for displaying the information about the mesh of sculpt body. This tool also repairs error star points and error T points. The procedure to use this tool is discussed next.

- Click on the **Repair Body** tool from **SYMMETRY** drop-down; refer to Figure-28. The **REPAIR BODY** dialog box will be displayed; refer to Figure-29.

Figure-28. Repair Body tool

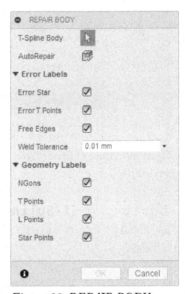

Figure-29. REPAIR BODY dialog box

- The **Select** button of **T-Spline Body** section is active by default. Click on the body to select.
- Click on the **AutoRepair** button from **REPAIR BODY** dialog box to repair error star points, error T points, and free edges.
- Select the **Error Star** check box of **Error Labels** section to display a red star on star points with an error.
- Select **Error T Points** check box of **Error Labels** section to display a red T on T point with an error.
- Select **Free Edges** check box to highlight the open edges on the body.
- Click in the **Weld Tolerance** edit box and specify the distance between edges to weld when using **AutoRepair** button.
- Select the **NGons** check box from **Geometry Labels** section to display the NGons with the number of edges on the selected model. NGons are faces which less than or more than 4 edges.

- Select the **T Points** check box to display a yellow T on T points of model.
- Select the **L Points** check box to display a yellow L on L Points of the selected model.
- Select the **Star Points** check box to display a yellow star on star points of model.
- After specifying the parameters, click on the **OK** button from **REPAIR BODY** dialog box.

Make Uniform

The **Make Uniform** tool is used to create uniform surface of the selected body. This tool is used for making all the knots interval of selected body uniform. The procedure to use this tool is discussed next.

- Click on the **Make Uniform** tool from **UTILITIES** drop-down; refer to Figure-30. The **MAKE UNIFORM** dialog box will be displayed; refer to Figure-31.

Figure-30. Make Uniform tool

Figure-31. MAKE UNIFORM dialog box

- The **Select** button of **T-Spline Body** section is active by default. Click on the body to select and click on the **OK** button. The tool will be applied.

Convert

The **Convert** tool is used to create a sculpt object into other forms. The type of body created depends on selected body. The procedure to use this tool is discussed next.

- Click on the **Convert** tool from **UTILITIES** drop-down; refer to Figure-32. The **CONVERT** dialog box will be displayed; refer to Figure-33.

Figure-32. Convert tool

Figure-33. CONVERT dialog box

- Select **T-Spline to BRep** option from **Convert Type** drop-down to convert a T-Spline body into solid body from.
- Select **BRep Face to T-Splines** option from **Convert Type** drop-down to convert a surface into sculpt.
- Select **Quad Mesh to T-Spline** option from **Convert Type** drop-down to convert a mesh body to a T-Spline body (sculpt).
- In our case, we are converting a T-Spline body to a solid body. The **Select** button of **Selection** section is active by default. Click on the sculpt body to convert; refer to Figure-34.

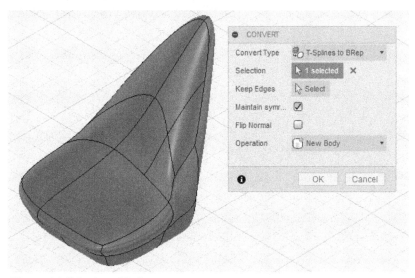

Figure-34. Selection of body to convert

- Click on the **Select** button from **Keep Edges** section and select the required edges to maintain the selected edge in converted body.
- Select the **Maintain symmetry** check box from **CONVERT** dialog box to maintain the symmetry of T-Spline body after conversion.
- Select the **Flip Normal** check box to changes the normal direction of the selected bodies.
- After specifying the parameters, click on the **OK** button from **CONVERT** dialog box; refer to Figure-35. The converted body will be displayed in **MODEL Workspace**.

Figure-35. Converted body

Maintain Better Performance/Display

The **Maintain Better Performance/Display** tool is used to toggle between better performance or better display. The better display shows the bodies at highest quality and better performance calculates modification by applying G0 conditions at star points.

FOR STUDENT NOTES

Chapter 14

Mesh Design

Topics Covered

The major topics covered in this chapter are:

- *BRep to Mesh*
- *Make Closed Mesh*
- *Erase and Fill*
- *Plane Cut*
- *Reverse Normal*
- *Separate and Merge Bodies*

INTRODUCTION

A model of mesh consists of vertices, edges, and faces that use polygonal representation, including triangles and quadrilaterals, to define a 3D shape. Mesh models has no mass properties but we can create primitive mesh forms.

OPENING THE MESH WORKSPACE

The **MESH Workspace** is used to create mesh models by using various tools of **MESH Workspace**. Generally, the **MESH Workspace** is not present in **Workspace** drop-down. The button to open the **MESH Workspace** is displayed in **MODEL Workspace**. The procedure is discussed next.

*   Click on the **MODEL Workspace** from **Workspace** drop-down. The **MODEL Workspace** will be displayed in **Autodesk Fusion 360** window.
*   Click on **Create Mesh** tool from **CREATE** drop-down in the **Toolbar**; refer to Figure-1. The **MESH Workspace** will be displayed; refer to Figure-2.

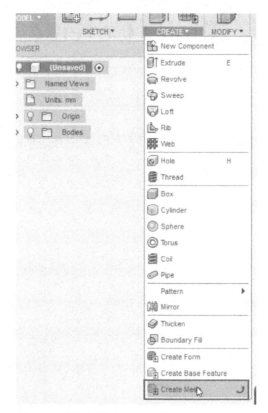

Figure-1. Mesh Workspace tool

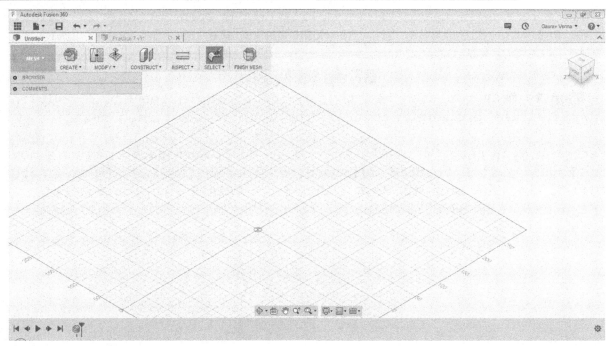

Figure-2. MESH Workspace window

INSERTING FILE

In this section, we will learn to insert a selected format file into Mesh Workspace and convert a solid model to mesh model.

Insert Mesh

The **Insert Mesh** tool is used for inserting a .OBJ or .STL mesh file into current design. The procedure to use this tool is discussed next.

- Click on the **Insert Mesh** tool from **Create** drop-down; refer to Figure-3. The **Open** dialog box will be displayed; refer to Figure-4.

Figure-3. Insert Mesh tool

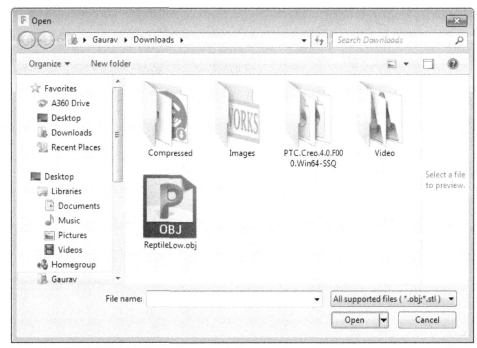

Figure-4. Open dialog box for inserting mesh file

• Select on the required file and click on the **Open** button. The selected file will be displayed in the **MESH Workspace** window.

BRep to Mesh

The **BRep to Mesh** tool is used to convert the selected solid body into mesh body. The procedure to use this tool is discussed next.

• At first, you need to create a model in **MODEL Workspace** or open a model and then switch to **MESH Workspace** to convert the body to mesh body.
• Click on the **BRep to Mesh** tool from **CREATE** drop-down; refer to Figure-5. The **BREP TO MESH** dialog box will be displayed; refer to Figure-6.

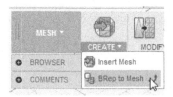

Figure-5. BRep to Mesh tool

• The **Select** button of **Body** section is active by default. Click on the body to select; refer to Figure-7.

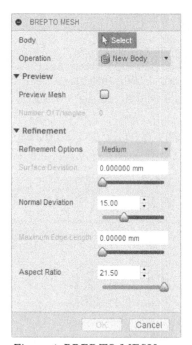

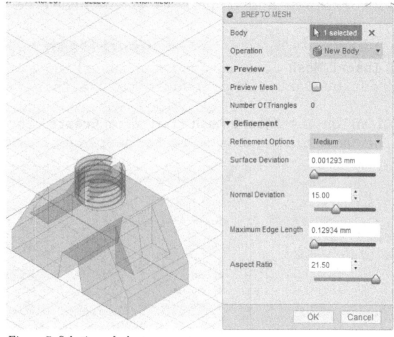

Figure-6. BREP TO MESH dialog box

Figure-7. Selecting a body

• Select the **Preview Mesh** check box under **Preview** node from **BREP TO MESH** dialog box to display the preview of mesh body of the selected solid body. The preview of mesh body will be displayed along with **Number of Triangles**; refer to Figure-8.

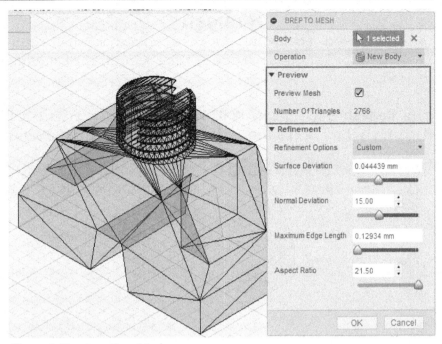

Figure-8. Preview of mesh body

- Select the **High**, **Medium**, or **Low** option from **Refinement Options** drop-down to set the refinement of mesh body automatically.
- If you want to set the refinement manually then click on the **Custom** option from **Refinement Options** drop-down.
- Move the **Surface Deviation**, **Normal Deviation**, **Maximum Edge Length**, and **Aspect Ratio** sliders to set the respective values.
- After specifying the parameters, click on the **OK** button from **BREP TO MESH** dialog box to complete the process; refer to Figure-9.

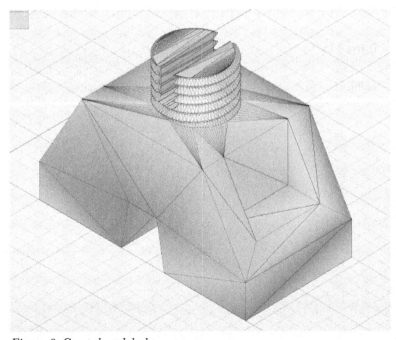

Figure-9. Created mesh body

MODIFICATION TOOLS

In this section, we will discuss various tools which are used to modify the mesh body.

Remesh

The **Remesh** tool is used to refine the selected mesh faces or body to form regular-shaped triangular faces. The procedure to use this tool is discussed next.

- Click on the **Remesh** tool from **MODIFY** drop-down; refer to Figure-10. The **REMESH** dialog box will be displayed; refer to Figure-11.

Figure-10. Remesh tool

Figure-11. REMESH dialog box

- The **Select** button of **Mesh Faces/Body** section is active by default. Click on the face to select.
- Select **Uniform** option from **Meshing Type** drop-down for creating the similar size face on the entire selection. This option is used for keeping the face sizes even.
- Select **Adaptive** option from **Meshing Type** drop-down for smaller faces in the region of high detail and larger faces in the region of low detail. This option is used for preserving details on the selected model.
- Click in the **Density** edit box and enter the desired value of density. You can also specify the value of density by moving the **Density** slider from **REMESH** dialog box; refer to Figure-12. It controls the number of faces created.

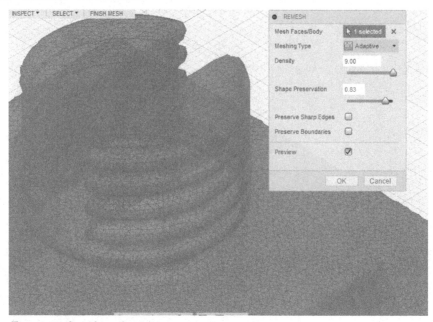

Figure-12. Specifying density

- Click in the **Shape Preservation** edit box and enter the value. You can also specify the value by moving the **Shape Preservation** slider. The value of **Shape Preservation** lies between 0 to 1.
- Select the **Preserve Sharp Edges** check box from **REMESH** dialog box to preserve the sharp edges from the input mesh.
- Select the **Preserve Boundaries** check box from **REMESH** dialog box to make sure that any open boundaries of the selected model do not change shape. This option is useful if you have two separate bodies meeting at open boundaries that you wish to merge later.
- Select the **Preview** check box to view the mesh preview before it is created.
- After specifying the parameters, click on the **OK** button from **REMESH** dialog box. The created mesh body will be displayed; refer to Figure-13.

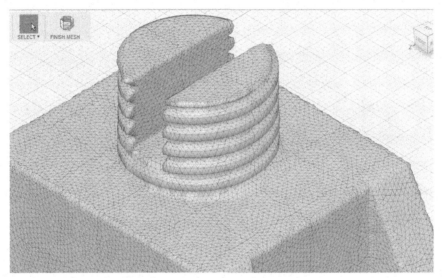

Figure-13. Created mesh body1

Reduce

The **Reduce** tool is used to reduce the number of faces on your model while trying to maintain its shape. The procedure to use this tool is discussed next.

- Click on the **Reduce** tool from **MODIFY** drop-down; refer to Figure-14. The **REDUCE** dialog box will be displayed; refer to Figure-15.

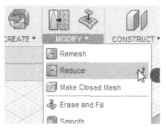

Figure-14. Reduce tool

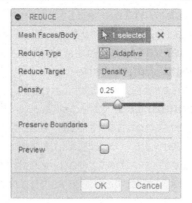

Figure-15. REDUCE dialog box

- The **Select** button of **Mesh Faces/Body** section is active by default. Click on the body to select.
- Click in the **Density** edit box and enter the desired value; refer to Figure-16.

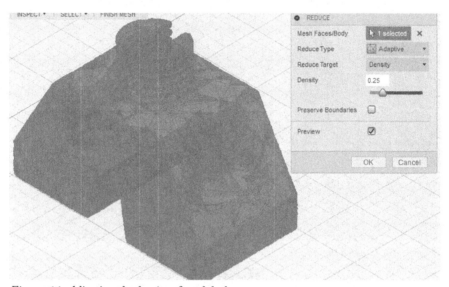

Figure-16. Adjusting the density of mesh body

- Select the **Preview** check box to view the mesh preview before it is created.
- After specifying the parameter, click on the **OK** button from **REDUCE** dialog box. The mesh body will be created; refer to Figure-17.

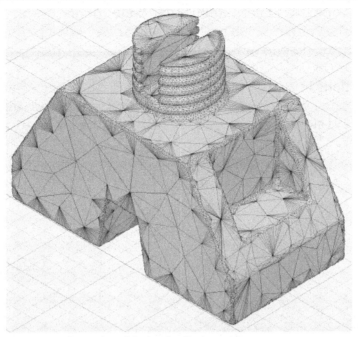

Figure-17. Created mesh body after Reduce tool

Make Closed Mesh

The **Make Closed Mesh** tool is used for rebuilding the selected mesh body as a new closed mesh. The procedure to use this tool is discussed next.

- Click on the **Make Closed Mesh** tool from **MODIFY** drop-down; refer to Figure-18. The **MAKE CLOSED MESH** dialog box will be displayed; refer to Figure-19.

Figure-18. Make Closed Mesh tool

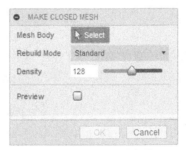

Figure-19. MAKE CLOSED MESH dialog box

- The **Select** button of **Mesh Body** section is active by default. Click on the body to select; refer to Figure-20.
- Select **Standard** option from **Rebuild Mode** drop-down to rebuilds the mesh with default behavior. This option provides a good balance of speed and accuracy, but sharp edges will become soft.
- Select **Preserve Sharp Edges** option from **Rebuild Mode** drop-down to rebuild the mesh similar to the **Standard** option but it will preserve sharp edges. In this option the mesh density is higher near the edge.
- Select **Accurate** option from **Rebuilds Mode** drop-down to create a closed mesh of the selected body in more advanced method. In this performance will be slower than standard but accuracy may be improved.

- Select **Blocky** option from **Rebuilds Mode** drop-down to rebuilds the model as simple cubes. This option does not provide an accurate approximation of the input shape, this is just intended to give your model a bulky aesthetic.
- Click in the **Density** edit box and enter the desired value. You can also specify the value of density by moving the **Density** slider.
- Select the **Preview** check box to view the mesh preview before it is created.
- After specifying the parameter, click on the **OK** button from **MAKE CLOSED MESH** dialog box. The mesh body will be created; refer to Figure-21.

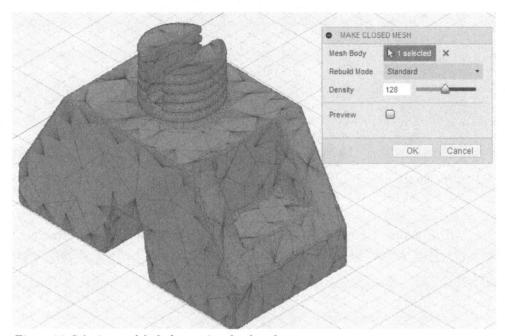

Figure-20. Selecting mesh body for creating closed mesh

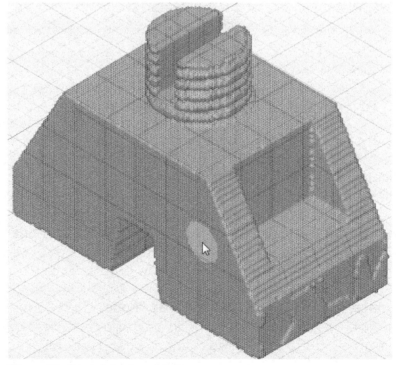

Figure-21. Created closed mesh body

Erase and Fill

The **Erase and Fill** tool is used to fill a hole or heal regions or defects on a mesh body. The procedure to use this tool is discussed next.

- Click on the **Erase and Fill** tool from **MODIFY** drop-down in the **Toolbar**; refer to Figure-22. The **ERASE AND FILL** dialog box will be displayed; refer to Figure-23.

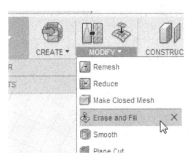

Figure-22. Erase and Fill tool

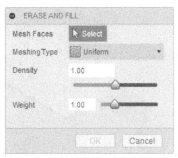

Figure-23. ERASE AND FILL
dialog box

- The **Select** button of **Mesh Faces** section is active by default. Click on the face to select. You can also use window selection to select the body; refer to Figure-24.

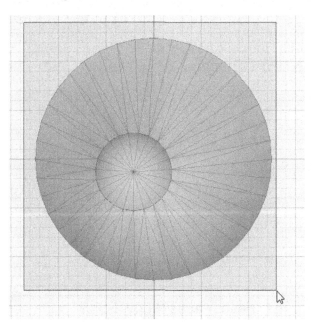

Figure-24. Selection of hole

- Select **Uniform** option from **Meshing Type** drop-down to fill the selected region with regular shaped triangles. This option gives the smoothest and most reliable results.
- Select **Minimal** option from **Meshing Type** drop-down to uses the minimum number of faces to fill the selected hole.
- Click in the **Density** edit box and enter the desired value. You can also specify the value of density by moving the **Density** slider.
- Click in the **Weight** edit box and enter the weight mesh after filling. You can also specify the weight by moving the **Weight** slider.

- After specifying the parameters, click on the **OK** button from **ERASE AND FILL** dialog box. The filled hole will be displayed; refer to Figure-25.

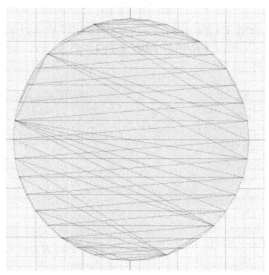

Figure-25. Filled hole

Smooth

The **Smooth** tool is used to smooth out uneven regions on the mesh. The procedure to use this tool is discussed next.

- Click on the **Smooth** tool from **MODIFY** drop-down; refer to Figure-26. The **SMOOTH** dialog box will be displayed; refer to Figure-27.

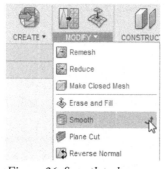

Figure-26. Smooth tool

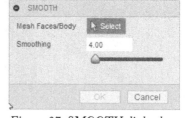

Figure-27. SMOOTH dialog box

- The **Select** button of **Mesh Faces/Body** section is active by default. Click on the body to select; refer to Figure-28.
- Click in the **Smoothing** edit box and enter the desired value for smoothing of mesh body. You can also move the **Smoothing** slider to specify the smoothing value.
- After specifying the parameters, click on the **OK** button from **SMOOTH** dialog box. The mesh body will be displayed; refer to Figure-29.

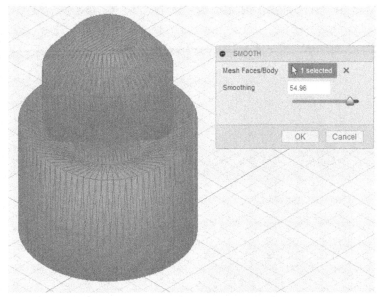

Figure-28. Selecting Mesh body for smooth

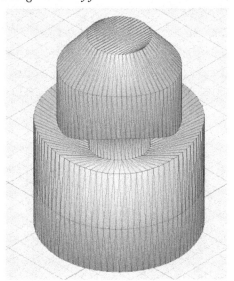

Figure-29. Smoothened mesh body

Plane Cut

The **Plane Cut** tool is used to splits the selected body with a user defined plane. The procedure to use this tool is discussed next.

- Click on the **Plane Cut** tool from **MODIFY** drop-down; refer to Figure-30. The **PLANE CUT** dialog box will be displayed; refer to Figure-31.

Figure-30. Plane Cut tool

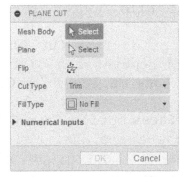

Figure-31. PLANE CUT dialog box

- The **Select** button of **Mesh Body** section is active by default. Click on the body to select. The manipulator will be displayed on the selected body; refer to Figure-32.

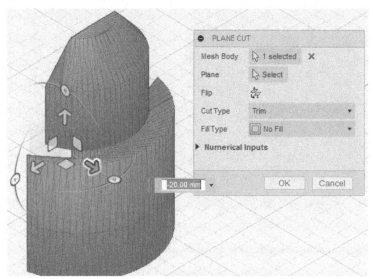

Figure-32. Manipulator displayed on the selected body

- Move the manipulator to split the body. If you want to manually split the selected body then click on the **Select** button of **Plane** section and select the required plane; refer to Figure-33.

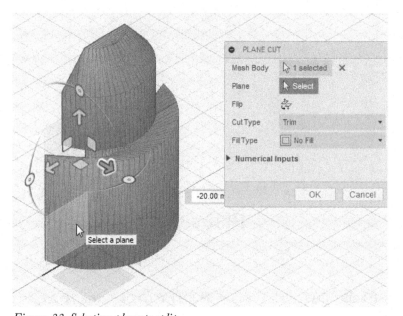

Figure-33. Selecting plane to split

- Click on the **Flip** button from **PLANE CUT** dialog box to flip the direction of split.
- Select **Trim** option from **Cut Type** drop-down to split the body into two sides and removes one of the sides.
- Select **Split Body** option from **Cut Type** drop-down to split the body to create two separate mesh body.
- Select **Split Faces** option from **Cut Type** drop-down to splits the faces that intersect the plane, but keeps the body intact. This option will create a new face group on one side of the split.

- Select **No Fill** option from **Fill Type** drop-down if you want to leave the open boundary at the cut.
- Select **Uniform** option from **Fill Type** drop-down if you want to fill the hole with new faces of regular shape.
- Select **Minimal** option from **Fill Type** drop-down of you want to fill the hole with minimal number of possible faces.
- Click on the **Numerical Inputs** node from **PLANE CUT** dialog box to manually enter the value of manipulator in respective edit box.
- After specifying the parameters, click on the **OK** button from **PLANE CUT** dialog box to split the model. The model will be displayed; refer to Figure-34.

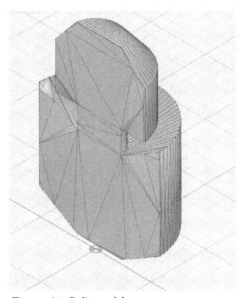

Figure-34. Split model

Reverse Normal

The **Reverse Normal** tool is used to flip the normal direction of the selected face. The procedure to use this tool is discussed next.

- Click on the **Reverse Normal** tool from **MODIFY** drop-down; refer to Figure-35. The **REVERSE NORMAL** dialog box will be displayed; refer to Figure-36.

Figure-35. Reverse Normal tool

Figure-36. REVERSE NOR-MAL dialog box

- The **Select** button of **Mesh Face** section is active by default. Click on the face of mesh body to select; refer to Figure-37

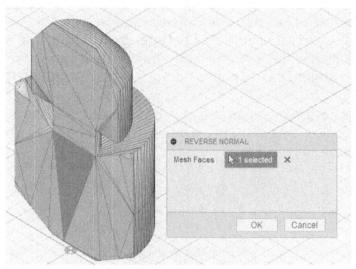

Figure-37. Selecting face for mesh body

- You can also select multiple faces by clicking on them.
- After selecting the required faces, click on the **OK** button from **REVERSE NORMAL** dialog box. The normal direction of selected face will be flipped; refer to Figure-38.

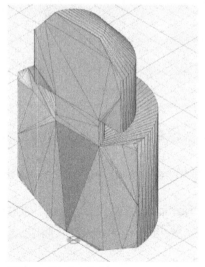

Figure-38. Applied reverse normal on Selected face

Delete Faces

The **Delete Faces** tool is used to remove the selected face from the body. The procedure to use this tool is discussed next.

- Click on the **Delete Faces** tool from **MODIFY** drop-down; refer to Figure-39. The **DELETE FACES** dialog box will be displayed; refer to Figure-40.

Figure-39. Delete Faces tool

Figure-40. DELETE FACES dialog box

- The **Select** button of **Mesh Faces** section is active by default. Click on the required face to select; refer to Figure-41.

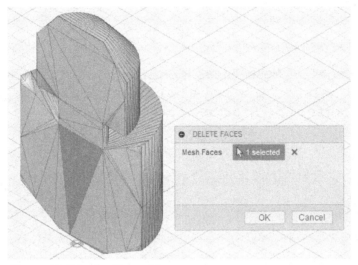

Figure-41. Selecting face to delete

- Click on the **OK** button from **DELETE FACES** dialog box to delete the selected face. The face will be deleted; refer to Figure-42.

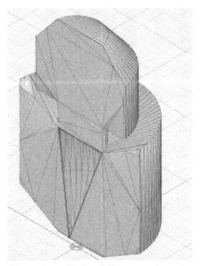

Figure-42. Deleted Face

Separate

The **Separate** tool is used to create a new mesh body from selected set of faces. The procedure to use this tool is discussed next.

- Click on the **Separate** tool from **MODIFY** drop-down; refer to Figure-43. The **SEPARATE** dialog box will be displayed; refer to Figure-44.

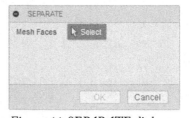

Figure-44. SEPARATE dialog box

Figure-43. Separate tool

- The **Select** button of **Mesh Faces** section is active by default. Click on the face to select. You can also select multiple faces by holding the **CTRL** key; refer to Figure-45.

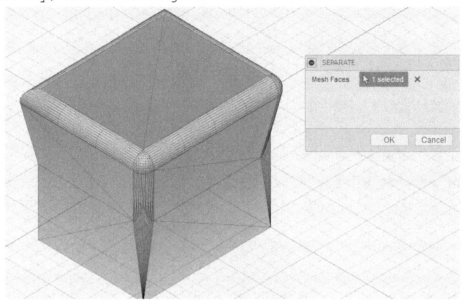

Figure-45. Selecting face for separate

- After selection of required faces, click on the **OK** button from **SEPARATE** dialog box. The selected face will be separated; refer to Figure-46.

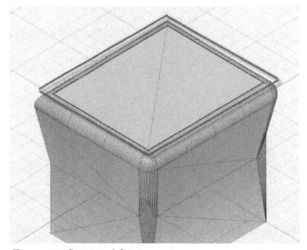

Figure-46. Separated face

Merge Bodies

The **Merge Bodies** tool is used to create a single body from multiple input bodies. If the input bodies have touching boundary edges then these edges will be stitched together. The procedure to use this tool is discussed next.

- Click on the **Merge Bodies** tool from **MODIFY** drop-down; refer to Figure-47. The **MERGE BODIES** dialog box will be displayed; refer to Figure-48

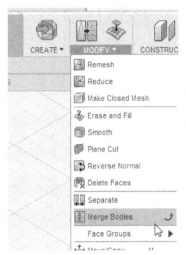

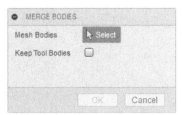

Figure-48. MERGE BODIES
dialog box

Figure-47. Merge Bodies tool

- The **Select** button of **Mesh Bodies** section is active by default. Click on the body to select; refer to Figure-49.

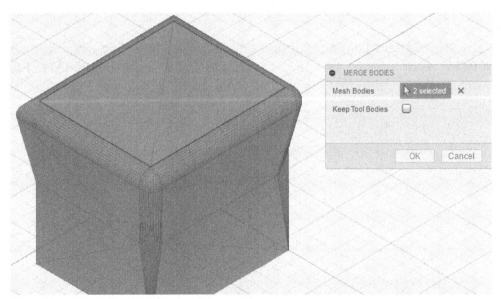

Figure-49. Selecting bodies to merge

- Select the **Keep Tool Bodies** check box from **MERGE BODIES** dialog box to keep the input bodies retained after operation.
- After specifying the parameter, click on the **OK** button from **MERGE BODIES** dialog box. The selected bodies will be merged; refer to Figure-50

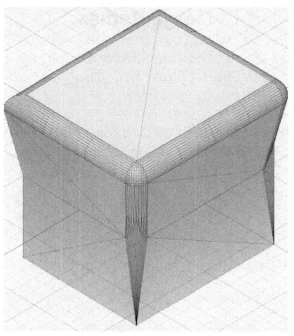

Figure-50. Merged bodies

Face Groups

The tools of Face Groups cascading menu are used to segment a mesh body into logical regions. In this, we can generate, clear, or create the face groups.

Generate Face Groups

The **Generate Face Groups** tool is used to create face groups on a selected body, using normal angles between faces to determine the boundaries of each face group. The procedure to use this tool is discussed next.

- Click on the **Generate Face Groups** tool of **Face Groups** cascading menu from **MODIFY** drop-down; refer to Figure-51. The **GENERATE FACE GROUPS** dialog box will be displayed; refer to Figure-52.

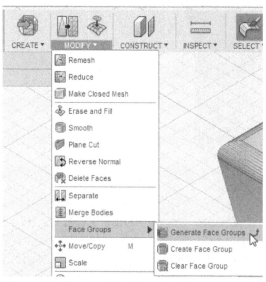

Figure-52. GENERATE FACE GROUPS dialog box

Figure-51. Generate Face Groups tool

- The **Select** button of **Mesh Body** section is active by default. Click on the mesh body to select; refer to Figure-53.

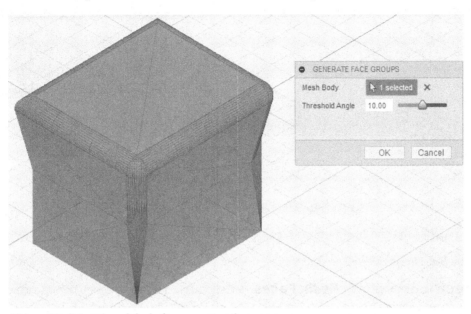

Figure-53. Selected mesh body for generating face groups

- Click in the **Threshold Angle** edit box and enter the value of angle between the adjacent faces where face group boundaries will be drawn on exceeding the specified values.
- After specifying the parameters, click on the **OK** button from **GENERATE FACE GROUPS** dialog box. The generated face groups will be displayed; refer to Figure-54.

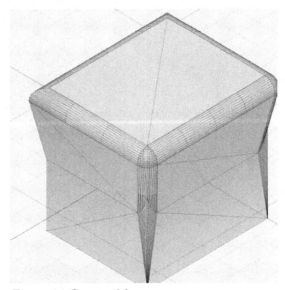

Figure-54. Generated face groups

Create Face Group

The **Create Face Group** tool is used to create a new face group from a selected set of faces. The procedure to use this tool is discussed next.

- Click on the **Create Face Group** tool of **Face Groups** cascading menu from **MODIFY** drop-down; refer to Figure-55. The **CREATE FACE GROUP** dialog box will be displayed; refer to Figure-56.

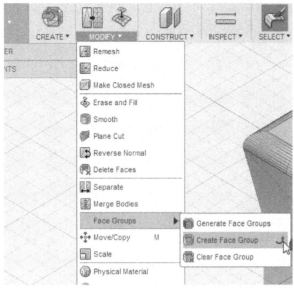

Figure-55. *Create Face Groups tool*

Figure-56. *CREATE FACE GROUP dialog box*

- The **Select** button of **Mesh Faces** section is active by default. Click on the required face to select; refer to Figure-57.

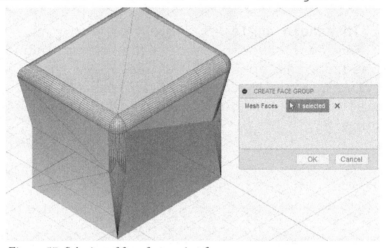

Figure-57. *Selection of faces for creating face group*

- After selection of required faces, click on the **OK** button from **CREATE FACE GROUP** dialog box. The created face group will be displayed; refer to Figure-58.

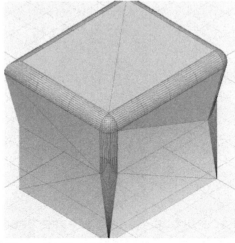

Figure-58. *Created face group*

Clear Face Group

The **Clear Face Group** tool is used to clear any existing face group from selected body. The procedure to use this tool is discussed next.

- Click on the **Clear Face Group** tool of **Face Groups** cascading menu from **MODIFY** drop-down; refer to Figure-59. The **CLEAR FACE GROUP** dialog box will be displayed; refer to Figure-60.

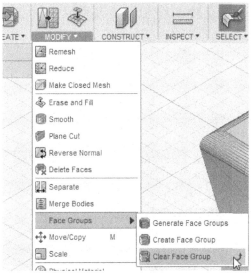

Figure-60. CLEAR FACE GROUP dialog box

Figure-59. Clear Face Group tool

- The **Select** button of **Mesh Faces/Body** section is active by default. Click on the existing face group to clear; refer to Figure-61.

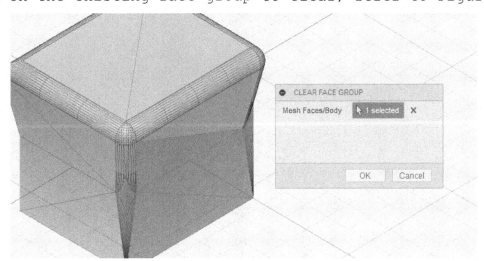

Figure-61. Selectd Faces to clear

- After selection of faces, click on the **OK** button from **CLEAR FACE GROUP** dialog box. The selected face will be cleared; refer to Figure-62.

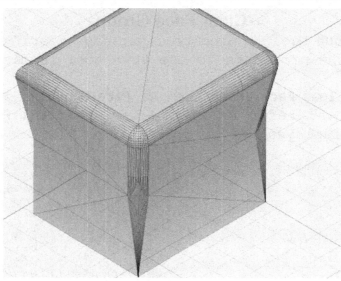

Figure-62. Cleared selected face group

- Others tools of **MODIFY** drop-down are same as discussed earlier in this book.
- After creating or modifying the mesh body, click on the **FINISH MESH** button from **Toolbar** to exit the **MESH Workspace**.

PRACTICAL

Create the mesh model as shown in Figure-63.

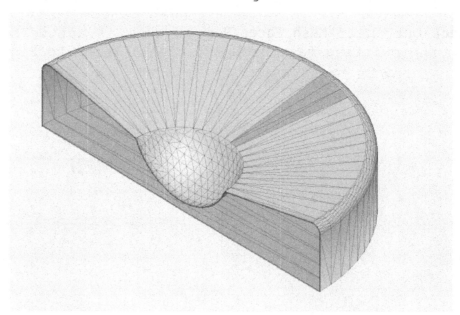

Figure-63. final model

Converting model into mesh

Before modifying the mesh file, we need to convert the model into mesh file.

- Create or open the model of this practical in the **MODEL Workspace**. The file is available in the resource kit of this book.
- Click on the **Create Mesh** tool of **CREATE** drop-down from **Toolbar**. The **MESH Workspace** will be displayed along with the model; refer to Figure-64.

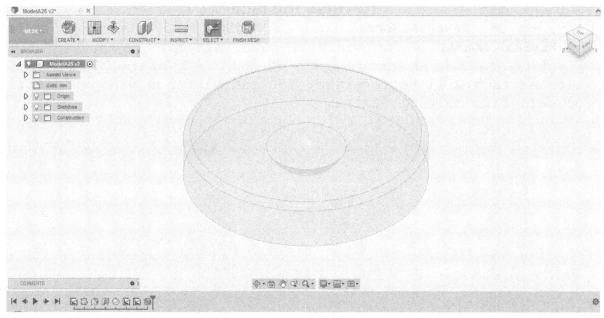

Figure-64. Added model into MESH Workspace

- Click on the **BRep to Mesh** tool of **CREATE** drop-down from **Toolbar**. The **BREP TO MESH** dialog box will be displayed.
- The **Select** button of **Body** option is active by default. You need to select the Brep model to convert it into mesh model and specify the parameters as displayed in Figure-65.

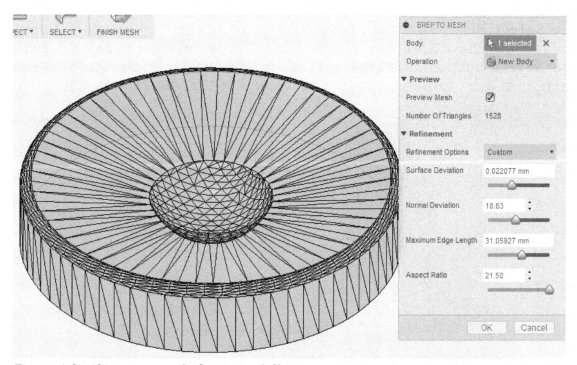

Figure-65. Specifying parameters for Creating mesh file

- After specifying the parameters, click on the **OK** button from **BREP TO MESH** dialog box. The mesh body will be created and displayed on the **MESH Workspace**.

Reverse the face

- Click on the **Reverse Normal** tool of **Modify** drop-down from **Toolbar**. The **REVERSE NORMAL** dialog box will be displayed.
- Select the face as shown in Figure-66 and click on the **OK** button. The selected face will be reversed. You can also use window selection for selecting multiple faces.

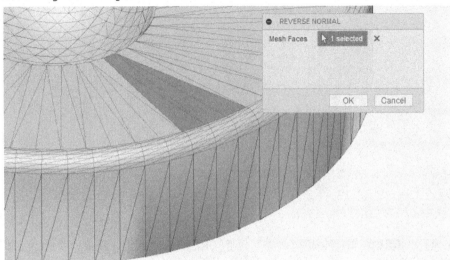

Figure-66. Selection for reverse normal

Plane Cut

- Click on the **Plane Cut** tool of **MODIFY** drop-down from **Toolbar**. The **PLANE CUT** dialog box will be displayed.
- The **Select** button of **Mesh Body** option is active by default. You need to select the model for plane cut. Click on the model to select.
- The **Select** button of **Plane** option is active by default. You need to select the YZ plane to cut the body; refer to Figure-67.

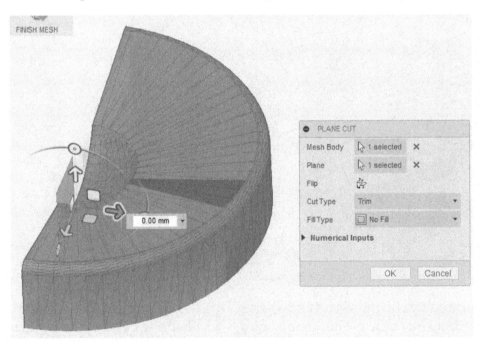

Figure-67. Selecting plane for plane cut

- Specify the parameters as shown in above figure and click on the **OK** button from **PLANE CUT** dialog box.
- After following all the steps discussed above, the model will be displayed as shown in Figure-68.

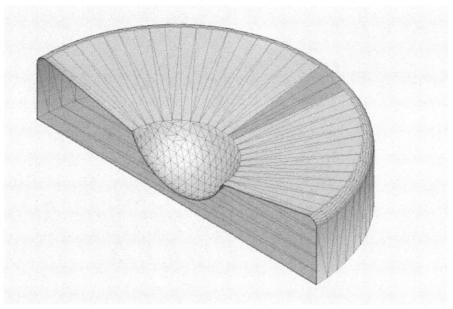

Figure-68. final model

PRACTICE

Create the model as displayed in the Figure-69.

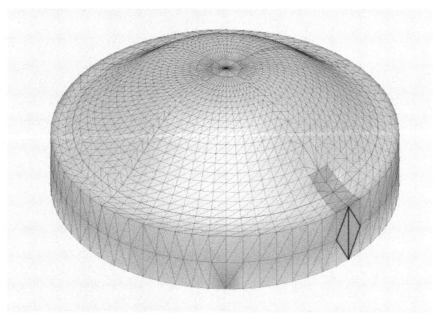

Figure-69. Practice1

FOR STUDENTS NOTES

Chapter 15

CAM

Topics Covered

The major topics covered in this chapter are:

- *New Setup*
- *Milling Machine Setup*
- *Turning Machine Setup*
- *Milling and Turning Tools*
- *Creating new Mill and Turning tool*

INTRODUCTION

CAM stands for Computer Aided Manufacturing. CAM is a mode where you can convert the model into a machine readable language that can be use for the manufacturing process(usually G code). The most common manufacturing process that are studied in CAM are Milling, Turning, Drilling, and Laser Cutting. In the CAM Workspace, we will be able to generate high-quality toolpaths within minutes. Depending on the Fusion 360 version, You can create high quality 2D, 3D, 5-Axis milling, and turning toolpaths for high .speed machining (HSM).

STARTING THE CAM WORKSPACE

The **CAM Workspace** is used for generating tool-paths for machining a part. The procedure to start **CAM Workspace** is discussed next.

- Click on the **CAM** option from **Workspace** drop-down from current workspace; refer to Figure-1. The **CAM Workspace** environment will be displayed; refer to Figure-2.

Figure-1. CAM workspace

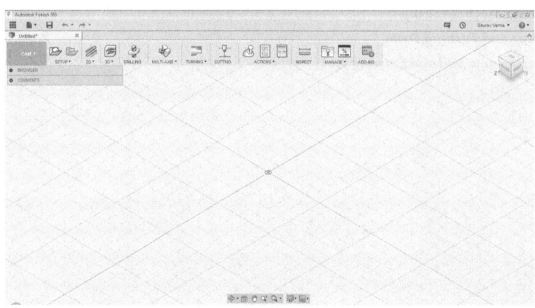

Figure-2. CAM Workspace

JOB SETUP

In this section, we will discuss the procedure of setting up the workpiece and create the required material stock for machining process. Job setup lets you define your stock for machining and machine type to be used like Milling or Turning. Stock is the workpiece out of which the final product will be produced after machining. The shape and size of stock depends upon the part which is to be created by machining.

New Setup

Before starting any machining project, you need to define a job setup to tell Autodesk Fusion 360 that the toolpaths will be generated for Milling machine or Turning machine. You also need to set the part zero location to be used machining coordinates. You can also define any fixture components for machining. The procedure to setup the workpiece is discussed next.

- Click on the **New Setup** tool from **SETUP** drop-down; refer to Figure-3. The **SETUP** dialog box will be displayed along with the workpiece; refer to Figure-4.

Figure-3. New Setup tool

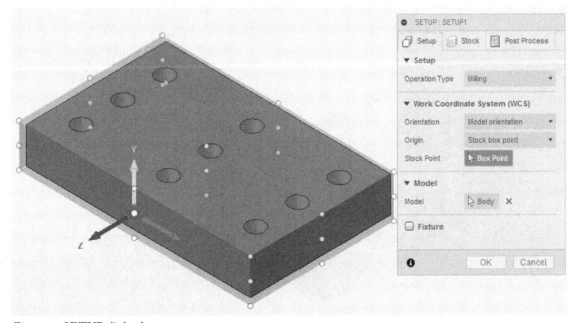

Figure-4. SETUP dialog box

Setting a Milling machine

In this section, we will discuss the procedure to set up a milling machine. The procedure is discussed next.

Setup

- Select the **Milling** option from the **Operation Type** drop-down in **Setup** section of **Setup** tab for setting up a milling operation.
- Select **Model Orientation** option in **Orientation** section from **Work Coordinate System (WCS)** to automatically set the orientation of coordinate system on workpiece for machining.
- Select the **Select Z axis/plane & X axis** option of **Orientation** section of **Work Coordinate System (WCS)** node to select the Z axis and X axis for setting the orientation of workpiece; refer to Figure-5. The updated **WCS** section will be displayed along with axis or face selection on part; refer to Figure-6

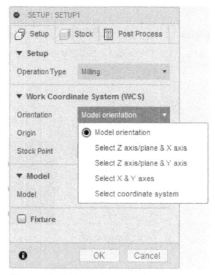

Figure-5. Orientation

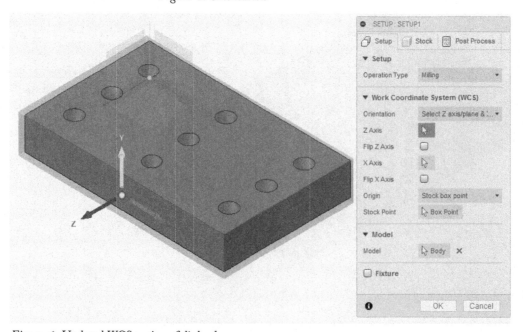

Figure-6. Updated WCS section of dialog box

- Click on the **Select Z axis/plane & Y axis** option of **Orientation** section of **Work Coordinate System (WCS)** to select the Z axis and Y axis for setting the orientation to reflect the machining plane.

- Click on the **Select X & Y axes** option of **Orientation** section of **Work Coordinate System (WCS)** to select the X axis and Y axis for setting the orientation to reflect the machining plane. You need to select a face or an edge of model to define the X and Y axis.

- Click on the **Select Coordinate System** option of **Orientation** section of **Work Coordinate System (WCS)** to define a user defined coordinate system in the model to set the WCS orientation.

- Click on the **Z Axis** button from **Work Coordinate System (WCS)** section and click on the desired axis or plane from Origin node in the **Browser** to define the Z axis. The Z axis should be perpendicular to machining plane.

- Select the **Flip Z Axis** check box to flip the selected direction of Z axis at 180 degree.

- The **X Axis** button is active by default. Click on the axis or plane to define the X axis; refer to Figure-7. The X axis will be defined perpendicular to the selected plane.

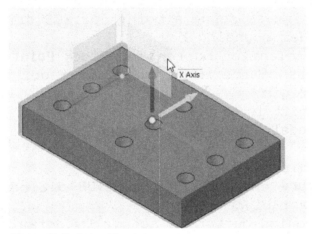

Figure-7. Defining X axis

- Select the **Flip X Axis** check box to flip the selected direction of X axis at 180 degree.

- Select **Model Origin** option of **Origin** drop-down from **Work Coordinate System (WCS)** section to uses the coordinate system (WCS) origin of the current part for the WCS origin.

- Select the **Selected Point** option of **Origin** drop-down to select a vertex or an edge for the WCS origin and click on the desired vertex or edge to define the WCS origin from model.

- Select **Stock box point** option of **Origin** drop-down from **Work Coordinate System (WCS)** section to define a point for WCS origin by selecting a point on the stock bounding box.

- The **Stock Point** button of **Origin** section is active by default. Click on the stock point from workpiece to define the point for WCS origin; refer to Figure-8.

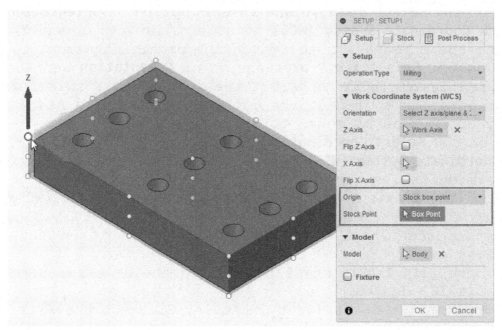

Figure-8. Defining stock box point

- Select **Model Box Point** option of **Origin** drop-down from **Work Coordinate System (WCS)** section to define a point for WCS origin by selecting a point on the model bounding box.
- The **Model Point** selection button for **Stock Point** section is active by default. Click on the model point from workpiece to define the point for WCS origin.
- The body/model which is considered for generating toolpaths is active by default in **Model** section from **Setup** tab. If there are multiple solid model in file then it is recommended to select the required model for machining process.
- Select the **Fixture** check box from **SETUP** dialog box to define the fixture for the workpiece.
- The **Fixture** selection button is active by default on selecting the **Fixture** check box. You need to select the component/body to define fixture; refer to Figure-9

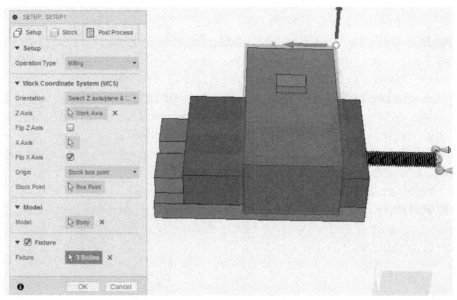

Figure-9. Selecting Fixture

Stock tab

- Click on the **Stock** tab of **SETUP** dialog box to define the workpiece dimensions. The **Stock** tab will be displayed; refer to Figure-10.

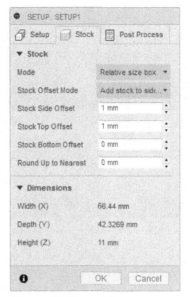

Figure-10. Setup tab

Fixed size box

- Select **Fixed size box** option from the **Mode** drop-down in **Setup** tab to create a rectangular stock body of defined parameter. The updated **Stock** tab will be displayed; refer to Figure-11.

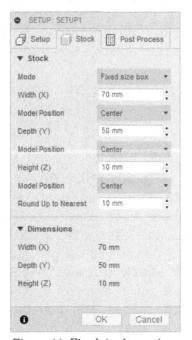

Figure-11. Fixed size box options

- Click in the **Width (X)** edit box and specify the width of the stock body.
- Select the **Offset from left side (-X)** option of **Model Position** drop-down from **Stock** section to offset the stock to the left side of model.
- Click in the **Offset** edit box and enter the required value; refer to Figure-12.

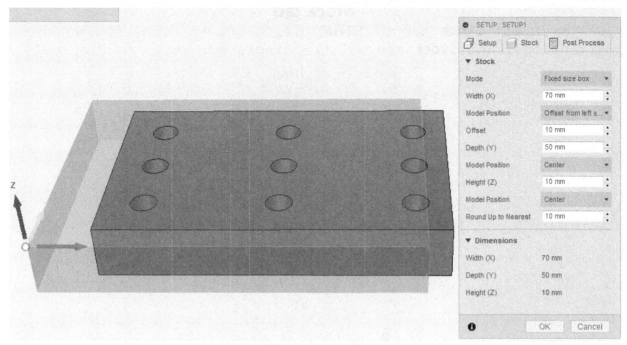

Figure-12. Offset from left side

- Select **Center** option of **Model Position** drop-down from **Stock** section to place the stock on the center of model.
- Select the **Offset from right side (+X)** option of **Model Position** drop-down from **Stock** section to offset the stock to the right side of model.
- Click in the **Offset** edit box and enter the required value. In our case we are selecting the **Center** option.
- Click in the **Depth (Y)** edit box and enter the required depth of the stock.
- Click in the **Height (Z)** edit box and enter the required height of stock.
- Click in the **Round Up to Nearest** edit box and enter the round of increment of the stock size.

Relative Size Box

- Select the **Relative Size Box** option from **Mode** drop-down in **Stock** section to create a rectangular stock body larger then the model by specifying the required values. The updated **SETUP** dialog box will be displayed; refer to Figure-13.
- Select the **No additional stock** option of **Stock Offset Mode** drop-down from **Stock** section to not add any offset value to the stock size.
- Select the **Add stock to the sides and top-button** option of **Stock Offset Mode** drop-down to add symmetric values to all side of stock and unique values to top and button offsets.
- Click in the respective offset edit box and enter the value as desired.
- Select the **Add stock to all sides** option of **Stock Offset Mode** drop-down if you want to enter the specific the values for all offset directions of stock.

- Click in the respective offset edit box and enter the value as desired.

Figure-13. Relative size box option

Fixed size cylinder

- Select the **Fixed size cylinder** option of **Mode** drop-down from **Stock** section to create a fixed size cylinder stock body. The updated **SETUP** dialog box will be displayed; refer to Figure-14.

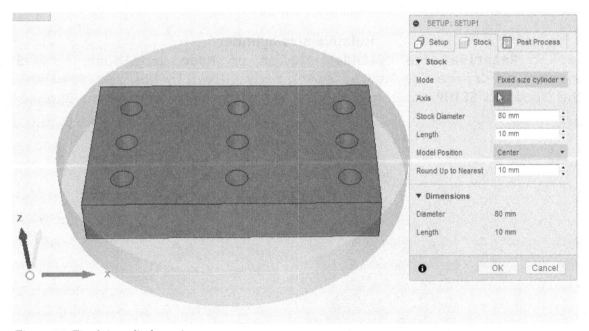

Figure-14. Fixed size cylinder option

- The **Axis** button of **Setup** section is active by default. You need to select the axis from model; refer to Figure-15.

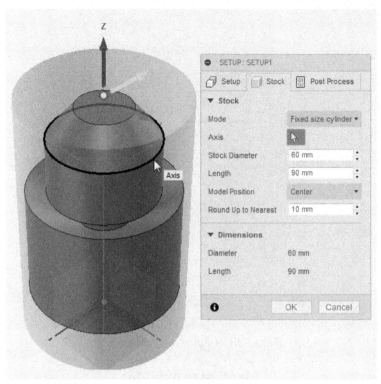

Figure-15. Selecting axis

- Click in the **Stock Diameter** edit box from **Stock** section and enter the desired diameter of stock.
- Click in the **Length** edit box and enter the desired length of stock.

Relative size cylinder

- Select **Relative size cylinder** option of **Mode** drop-down from **Stock** section to create the larger stock body. Specify the required values. The updated **SETUP** dialog box will be displayed; refer to Figure-16.

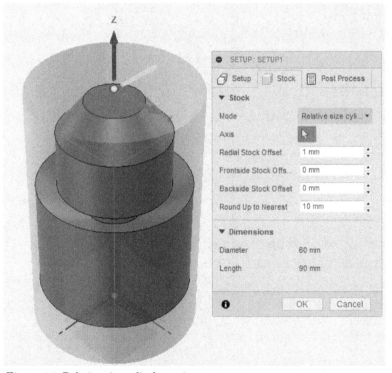

Figure-16. Relative size cylinder option

- The **Axis** button is active by default. Click on the axis from model to select.
- Click in the **Radial Stock Offset** edit box and enter the value of radial offset of the stock.
- Click **Frontside Stock Offset** edit box and specify the distance to machine beyond the front side of the model.
- Click in the **Backside Stock Offset** edit box and specify the distance to machine beyond the backside of the model.

Fixed size tube

- Select **Fixed size tube** option of **Mode** drop-down from **Stock** section to create a tube stock body of fixed size. The updated **SETUP** dialog box will be displayed; refer to Figure-17.

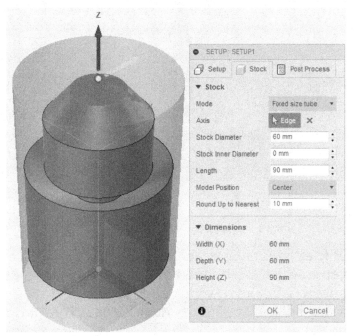

Figure-17. Fixed size tube option

- The **Axis** button is active by default. Click on the axis from model to select.
- Click in the **Stock Diameter** edit box from **Stock** section and enter the desired diameter of stock.
- Click in the **Stock Inner Diameter** edit box and enter the inner diameter of stock; refer to Figure-18.

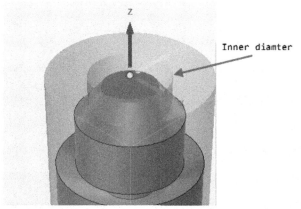

Figure-18. Inner diameter

- Click in the **Length** edit box and enter the desired length of stock.
- Select **Offset from front** option of **Model Position** drop-down from **Stock** section to offset the stock to the front side of model; refer to Figure-19.

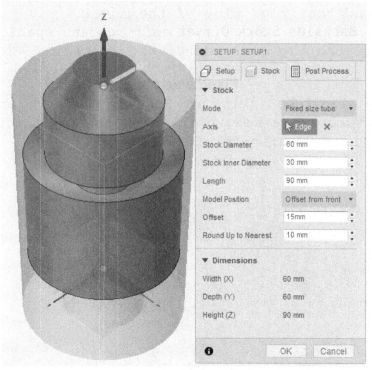

Figure-19. Offset from front option

- Click in the **Offset** edit box and enter the required value.
- Select the **Center** option of **Model Position** drop-down from **Stock** section to place the stock at the center of model.
- Select **Offset from back** option of **Model Position** drop-down from **Stock** section to offset the stock to the back side of model.

Relative size tube

- Select the **Relative size tube** option of **Mode** drop-down from **Stock** section to create the larger stock body then model by specifying the required values. The updated **SETUP** dialog box will be displayed; refer to Figure-20.

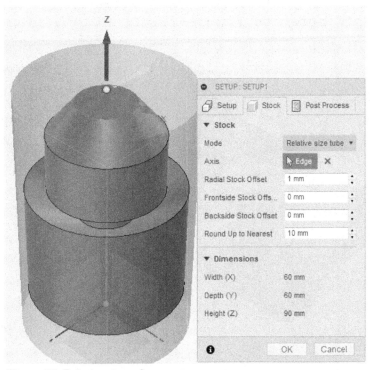

Figure-20. Relative size tube

- The edit boxes of **Relative size tube** were discussed in **Relative size cylinder** option of this unit.

From solid

- Select **From solid** option of **Mode** drop-down from **Stock** section to create a stock by selecting a solid body in a multi-body part or from a part file in an assembly. The updated **SETUP** dialog box will be displayed; refer to Figure-21.

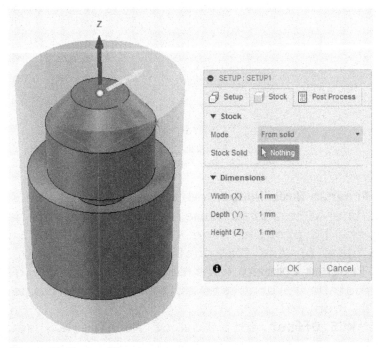

Figure-21. From solid option

- Click on the **Nothing** button of **Stock Solid** option and click on the body to select; refer to Figure-22.

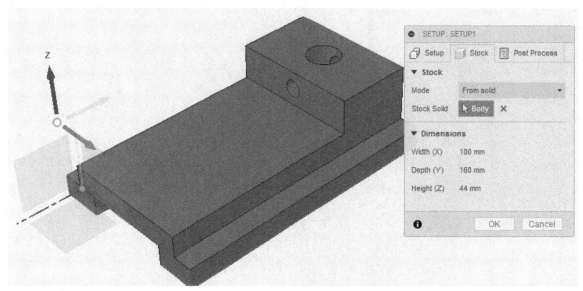

Figure-22. Selection of body for stock

- Click on the **Dimensions** node of **Stock** tab from **SETUP** dialog box to check the exact dimensions of the stock.

Post Process tab

Click on the **Post Process** tab from **SETUP** dialog box. The **Post Process** tab will be displayed; refer to Figure-23

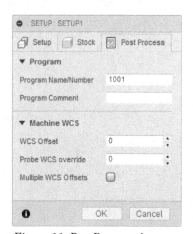

Figure-23. Post Process tab

- Click in the **Program Name/Number** edit box of **Setup** tab from **Post Process** tab to define the NC program name/number. This number is output at the start of the NC program code as the "O" number. It is also used as a storage name on the CNC controller.
- Click in the **Program Comment** edit box and enter the desired comment. The comment must be in brackets so that the contained text will not be read by CNC control.
- Click on the **WCS Offset** edit box of **Machine WCS** section from **Post Process** tab to select the work offset used in the NC machine to

provide tool compensation. The output in the NC program is generally produced by G54 through G59 codes, but will vary between various NC Controls/Machines. Zero (0) in WCS Offset will output the first available fixture offset and 1 will output the 2nd available offset.

- Click in the **Probe WCS override** edit box of **Machine WCS** section to enter the value of offset for probe while checking coordinates in CMM.
- Select the **Multiple WCS Offsets** check box to enter multiple offset values in respective edit box. This check box is used when there are multiple tools in the machine for machining.
- The **Operation Order** drop-down of **Multiple WCS Offsets** check box is used to specifies the ordering of the individual operations.
- Select **Preserve Order** option from **Operation Order** drop-down to the features which are machined in the order in which they are selected.
- Select **Order by Operation** option from **Operation Order** drop-down to specify the ordering of the individual operations.
- Select **Order by tool** option from **Operation Order** drop-down to specify the ordering of operations by tool.
- After specifying the parameters for creating the required stock, click on the **OK** button from **SETUP** dialog box. The **Setup1** will be added in **Browser**; refer to Figure-24.
- If you want to edit the earlier created stock then right click on the stock and click on **Edit** button from marking menu; refer to Figure-25.

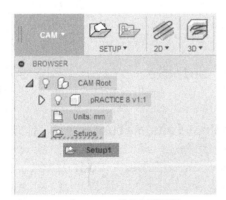

Figure-24. Setup1 in BROWSER

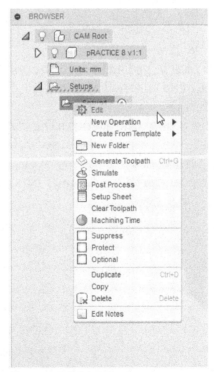

Figure-25. Edit stock

- Modify the required parameters from **SETUP** dialog box and click on **OK** button.

Turning

In this section, we will discuss about the procedure of setting up a **Turning** operation. The procedure is discussed next.

- Click on the **Turning or mill/turn** option from **Operation Type** drop-down in **SETUP** dialog box. The options used in turning process will be displayed along with the model; refer to Figure-26.

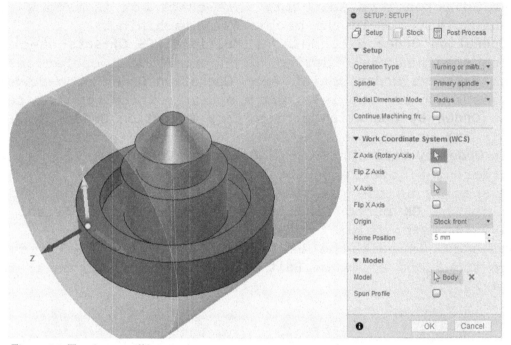

Figure-26. Turning or mill turn option

- Select **Primary spindle** or **Secondary spindle** option from **Spindle** drop-down in **Setup** section to specify the spindle to be used if your machine has two spindles.
- Select the **Continue Machining from Previous Setup** check box from **Setup** section if you want to use the machine setup earlier created.
- Click on the **Z Axis (Rotary Axis)** button of **Work Coordinate System (WCS)** section from **SETUP** dialog box and select the Z axis as required; refer to Figure-27.

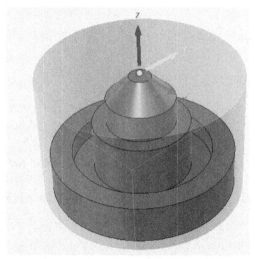

Figure-27. Selecting Z axis for turning

- Select the **Flip Z Axis** check box to flip the selected direction of Z axis at 180 degree.
- The **X Axis** button is active by default. Click on the axis or plane to define the X axis.
- Select the **Flip X Axis** check box to flip the selected direction of X axis at 180 degree.
- The **Origin** drop-down will define where zero will be located on the part. Click on the **Origin** drop-down from **Work Coordinate System (WCS)** section to select the required option.
- Click in the **Home Position** edit box and enter the value to specify the home position along the Z axis.
- Select the **Spun Profile** check box of **Model** section from **Setup** tab to generate profile for turning.
- Click in the **Spun Profile Tolerance** edit box of **Model** section and enter the desired value of tolerance
- Select the **Spun Profile Smoothing** check box to smooth the profile.

The tools of Stock and Post process tab have been discussed earlier in Milling section.

- After specifying the parameters, click on the **OK** button from **SETUP** dialog box to complete the process of creating stock.

Cutting

In this section, we will discuss about the procedure of setting up a **Cutting** operation. The procedure is discussed next.

- Click on the **Cutting** option of **Operation Type** drop-down from **SETUP** dialog box. The options used in cutting process will be displayed; refer to Figure-28.

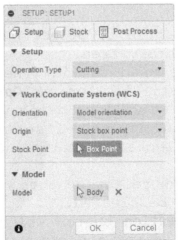

Figure-28. SETUP dialog box for Cutting operation

- Set the orientation and zero point for cutting operation from **SETUP** dialog box.
- The other options in the dialog box have been discussed earlier.
- After specifying the parameters, click on the **OK** button from **SETUP** dialog box.

TOOL SELECTION

Before proceeding towards the tools used to generate 2D and 3D path, you need to know various types of tools used in machining and their selection criteria. For this we need to discuss the **Select Tool** dialog box; refer to Figure-29 which is discussed next. To display this dialog box, click on the **2D Adaptive Clearing** tool from the **2D** drop-down in the Toolbar. A dialog box will be displayed. Click on the **Select** button for **Tool** section in the **Tool** node of the dialog box. The **Select Tool** dialog box will be displayed. Note that in this chapter, we will discuss about cutting tools only. In next chapter, we will discuss the toolpaths.

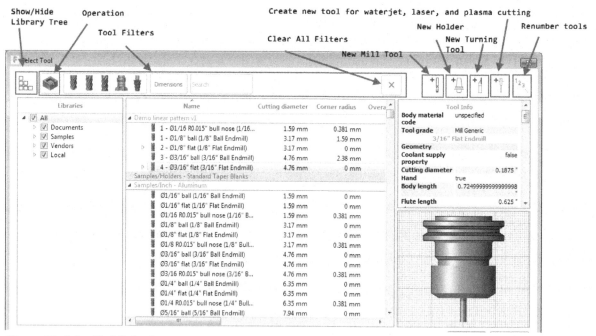

Figure-29. Select Tool dialog box

- The **Show/Hide Library** button of **Select Tool** dialog box is used to display or hide the library tree.
- The **Operation** button is used to view the selected types of tools from the mixed tool list. Click on the **Operation** button from **Select Tool** dialog box. The list will be displayed; refer to Figure-30
- Click on the desired tool from the list and click on the **OK** button. The tools used for selected operation will be displayed on the **Select Tool** dialog box.
- If you want to erase the selection then click on the **Clear** button. The selected tool will be cleared.
- Click on the **Tools selection** button which is placed right to the **Operation** button. The list of tools type will be displayed; refer to Figure-31

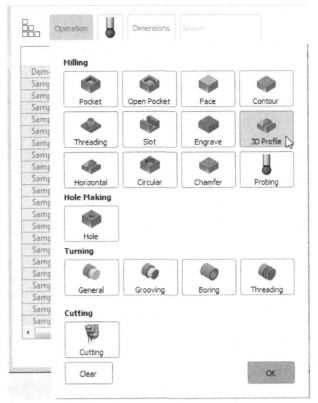

Figure-30. Tools under operation button

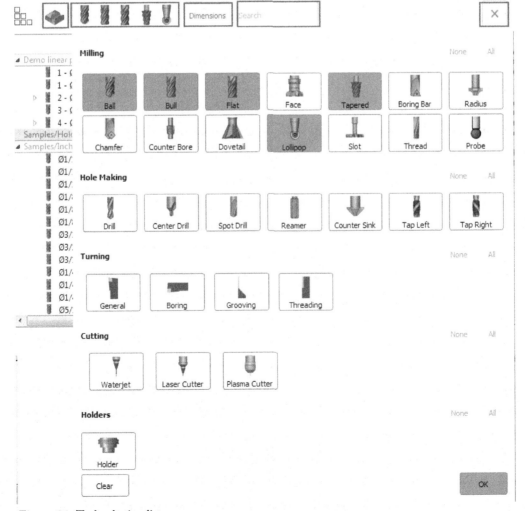

Figure-31. Tools selection list

- From the tools displayed, click on the required types of tools from different sections like Milling, Hole Making, Turning, Cutting, and Holders. After selection, click on the **OK** button. The tools related to your selection will be displayed in **Tool record** section of **Select Tool** dialog box.
- If you want to erase the selection then click on the **Clear** button. The entries will be cleared.
- If you want to filter the tools according to the parameter then click on the **Dimension** button from **Select Tool** dialog box. The **Dimension** box will be displayed; refer to Figure-32.

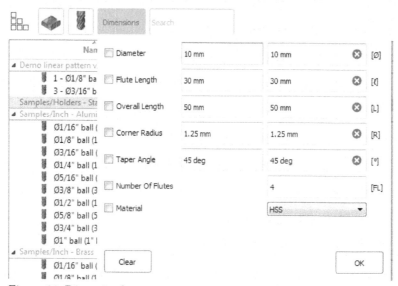

Figure-32. Dimension box

- Select the specific parameter check box from **Dimension** box and enter the desired value in respective edit box.
- After specifying the parameters, click on the **OK** button from **Dimension** box. The tools of selected parameter will be displayed.
- If you want to search the tool by its name then click on the **Search** box from **Select Tool** dialog box and enter the required keyword. The tools similar to the entered keywords will be displayed in **Tool record** section.
- If you want to erase all the applied filters for tool selection, click on the **Clear All Filter** button from **Select Tool** dialog box.

TOOLS USED IN CNC MILLING AND LATHE MACHINES

The tools used in CNC machines are made of cemented carbide, High Speed Steel, Tungsten Alloys, Ceramics, and many other hard materials. The shapes and sizes of tools used in Milling machines and Lathe machines are different from each other. These tools are discussed next.

Milling Tools

There are various type of milling tools for different applications. These tools are discussed next.

End Mill

End mills are used for producing precision shapes and holes on a Milling or Turning machine. The correct selection and use of end milling cutters is paramount with either machining centers or lathes. End mills are available in a variety of design styles and materials; refer to Figure-33.

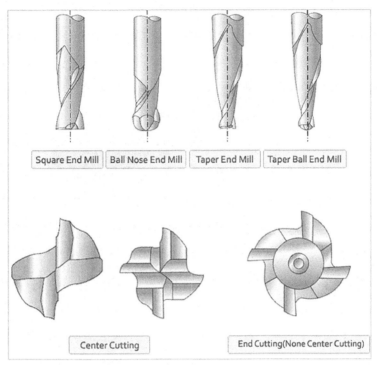

Figure-33. End Mill tool types

Titanium coated end mills are available for extended tool life requirements. The successful application of end milling depends on how well the tool is held (supported) by the tool holder. To achieve best results an end mill must be mounted concentric in a tool holder. The end mill can be selected for the following basic processes:

FACE MILLING - For small face areas, of relatively shallow depth of cut. The surface finish produced can be 'scratchy".
KEYWAY PRODUCTION - Normally two separate end mills are required to produce a quality keyway.
WOODRUFF KEYWAYS - Normally produced with a single cutter, in a straight plunge operation.
SPECIALTY CUTTING - Includes milling of tapered surfaces, "T" shaped slots & dovetail production.
FINISH PROFILING - To finish the inside/outside shape on a part with a parallel side wall.
CAVITY DIE WORK - Generally involves plunging and finish cutting of pockets in die steel. Cavity work requires the production of three dimensional shapes. A Ball type end mill is used for the finishing cutter with this application.

Roughing End Mills, also known as ripping cutters or hoggers, are designed to remove large amounts of metal quickly and more efficiently than standard end mills; refer to Figure-34. Coarse tooth roughing end mills remove large chips for heavy cuts, deep slotting and rapid stock removal on low to medium carbon steel and alloy steel prior to a finishing application. Fine tooth roughing end mills remove less material but the pressure is distributed over many more teeth, for longer tool life and a smoother finish on high temperature alloys and stainless steel.

Figure-34. Roughing End Mill

Bull Nose Mill

Bull nose mill look alike end mill but they have radius at the corners. Using this tool, you can cut round corners in the die or mold steels. Shape of bull nose mill tool is given in Figure-35.

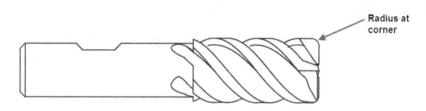

Figure-35. Bull Nose Mill cutter

Ball Nose Mill

Ball nose cutters or ball end mills has the end shape hemispherical; refer to Figure-33. They are ideal for machining 3-dimensional contoured shapes in machining centres, for example in moulds and dies. They are sometimes called ball mills in shop-floor slang. They are also used to add a radius between perpendicular faces to reduce stress concentrations.

Face Mill

The Face mill tool or face mill cutter is used to remove material from the face of workpiece and make it plane; refer to Figure-36.

Figure-36. Face milling tool

Radius Mill and Chamfer Mill

The Radius mill tool is used to apply round (fillet) at the edges of the part. The Chamfer mill tool is used to apply chamfer at the edges of the part. Figure-37 shows the radius mill tool and chamfer mill tool.

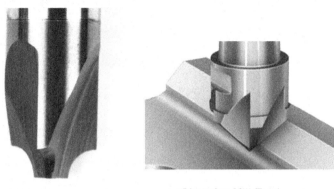

Radius Mill Tool   **Chamfer Mill Tool**

Figure-37. Radius mill and Chamfer mill tool

Slot Mill

The Slot mill tool is used to create slot or groove in the part metal. Figure-38 shows the shape of slot mill tool.

Figure-38. Slot mill tool

Taper Mill

In CNC machining, taper end mills are used in many industries for a large number of applications, such as walls with draft or clearance angle, tool and die work, mold work, even for reaming holes to make them conical. There are mainly two types of taper mills, Taper End Mill and Taper Ball Mill; refer to Figure-33.

Dove Mill

Dove mill or Dovetail cutters are designed for cutting dovetails in a wide variety of materials. Dovetail cutters can also be used for chamfering or milling angles on the bottom surface of a part. Dovetail cutters are available in a wide variety of diameters and in 45 degree or 60 degree angles; refer to Figure-39.

Figure-39. Dovetail milling cutters

Lollipop Mill

The Lollipop mill tool is used to cut round slot or undercuts in workpiece. Some tool suppliers use a name Undercut mill tool in place of Lollipop mill in their catalog. The shape of lollipop mill tool is given in Figure-40.

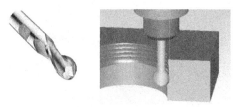

Figure-40. Lollipop mill tool

Engrave Mill

The Engrave mill tool is used to perform engraving on the surface of workpiece. Engraving has always been an art and it is also true for CNC machinist. You can find various shapes of engraving tool that are single flute or multi-flute; refer to Figure-41. You can use ball mill/ end mill for engraving or you can use specialized engrave mill tool for engraving. This all depends on your requirement. If you want to perform engraving on softer materials or plastics then it is better to use ball end mill but if you want an artistic shade on the surface then use the respective engrave mill tool. Keep a note of maximum depth and spindle speed mentioned by your engrave mill tool supplier.

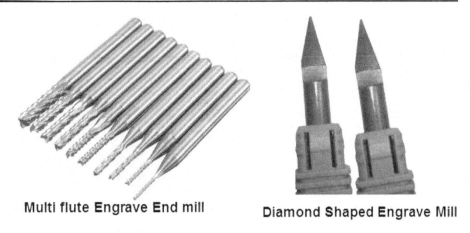

Multi flute Engrave End mill   **Diamond Shaped Engrave Mill**

Figure-41. Engrave mill tools

Thread Mill

The Thread mill tool is used to generate internal or external threads in the workpiece. The most common question here is if we have Taps to create thread then why is there need of Thread mill tool. The answer is less machining time on CNC, tool cost saving, more parts per tool, and better thread finish. Now, you will ask why to use tapping. The answer is low machine cost. Figure-42 shows thread mill tools.

Figure-42. Thread Mill

Barrel Mill

Barrel Mill tool is the tool recently being highly used in machining turbine/impeller blades and other 5-axis milling operations. Barrel Mill has conical shape with radius at its end; refer toFigure-43. Note that earlier Ball mill tools were used for irregular surface contouring but Barrel Mill tools give much better surface finish so they are highly in demand for 5-axis milling now a days.

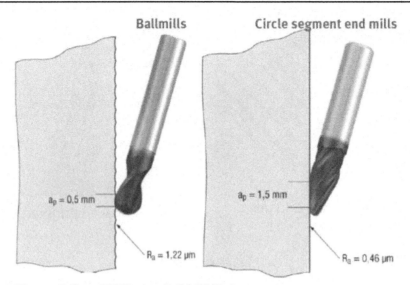

Figure-43. Barrel Mill versus Ball Mill Tool

Drill Bit

Drill bit is used to make a hole in the workpiece. The hole shape depends on the shape of drill bit. Drill bits for various purposes are shown in Figure-44. Note that drill is the machine or holder in which drill bit is installed to make cylindrical holes. There are mainly four categories of drill bit; Twist drill bit, Step drill bit, Unibit (or conical bit), and Hole Saw bit (Refer to Figure-45). Twist drill bits are used for drilling holes in wood, metal, plastic and other materials. For soft materials the point angle is 90 degree, for hard materials the point angle is 150 degree and general purpose twist drill bits have angle of 150 degree at end point. The Step drill bits are used to make counter bore or countersunk holes. The Unibits are generally used for drilling holes in sheetmetal but they can also be used for drilling plastic, plywood, aluminium and thin steel sheets. One unibit can give holes of different sizes. The Hole saw bit is used to cut a large hole from the workpiece. They remove material only from the edge of the hole, cutting out an intact disc of material, unlike many drills which remove all material in the interior of the hole. They can be used to make large holes in wood, sheet metal and other materials.

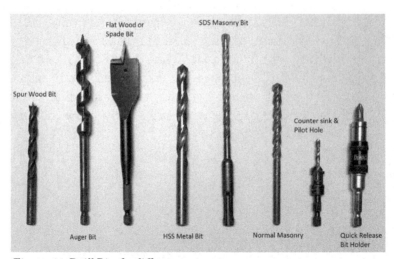

Figure-44. Drill Bits for different purposes

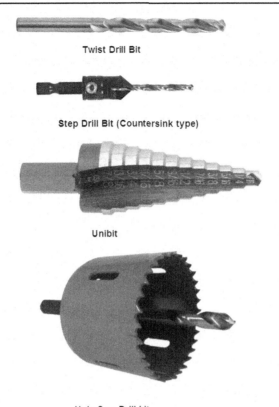

Twist Drill Bit

Step Drill Bit (Countersink type)

Unibit

Hole Saw Drill bit

Figure-45. Types of drill bits

Reamer

Reamer is a tool similar to drill bit but its purpose is to finish the hole or increase the size of hole precisely. Figure-46 shows the shape of a reamer.

Bore Bar

Bore Bar or Boring Bar is used to increase the size of hole; refer to Figure-47. One common question is why to use bore bar if we can perform reaming or why to perform reaming when we have bore bar. The answer is accuracy. A reamer does not give tight tolerance in location but gives good finish in hole diameter. A bore bar gives tight tolerance in location but takes more time to machine hole as compared to reamer. The decision to choose the process is on machinist. If you need a highly accurate hole then perform drilling, then boring and then reaming to get best result.

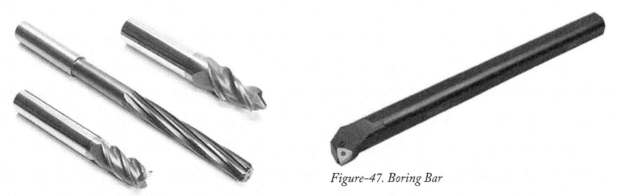

Figure-47. Boring Bar

Figure-46. Reamer tool

Lathe Tools or Turning Tools

The tools used in CNC lathe machines use a different nomenclature. In CNC lathe machines, we use insert for cutting material. The Insert Holder and Inserts have a special nomenclature scheme to define their shapes. First we will discuss the nomenclature of Insert holder and then we will discuss the nomenclature of Inserts.

Insert Holders

Turning holder names follow an ISO nomenclature standard. If you are working on a CNC shop floor with lathes, knowing the ISO nomenclature is a must. The name looks complicated, but is actually very easy to interpret; refer to Figure-48.

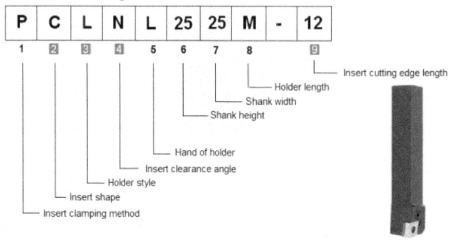

Figure-48. CNC Lathe Insert Holder nomenclature

When selecting a holder for an application, you mainly have to concentrate on the numbers marked in red in the above nomenclature. The others are decided automatically (e.g., the shank width and height are decided by the machine), or require less effort. In Figure-49, the rows with the question mark indicate the parameters that require the decision by machinist based on job.

	Parameter		How is this decided ?
1	Insert clamping method		Select based on cutting forces. Top clamping is the most sturdy, screw clamping the least.
2	Insert shape	?	Decided by the contour that you want to turn.
3	Holder style	?	Decided by the contour that you want to turn.
4	Insert clearance angle	?	Positive / Negative, based on application.
5	Hand of holder		Decided based on whether you want to cut towards the chuck or away from the chuck, and on turret position - turret front / rear
6	Shank height		Decided by holder size.
7	Shank width		Decided by machine.
8	Holder length		Decided by machine.
9	Insert cutting edge length	?	Decide based on depth of cut you want to use.

Figure-49. CNC Lathe Insert Holder nomenclature parameters

Figure-50 and Figure-51 show the options available for each of the parameters.

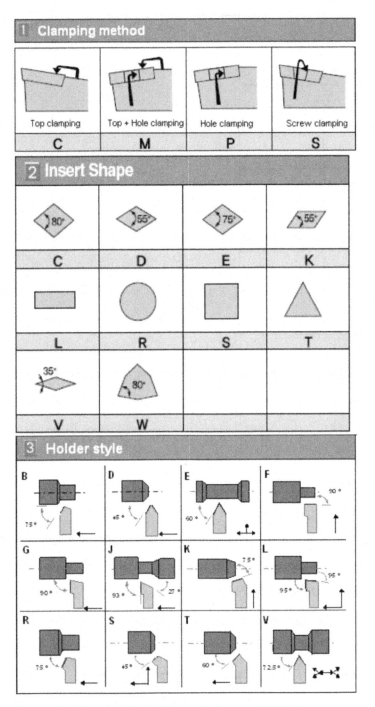

Figure-50. Clamping Method, Insert Shapes, and Holder Style

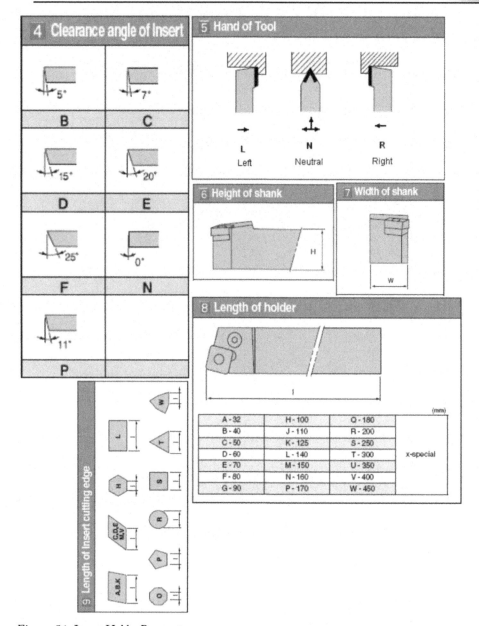

Figure-51. Insert Holder Parameters

CNC Lathe Insert Nomenclature

General CNC Insert name is given as

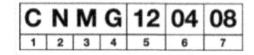

Meaning of each box in nomenclature is given next.

1 = Turning Insert Shape

The first letter in general turning insert nomenclature tells us about the general turning insert shape, turning inserts shape codes are like C, D, K, R, S, T, V, W. Most of these codes surely express the turning insert shape like

C = C Shape Turning Insert

D = D Shape Turning Insert
K = K Shape Turning Insert
R = Round Turning Insert
S = Square Turning Insert
T = Triangle Turning Insert
V = V Shape Turning Insert
W = W Shape Turning Insert

Figure-52 shows the turning inserts shapes.

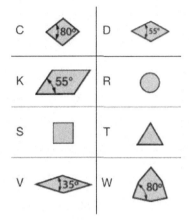

Figure-52. Turning Insert Shapes

The general turning insert shape play a very important role when we choose an insert for machining. Not every turning insert with one shape can be replaced with the other for a machining operation. As C, D, W type turning inserts are normally used for roughing or rough machining.

2 = Turning Insert Clearance Angle

The second letter in general turning insert nomenclature tells us about the turning insert clearance angle.

The clearance angle for a turning insert is shown in Figure-53.

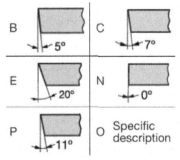

Figure-53. Turning insert clearance angle

Turning insert clearance angle plays a big role while choosing an insert for internal machining or boring small components, because if not properly chosen the insert bottom corner might rub with the component which will give poor machining. On the other hand a turning insert with 0° clearance angle is mostly used for rough machining.

3 = Turning Insert Tolerances

The third letter of general turning insert nomenclature tells us about the turning insert tolerances. Figure-54 shows the tolerance chart.

Code Letter	Cornerpoint (inches)	Thickness (inches)	Inscribed Circle (in)	Cornerpoint (mm)	Thickness (mm)	Inscribed Circle (mm)
A	.0002"	.001"	.001"	.005mm	.025mm	.025mm
C	.0005"	.001"	.001"	.013mm	.025mm	.025mm
E	.001"	.001"	.001"	.025mm	.025mm	.025mm
F	.0002"	.001"	.0005"	.005mm	.025mm	.013mm
G	.001"	.005"	.001"	.025mm	.13mm	.025mm
H	.0005"	.001"	.0005"	.013mm	.025mm	.013mm
J	.002"	.001"	.002-.005"	.005mm	.025mm	.05-.13mm
K	.0005"	.001"	.002-.005"	.013mm	.025mm	.05-.13mm
L	.001"	.001"	.002-.005"	.025mm	.025mm	.05-.13mm
M	.002-.005"	.005"	.002-.005"	.05-.13mm	.13mm	.05-.15mm
U	.005-.012"	.005"	.005-.010"	.06-.25mm	.13mm	.08-.25mm

Figure-54. Insert tolerance chart

4 = Turning Insert Type

The fourth letter of general turning insert nomenclature tells us about the turning insert hole shape and chip breaker type; refer to Figure-55.

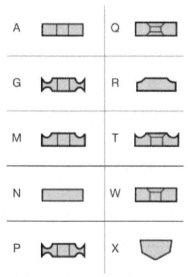

Figure-55. Turning Insert hole shape and chip breaker

5 = Turning Insert Size

This numeric value of general turning insert tells us the cutting edge length of the turning insert; refer to Figure-56.

Figure-56. Turning Insert Cutting Edge Length

6 = Turning Insert Thickness

This numeric value of general turning insert tells us about the thickness of the turning insert.

7 = Turning Insert Nose Radius

This numeric value of general turning insert tells us about the nose radius of the turning insert.

Code	=	Radius Value
04	=	0.4
08	=	0.8
12	=	1.2
16	=	1.6

You can know more about tooling from your tool supplier manual.

Creating New Mill Tool

In this topic we will learn about the procedure to create a new mill tool according to the part requirement. The procedure is discussed next.

• Click on the **New Mill** Tool button from **Select Tool** dialog box. The **Library** dialog box will be displayed; refer to Figure-57

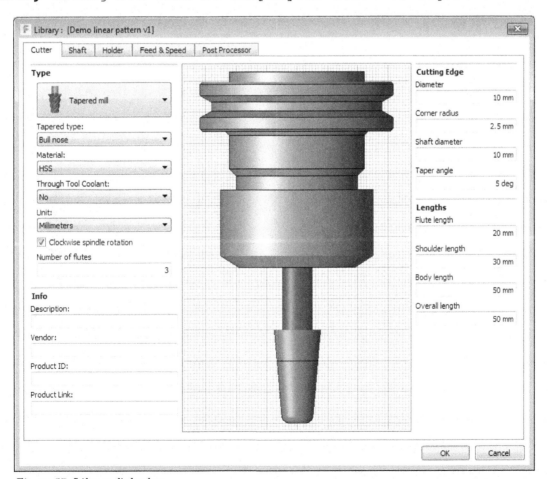

Figure-57. Library dialog box

• Click on the **Type** drop-down of **Cutter** tab from **Library** dialog box and select the required mill type. The selected mill will be displayed in graphics section.

- Click on the **Material** drop-down and select the required material for tool.
- Click on the **Through Tool Coolant** drop-down from **Cutter** tab and select the desired option.
- Click on the **Unit** drop-down from **Cutter** tab and select the desired unit.
- Select the **Clockwise spindle rotation** check box to set the rotation of spindle to clockwise direction.
- Click in the **Number of flutes** edit box and enter the enter the desired value.
- Specify the information about the tool in **Info** section of **Cutter** tab in their respective edit box.
- Click in the **Diameter** edit box of **Cutting Edge** section and specify the diameter of tool.
- Click in the **Shaft diameter** edit box of **Cutting Edge** section and specify the diameter of shaft.
- Similarly, Click on the respective edit box and enter the required parameters in **Lengths** section of **Cutter** tab.
- Click on the **Shaft** tab from **Library** dialog box. The **Shaft** tab will be displayed; refer to Figure-58

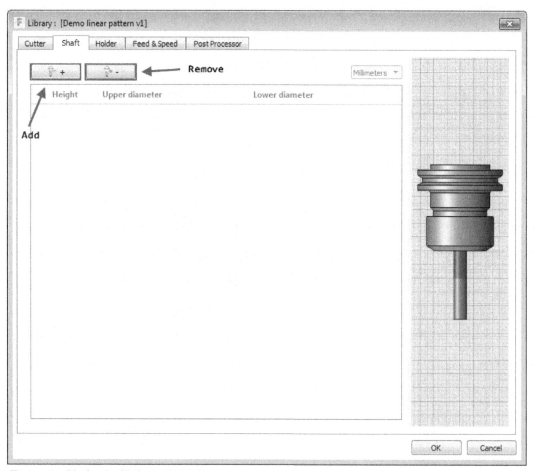

Figure-58. Shaft tab of Library dialog Box

- To add the shaft, click on the **Add** button and if you want to delete the created shaft then click on the **Remove** button from Shaft tab.

- Double-click on the shaft parameter and enter the desired value to edit the dimension of created shaft; refer to Figure-59.

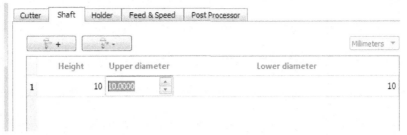

Figure-59. Editing the value of shaft

- Click on the **Holder** tab from **Library** dialog box to select the desired tool holder. The **Holder** tab will be displayed; refer to Figure-60.

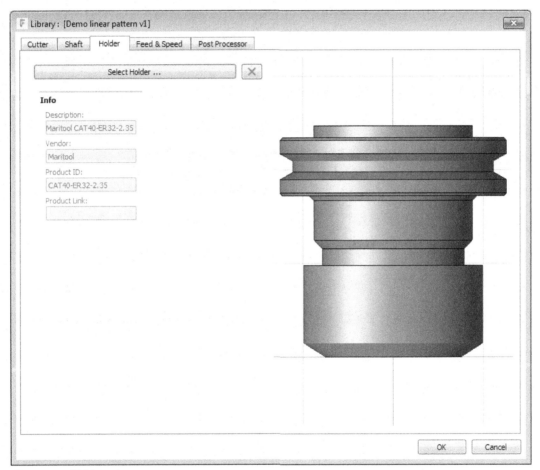

Figure-60. Holder tab

- Click on the **Select Holder** button from **Holder** tab. The **Select Holder** dialog box will be displayed; refer to Figure-61.

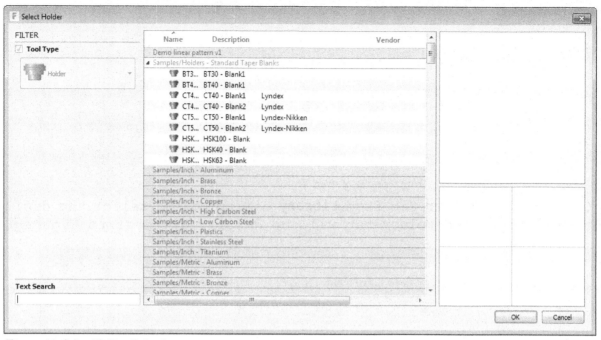

Figure-61. Select Holder dialog box

- Click on the desired holder from **Holder** record section and click on the **OK** button from **Select Holder** dialog box. The selected holder will be displayed in **Holder** tab of **Library** dialog box.
- Click on the **Feed & Speed** tab of **Library** dialog box to set the feed and speed of tool. The **Feed & Speed** tab will be displayed; refer to Figure-62.

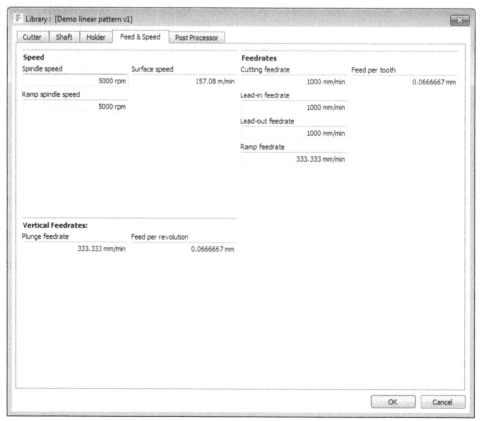

Figure-62. Feed & Speed tab

- Click in the edit boxes and enter the desired value of the respective parameter.
- Click in the **Post Processor** tab of **Library** dialog box to specify the post processor values of NC machining. The **Post Processor** tab will be displayed; refer to Figure-63.

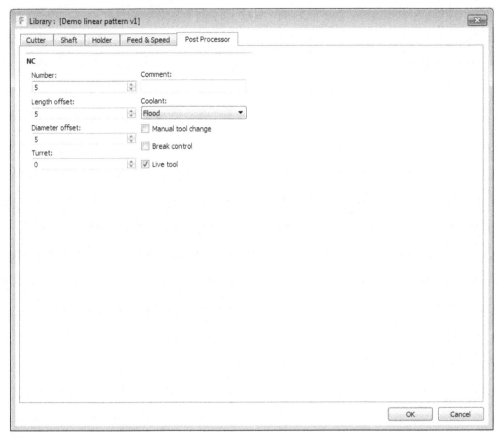

Figure-63. Post Processor tab

- Click in the specific edit box and enter the desired value of the respective parameter.
- Click on the **Coolant** drop-down and select the required coolant type.
- After specifying the parameters from **Library** dialog box, click on the **OK** button. The mill tool will be added in **Tool record** section.

Creating New Holder

In this section, we will discuss about the procedure of creating new tool holder. The procedure is discussed next.

- Click on the **New Holder** button from **Select Tool** dialog box. The **Library** dialog box will be displayed; refer to Figure-46.

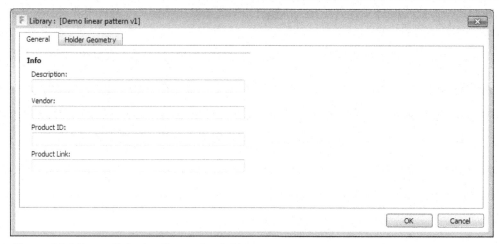

Figure-64. Library dialog box for creating holder

- Click in the specific edit box of **General** tab from **Library** dialog box and enter the information of holder as desired.
- Click in the **Holder Geometry** tab from **Library** dialog box to view or change the geometry of holder; refer to Figure-65.

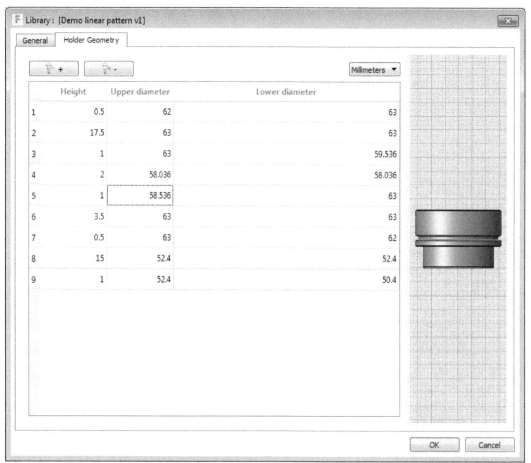

Figure-65. Holder Geometry tab of Library dialog box

- To change the parameter, double-click on the specific parameter and enter the desired value. The holder in graphic window will be updated.
- After specifying the parameter, click on the **OK** button from **Library** dialog box. The created tool holder will be displayed in **Select Tool** dialog box.

Creating New Turning tool

In this section, we will discuss about the procedure of creating a new turning tool. The procedure is discussed next.

- Click on the **New Turning** tool from **Select Tool** dialog box. The **Library** dialog box will be displayed; refer to Figure-66.

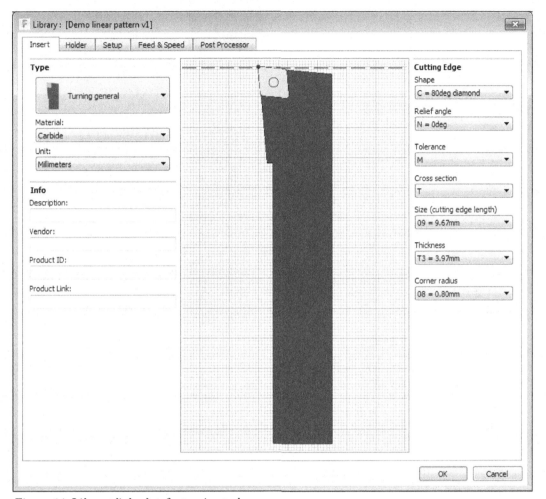

Figure-66. Library dialog box for turning tool

- Click on the **Type** drop-down from **Insert** tab and select the required turning tool type.
- Enter the required information about the tool in **Info** section of **Insert** tab in their respective edit box.
- Click on the **Shape** drop-down of **Cutting Edge** section from **Insert** tab and select the desired option.
- Click on the **Relief angle** drop-down of **Cutting Edge** section from **Insert** tab and select the desired option.
- Similarly, select the desired options from respective drop-down of **Cutting Edge** section from **Insert** tab.
- Enter the required parameter of **Holder** tab of **Library** dialog box as desired.
- Click in the **Setup** tab from **Library** dialog box to setup the tool. The **Setup** tab will be displayed; refer to Figure-67

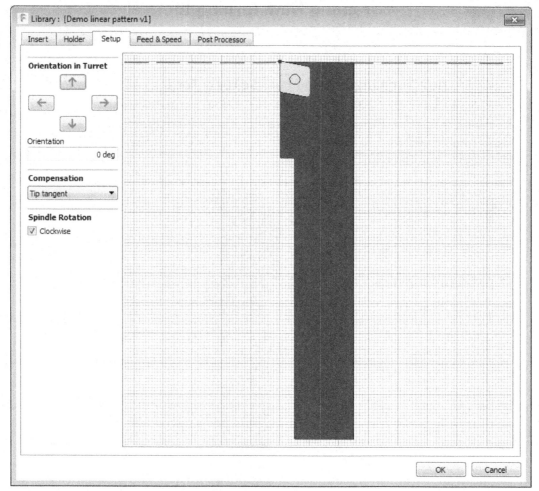

Figure-67. Setup tab for turning

- These four arrow key is used to adjust the orientation of the tool in turret. These orientation is based on the cutting part of the tool.
- Click on the upper key to orient the tool's tip upward, left key to orient the tool to left side, right key to orient the tool to right side, and down key to orient the tool's tip downward. If you want to orient the tool at a specified angle then click on the **Orientation** edit box and enter the specific angle.
- Click on the **Compensation** drop-down from **Setup** tab and select the desired option.
- Select the **Clockwise** check box from **Spindle Rotation** section to rotate the direction of tool.
- After specifying the required parameters from all tabs of Library dialog box, click on the **OK** button. The turning tool will be displayed on **Select Tool** dialog box.

Creating new tool for Waterjet/ Plasma Cutter/ Laser Cutter

In this section, we will discuss about the procedure of creating the Waterjet/ Plasma Cutter/ Laser Cutter tool. The procedure is discussed next.

• Click on the **Create new tool for waterjet, laser, and plasma cutting** tool from **Select Tool** dialog box. The **Library** dialog box will be displayed; refer to Figure-68.

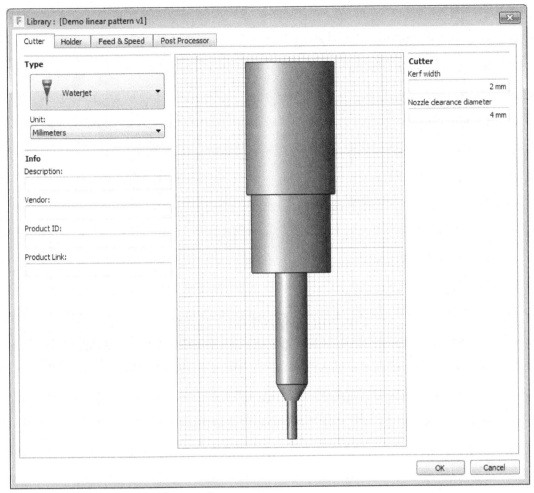

Figure-68. Library dialog box for waterjet

• The options of **Library** dialog box were discussed earlier in **Creating New Mill tool** section of this chapter.
• After specifying the parameters, click on the **OK** button from **Library** dialog box. The created tool will be displayed in **Select Tool** dialog box.
• After selecting the required tool from **Select Tool** dialog box, click on the **OK** button. The tool will be selected for the required operation.

PRACTICAL

Create the stock of the given model Figure-69 and create a required milling tool.

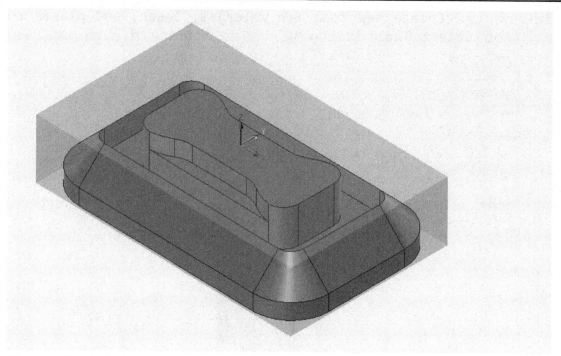

Figure-69. Practical 1.1

Opening Model in CAM

- Firstly, create and save the part in **MODEL Workspace**. The part file is available in the respective chapter folder of **Autodesk Fusion 360** folder.

- Click on the **CAM Workspace** from **Workspace** drop-down. The model will be displayed in the **CAM Workspace**; refer to Figure-70.

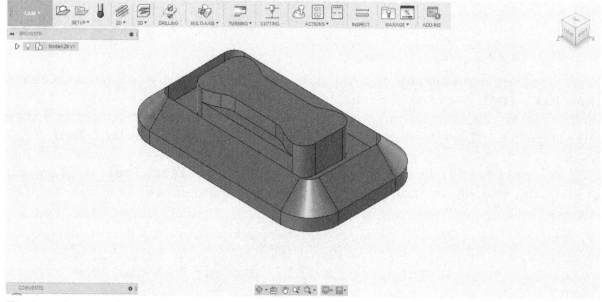

Figure-70. Model displayed in CAM Workspace

Creating Setup

- Click on the **New Setup** tool of **SETUP** drop-down from **Toolbar**. The **SETUP** dialog box will be displayed along with the setup; refer to Figure-71.

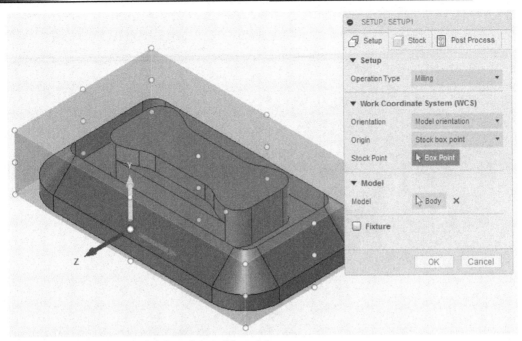

Figure-71. SETUP DIALOG box along with model

- Click on the **Milling** option of **Operation Type** drop-down from **Setup** tab to specify the type of operation.
- Click on the **Select Z axis/plane & X axis** option of **Orientation** drop-down from **Setup** tab and select the Z-axis for milling; refer to Figure-72.

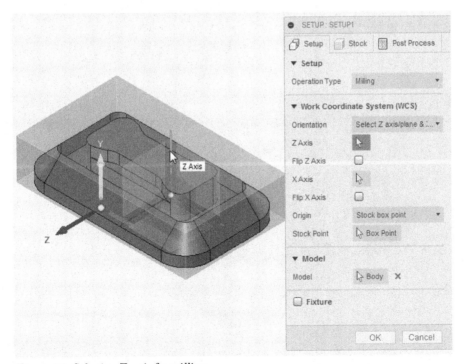

Figure-72. Selecting Z axis for milling

- The **Arrow** button of **X Axis** option is active by default. You need to click on the Z axis from origin to select as X axis.
- Click on the **Box Point** button of **Stock Point** option from **SETUP** tab and select the box point as displayed in the Figure-73.

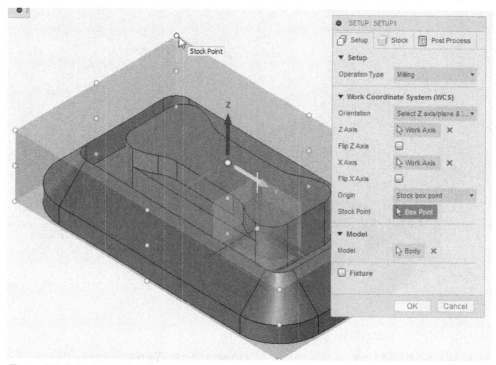

Figure-73. Selecting stock point

- Click on the **Relative Size Box** option of **Mode** drop-down from **Stock** tab and enter the parameters as shown in Figure-74.

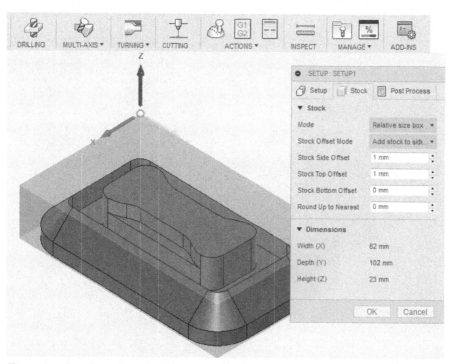

Figure-74. Specifying the parameters for creating stock

- After specifying the parameters of **SETUP** dialog box, click on the **OK** button from **SETUP** dialog box. The stock of model will be created.

Creating a new Milling tool

- Before creating a new tool, you need to know the inner and outer dimension of the model like height of pocket, radius, etc.

- Click on the **Tool Library** tool of **MANAGE** drop-down from **Toolbar**. The **CAM Tool Library** dialog box will be displayed.
- Click on the **New Mill Tool** button of **CAM Tool Library** dialog box; refer to Figure-75. The **Library** dialog box will be displayed.

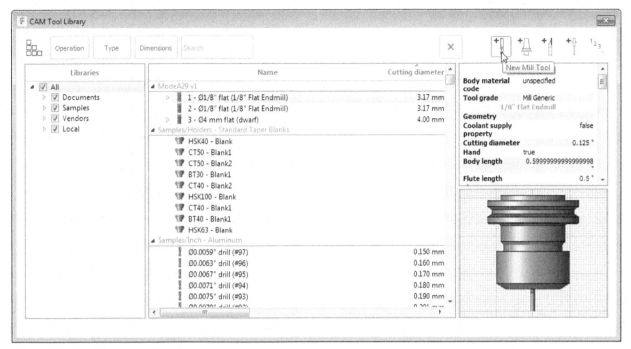

Figure-75. Creating new mill tool

- Enter the parameters in the **Cutter** tab of **Library** dialog box; refer to Figure-76.

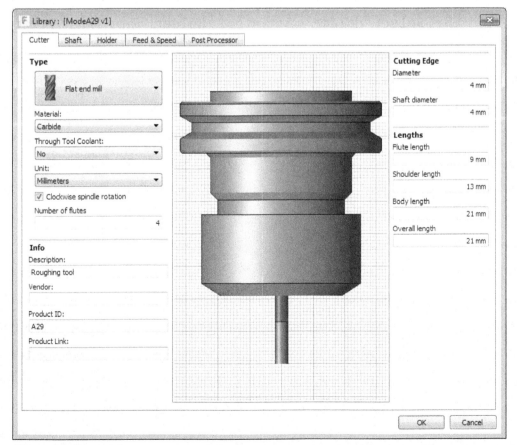

Figure-76. Cutter tab

- Click on the **Shaft** tab of **Library** dialog box and specify the parameters as displayed in Figure-77.

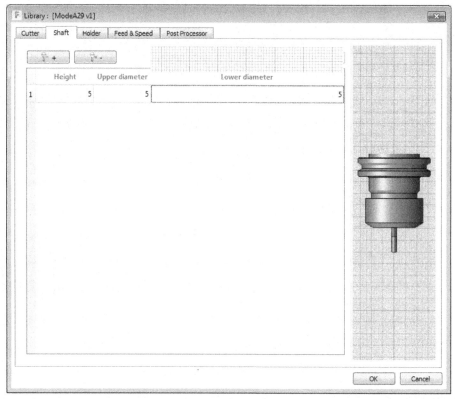

Figure-77. Shaft tab

- Click in the **Holder** tab of **Library** dialog box. The **Holder** tab will be displayed.
- Click on the **Select Holder** button of **Holder** tab. The **Select Holder** dialog box will be displayed.
- Select the holder as displayed in Figure-78 and click on **OK** button from **Select Holder** dialog box.

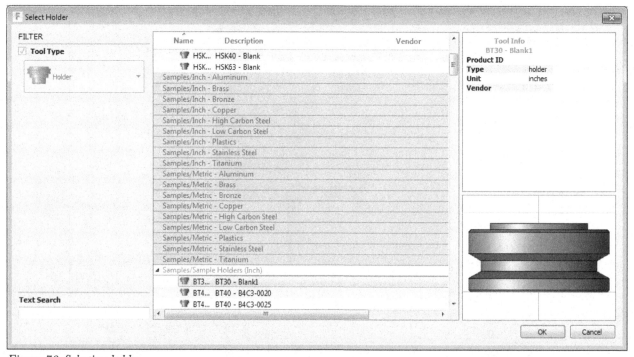

Figure-78. Selecting holder

- Click on the **Feed & Speed** tab of **Library** dialog box and specify the parameters as shown in Figure-79.

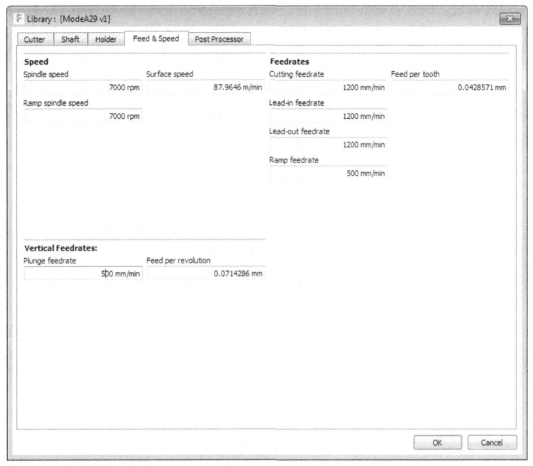

Figure-79. Feed & Speed tab1

- After specifying the parameters, click on the **OK** button from **Library** dialog box. The tool will be created and displayed in the tool library.

PRACTICAL 2

Create the stock for lathe machining as shown in Figure-80.

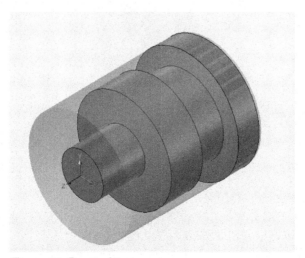

Figure-80. Practical 2

Adding model in CAM Workspace

- Create and save the part in **MODEL Workspace**. The part file is also available in the respective chapter folder of **Autodesk Fusion 360 Resources** folder.

- Click on the **CAM Workspace** from **Workspace** drop-down. The model will be displayed in the **CAM Workspace**; refer to Figure-81.

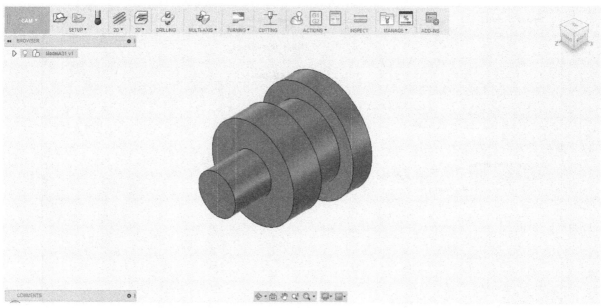

Figure-81. Model in CAM Workspace

Creating Setup

- Click on the **New Setup** tool of **SETUP** drop-down from **Toolbar**. The **SETUP** dialog box will be displayed along with the stock of model; refer to Figure-82.

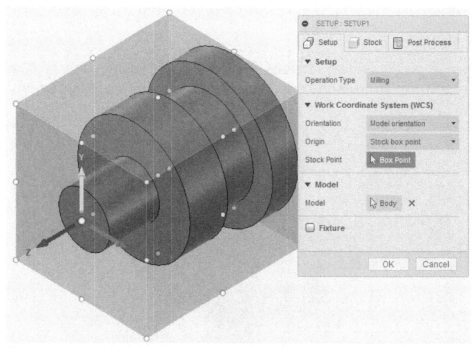

Figure-82. Model along with SETUP dialog box

- Click on the **Turning or mill/turn** option of **Operation Type** drop-down from **Setup** tab to specify the machining type.
- The **Arrow** button of **Z Axis(Rotary Axis)** option is active by default. Select the edge as displayed in Figure-83.

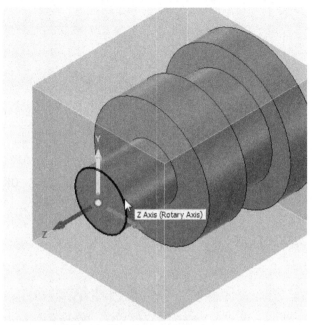

Figure-83. Selecting Z axis

- Click on the **Stock** tab of **SETUP** dialog box and specify the parameters as displayed in Figure-84.

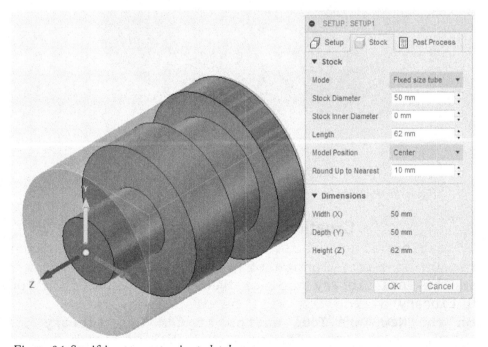

Figure-84. Specifying parameters in stock tab

- Click in the **Post Process** tab of **SETUP** dialog box and specify the parameters as displayed in Figure-85.

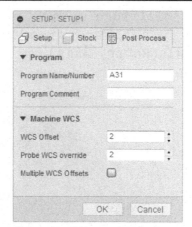

Figure-85. Post process tab of turning

- After specifying the parameters, click on the **OK** button from **SETUP** dialog box. The stock will be created and displayed on the model; refer to Figure-86.

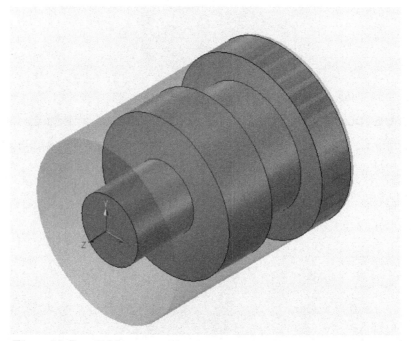

Figure-86. Practical 2

Creating new Turning tool

- Before creating a new tool, you need to know the dimensions of slot, radius of part, length of part, etc.
- Click on the **Tool Library** tool of **MANAGE** drop-down from **Toolbar**. The **CAM Tool Library** dialog box will be displayed.
- Click on the **New Turn Tool** button of **CAM Tool Library** dialog box; refer to Figure-87. The **Library** dialog box will be displayed.

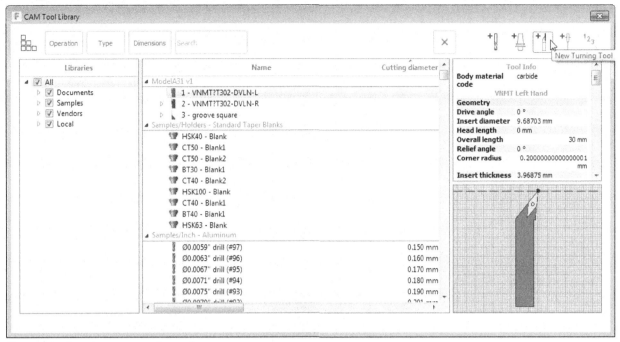

Figure-87. Creating new turning tool

- Specify the parameters of **Insert** tab from **Library** dialog box as displayed in Figure-88.

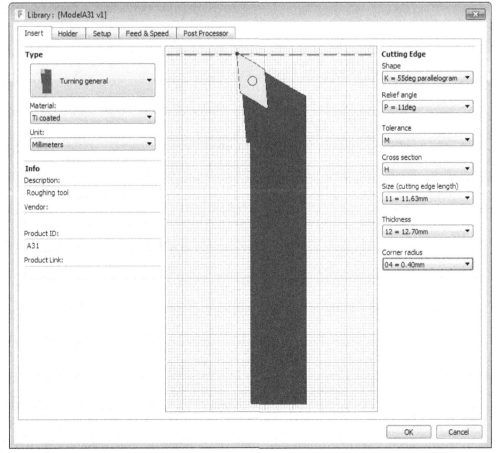

Figure-88. Insert tab

- Click on the **Holder** tab of **Library** dialog box and specify the parameters as displayed in Figure-89.

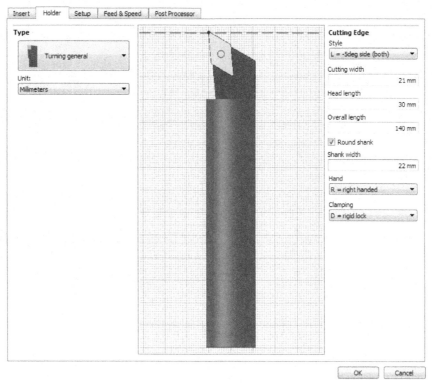

Figure-89. Hholder tab

- Click on the **Feed & Speed** tab of **Library** dialog box and enter the parameters as displayed in Figure-90.

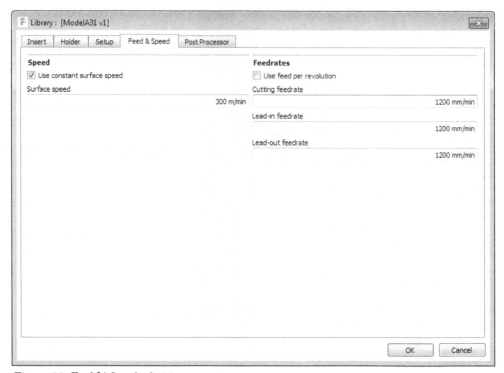

Figure-90. Feed & Speed tab132

- After specifying the parameters, click on the **OK** button from **Library** dialog box. The tool will be created and displayed in the tool library.

PRACTICE

Create the stock of given part; refer to Figure-91.

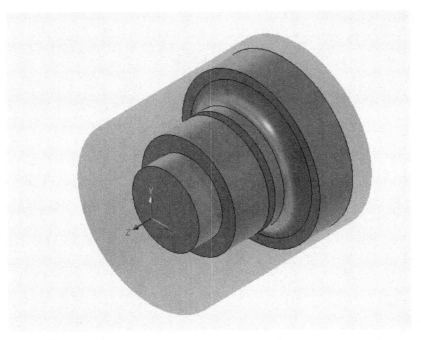

Figure-91. Practice

FOR STUDENT NOTES

Chapter 16

Generating Milling
Toolpaths - 1

The major topics covered in this chapter are:

- *2D Pocket Toolpath*
- *2D Contour Toolpath*
- *Trace Toolpath*
- *Bore Toolpath*
- *Engrave Toolpath*
- *Adaptive Clearing Toolpath*
- *Parallel Toolpath*
- *Contour Toolpath*
- *Horizontal Toolpath*
- *Scallop Toolpath*

GENERATING 2D TOOLPATHS

Till now, we have discussed the procedure of creating the stock and placing the tool on the stock. In this section, we will create the 2D toolpaths using tools in 2D drop-down of Toolbar.

2D Adaptive Clearing

The **2D Adaptive Clearing** tool is used to create a machining operation on the part which uses the adaptive path as per the parts curvature to avoid abrupt direction changes. The procedure to use this tool is discussed next.

*   Click on the **2D Adaptive Clearing** tool of **2D** drop-down from **Toolbar**; refer to Figure-1. The **2D ADAPTIVE** dialog box will be displayed; refer to Figure-2.

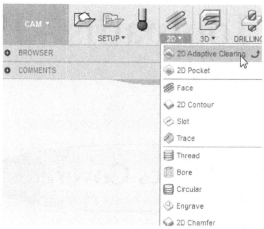

Figure-1. 2D Adaptive Clearing tool

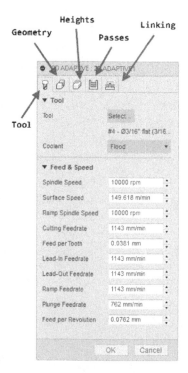

Figure-2. 2D ADAPTIVE dialog box

- Click on the **Select** button of **Tool** section from **2D ADAPTIVE** dialog box. The **Select Tool** dialog box will be displayed.
- Specify the feed rate and speed of tool in their respective edit boxes.

Geometry

In this section, we will discuss the procedure of selecting the geometry for removal of the extra material from workpiece. The procedure is discussed next.

- Click in the **Geometry** tab from **2D ADAPTIVE** dialog box. The geometry tab will be displayed; refer to Figure-3.

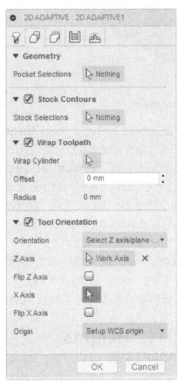

Figure-3. Geometry tab for 2D ADAPTIVE dialog box

- The **Nothing** selection button of **Pocket Sections** section in **Geometry** node is active by default. Click on the desired geometry from model to be machined; refer to Figure-4.

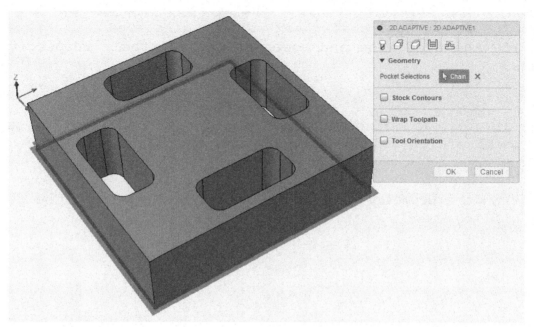

Figure-4. Selected geometry for 2D adaptive

- Select the **Stock Contours** check box of **Geometry** tab from **2D ADAPTIVE** dialog box to specify the parameter of stock that need to be faced.
- The **Nothing** selection button of **Stock Selection** section is active by default. You need to click on the edge or faces to select.
- Select the **Wrap Toolpath** check box of **Geometry** tab from **2D ADAPTIVE** dialog box to wrap the toolpath around a cylinder. This check box is generally used in creating a toolpath for round shaped body.
- Select the **Tool Orientation** check box of **Geometry** tab from **2D ADAPTIVE** dialog box to override the orientation of tool earlier defined in setup. The procedure to orient the tool has been discussed earlier.

Heights

The **Heights** tab of the **2D ADAPTIVE** dialog box is used to set different parameters of height of model. The procedure to define these heights are discussed next.

- Click on the **Heights** tab of **2D ADAPTIVE** dialog box. The **Heights** tab will be displayed along with model; refer to Figure-5.

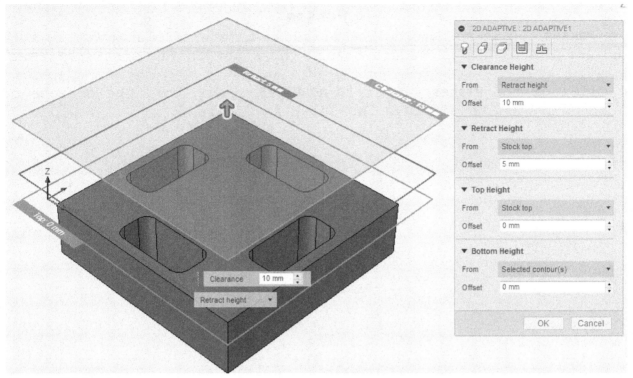

Figure-5. Heights tab

- The Clearance height is the distance between tool and workpiece at which the tool rapidly moves before the start of the toolpath for machining process.
- Click on the **From** drop-down of **Clearance Height** section from **Heights** tab and select the desired reference from which height will be measured.
- Click in the **Offset** edit box of **Clearance Height** section and enter the value of selected parameter.
- The Retract height is the distance in which tool moves upward after a cutting pass and before performing the next pass of machining or cutting. The Retract height should be set above the Feed Height and Top of workpiece.
- Click on the **From** drop-down of **Retract Height** section from **Heights** tab and select the desired parameter.
- Click in the **Offset** edit box of **Retract Height** section and enter the value of selected parameter.
- The Top height is the term used to define the top of the cut.
- Click on the **From** drop-down of **Top Height** section from **Heights** tab and select the desired parameter.
- Click in the **Offset** edit box of **Top Height** section and enter the value of selected parameter.
- The Bottom Height is the term used to define the depth of the cut in a workpiece. If it is not defined in the dialog box then system will automatically assume the stock depth.
- Click on the **From** drop-down of **Bottom Height** section from **Heights** tab and select the desired parameter.
- Click in the **Offset** edit box of **Bottom Height** section and enter the value of selected parameter.

Passes

The **Passes** tab is used to define various parameters related to the tool and workpiece. The procedure to use these options is discussed next.

- Click on the **Passes** tab of **2D ADAPTIVE** dialog box. The options of the tab will be displayed as shown in Figure-6.

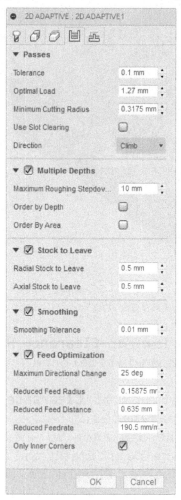

Figure-6. Passes tab

- Click in the **Tolerance** edit box of **Passes** section from **Passes** tab and enter the desired value of tolerance.
- Click in the **Optimal Load** edit box of **Passes** section from **Passes** tab and enter the desired value to specify the amount of engagement the adaptive strategies should maintain.
- Click in the **Minimum Cutting Radius** edit box and enter the desired value for applying minimum cutting radius to the sharp corners of workpiece while machining.
- Select the **Use Slot Clearing** check box to enable the pocket clearing with a slot along its middle, before continuing a spiral motion on the selected workpiece.
- Click in the **Slot Clearing Width** edit box and enter the desired value.
- Select **Conventional** option from **Direction** drop-down of **Passes** section to specify the material removal process by conventional milling. In **Conventional** milling, the cutting process is performed in opposite direction of tool movement which causes the tool to scoop up the

material. The cutting process is starts at zero thickness and increases up to maximum.

- Click on the **Climb** option of **Direction** drop-down from **Passes** section to specify the material removal process by climb milling. In Climb milling the cutting process starts by machining maximum thickness and then decreases to zero.

- Select the **Multiple Depths** check box from **Passes** tab to cut the material in multiple depths instead of cutting in one go.

- Click in the **Maximum Roughing Stepdown** edit box and enter the maximum value of step-down distance between two roughing depths.

- Select the **Order By Depth** check box to specify the machining of various depth from top to down.

- Select the **Order By Area** check box to specify the machining of multiple depth from larger area to less area.

- Select the **Stock to leave** check box from **Passes** tab to define the amount of stock to leave on the workpiece after roughing.

- Click in the **Radial Stock to Leave** edit box of **Stock to Leave** section and enter the value of stock to remain along round face of part.

- Click in the **Axial Stock to Leave** check box edit box of **Stock to Leave** section and enter the value of stock to remain along z axis or flat surface/body.

- Select the **Smoothing** check box from **Passes** tab to smooth the toolpath by removing excessive points and fitting arcs within a specified tolerance.

- Click in the **Smoothing Tolerance** edit box of **Smoothing** section and enter the desired value of tolerance.

- Select the **Feed Optimization** check box of **Passes** tab to specify the reduced feed at corners and curves.

- Click in the **Maximum Directional Change** check box of **Feed Optimization** section and specify the maximum value of angle change allowed.

- Click in the **Reduced Feed Radius** edit box and enter the value of minimum radius allowed before the feed is reduced.

- Click in the **Reduced Feed Distance** edit box and enter the value of distance before a corner to reduce the feed rate.

- Click in the **Reduced Feedrate** edit box and enter the value to specify the slower feedrate to be used when the tool is moving in X direction while cutting material.

- Select the **Only Inner Corners** check box of **Feed Optimization** section to reduce the feedrate at inner corners.

Linking

The options of **Linking** tab is used to define, how toolpath passes should be linked. The procedure to use this options is discussed next.

- Click on the **Linking** tab from **2D ADAPTIVE** dialog box. The **Linking** tab will be displayed; refer to Figure-7.

Figure-7. Linking tab

- Select **Full retraction** option of **Retraction Policy** drop-down to retract the tool up to retract height at the end of cutting process before moving to the start of next machining pass.
- Select **Minimum retraction** option of **Retraction Policy** drop-down to move the tool up to the minimum retraction height from one cutting pass to other pass where the tool clears the workpiece.
- The **High Feedrate Mode** drop-down is used to specify the tool rapid movements which should be output as true rapid movement. Click on the **High Feedrate Mode** drop-down and select the desired option.
- Select the **Allow Rapid Retract** check box from **Linking** tab to do the retracts as rapid movements. Clear the **Allow Rapid Retract** check box to force retract at lead-out feedrate.
- Click in the **Maximum Stay-Down Distance** edit box and enter the maximum value of distance between different cutting passes for which the tool stays down and does not retract.
- Click in the **Minimum Stay-Down Clearance** edit box and enter the radial clearance distance values for the stay down moves.
- Click on the **Stay-Down Level** drop-down of **Linking** section and select the desired option to control how much the tool will try to stay down around obstacles during cutting process.

- Click in the **Lift Height** edit box and enter the value of distance lift during reposition moves of the tool.
- Click in the **No-Engagement Feedrate** edit box of **Linking** section to specify the feedrate used for rapid movements where the tool is not in engagement with the material of workpiece during cutting pass.
- The options of **Leads & Transitions** section of **Linking** tab is used to specify how leads and transitions should be generated at the time of tool entering or exiting the material.
- Click in the **Horizontal Lead In/Out Radius** edit box and enter the value of radius as desired for horizontal entry/exit.
- Click in the **Vertical Lead In/Out Radius** edit box and enter the value of radius as desired for vertical entry/exit.
- The options in **Ramps** section of **Linking** tab are used to define how tool will entry into the material during its first cutting pass.
- Hover the cursor on the options of **Ramp Type** drop-down, the explanation of these options will be displayed on the screen. Select the option as required.
- The options of **Positions** section of **Linking** tab are used to specify the tool entry point.
- The **Nothing** selection button of **Predrill Positions** section is active by default. Click on the point in workpiece where drill has been done earlier so that the cutting tool can enter the material.
- Click on the **Nothing** selection button of **Active Positions** section and click on the geometry near the location of workpiece where you want to enter the tool.
- After specifying the parameter for **2D ADAPTIVE** dialog box, click on the **OK** button. The toolpath will be created and displayed on the workpiece; refer to Figure-8.

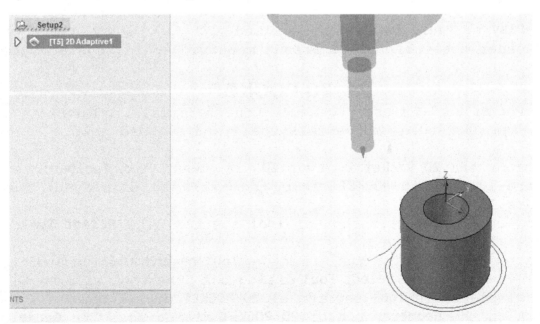

Figure-8. Toolpath created by adaptive clearing

- The option of the created toolpath will be added in **Browser Tree.** Right-click on the tool. A shortcut menu will be displayed; refer to Figure-9.

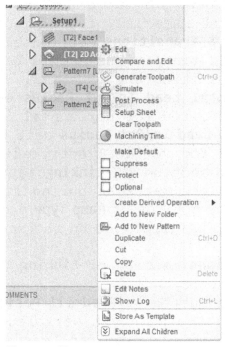

Figure-9. Shortcut menu

- If you want to edit the parameters then click on the **Edit** button from shortcut menu. The dialog box will be displayed of the selected toolpath.
- If you want to check the animation of material cutting by recently created toolpath then click on the **Simulate** button. The **SIMULATE** dialog box will be displayed along with simulation keys. Click on the **Play** or **Pause** button as required.
- The other tools of this shortcut menu will be discussed later.

2D POCKET

The **2D Pocket** tool is used to remove the material from pockets in the model. The procedure to use this tool is discussed next.

- Click on the **2D Pocket** tool of **2D** drop-down from **Toolbar;** refer to Figure-10. The **2D POCKET** dialog box will be displayed; refer to Figure-11.
- Click on the **Select** button of **Tool** section. The **Select Tool** dialog box will be displayed.
- Click on the required tool from the **Tool record** section and click on the **OK** button from **Select Tool** dialog box. The tool will be selected and displayed in **Tool** section of **2D POCKET** dialog box.
- Click on the **Geometry** tab of **2D POCKET** dialog box. The **Geometry** tab will be displayed.
- The **Nothing** button of **Pocket Selections** section from **Geometry** tab is active by default. Click on the surface/edge of the model to select; refer to Figure-12.

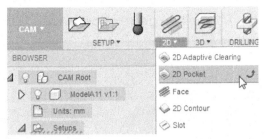

Figure-10. 2D Pocket tool

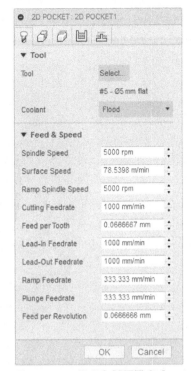

Figure-11. 2D POCKET dialog box

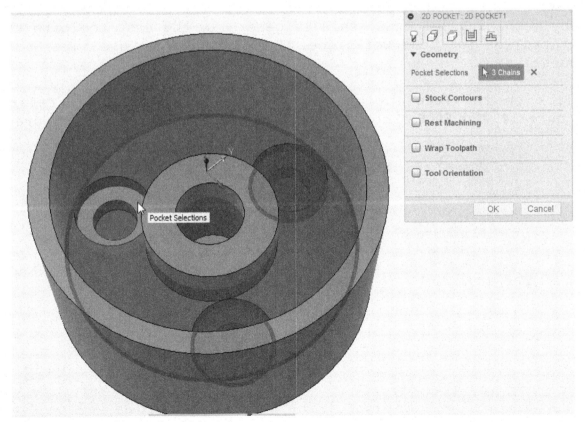

Figure-12. Pocket selections of model

- For multiple depths pocket, the **Multiple Depths** check should be selected from **Passes** tab.
- The other parameters have been discussed earlier in this chapter.

- After specifying the parameters, click on the **OK** button from **2D POCKET** dialog box. The toolpath will be generated and displayed on the model; refer to Figure-13.

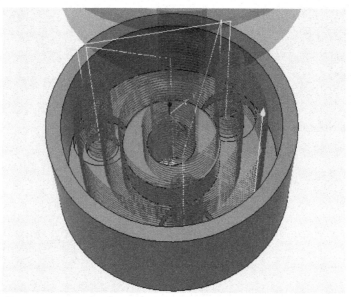

Figure-13. Created toolpath for 2D Pocket

Face

The **Face** tool is used to remove material from the face of the workpiece. In any machining sequence, this is generally the first toolpath to be generated for flat head workpiece. The procedure to use this tool is discussed next.

- Click on the **Face** tool of **2D** drop-down from **Toolbar**; refer to Figure-14. The **FACE** dialog box will be displayed; refer to Figure-15.

Figure-14. Face tool

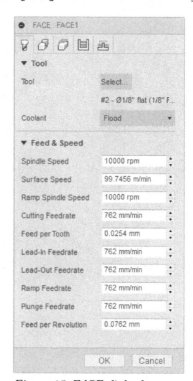

Figure-15. FACE dialog box

- Click on the **Select** button from **Tool** tab and select the required tool for facing according to your model from **Select Tool** dialog box.
- Click on the **Geometry** tab of **FACE** dialog box. The **Geometry** tab will be displayed.
- The **Nothing** selection button of **Stock Selections** section is active by default. Click on the edge of the model to select; refer to Figure-16.

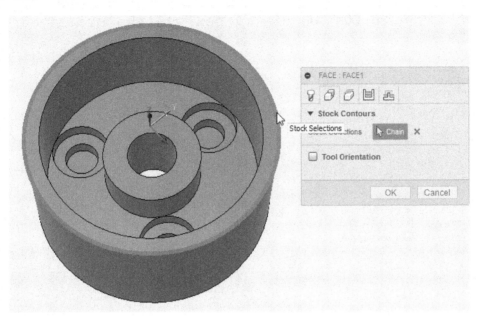

Figure-16. Selection for FACE dialog box

- The other tools of the dialog box have been discussed earlier.
- After specifying the parameters, click on the **OK** button from **FACE** dialog box. The toolpath will be created and displayed on workpiece; refer to Figure-17.

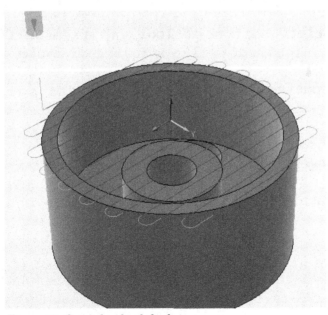

Figure-17. Created toolpath for facing

- Right-click on the recently created **Face** toolpath and click on the **Simulate** button to check the animation of facing operation.

2D Contour

The **2D Contour** tool is used to remove material by following the contour of the model. This tool is generally used to remove the material from outer/inner walls of the selected workpiece. The procedure to use this tool is discussed next.

- Click on the **2D Contour** tool of **2D** drop-down from **Toolbar**; refer to Figure-18. The **2D CONTOUR** dialog box will be displayed; refer to Figure-19.

Figure-18. 2D Contour tol

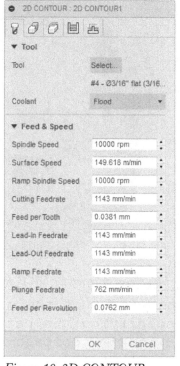

Figure-19. 2D CONTOUR dialog box

- Click on the **Select** button of **Tool** tab from **2D CONTOUR** dialog box and select the required tool according to your model from **Select Tool** dialog box.
- Click on the **Geometry** tab of **2D CONTOUR** dialog box. The **Geometry** tab will be displayed.
- The **Nothing** button of **Contour Selection** section is active by default. Click on the edge of model to select; refer to Figure-20.
- Click in the **Tangential Extension Distance** edit box of **Geometry** tab and enter the value of distance to extend the open contours tangentially.
- Select the **Separate Tangentially End Extension** check box to enable tangential extension at the end of cutting pass.

Tabs

The **Tabs** check box is used to hold the workpiece when you are cutting a part from sheet. This option is generally used when you are working on a sheet and after operation the part will be detached. The procedure to use the tab is discussed next.

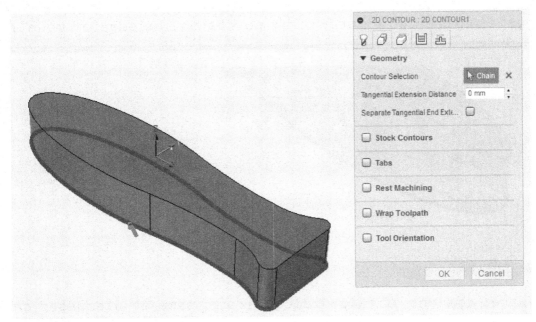

Figure-20. Selection of edge for contour

- Click on the **Tabs** check box to enable options related to tab.
- Select on the required shape of tab from **Tab Shape** drop-down.
- Click in the **Tab Width** edit box and enter the value of width of tab.
- Click in the **Tab Height** edit box and enter the value of height of tab.
- Select the **By distance** option of **Tab Positioning** drop-down from **Tabs** section to position the tab around the workpiece by distance entered; refer to Figure-21

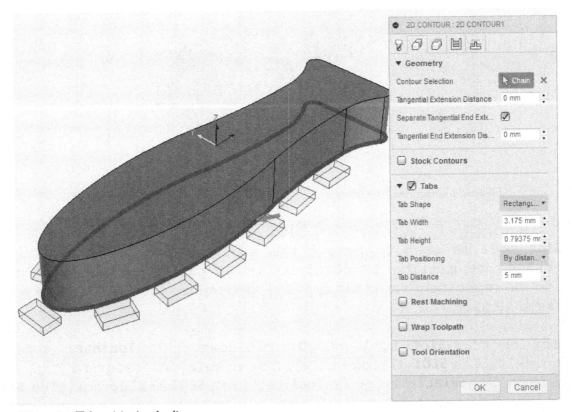

Figure-21. Tab positioning by distance

- Select the **At points** option of **Tab Positioning** option from **Tabs** section to position the tab by selecting points on the edge of model; refer to Figure-22.

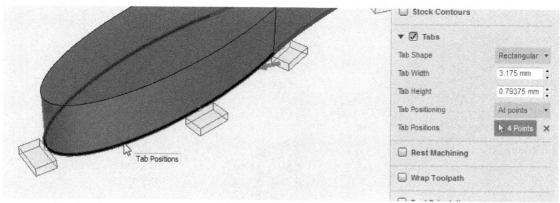

Figure-22. Positioning by points

- The other options of this dialog box are same as discussed earlier.
- After specifying the parameters, click on the **OK** button from **2D CONTOUR** dialog box. The toolpath will be generated and displayed on the model; refer to Figure-23.

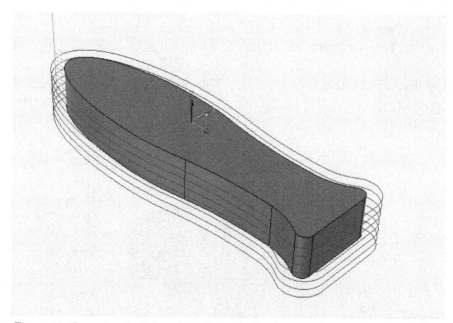

Figure-23. Created toolpath for solid

Slot

The **Slot** tool is used to remove the slot material from model. Note that this slot milling is governed in 2D plane sothe depth of cut need to be specified explicitly. The procedure to use this tool is discussed next.

- Click on the **Slot** tool of **2D** drop-down from **Toolbar**; refer to Figure-24. The **SLOT** dialog box will be displayed; refer to Figure-25.
- Click on the **Select** button of **Tool** tab from **SLOT** dialog box. The **Select Tool** dialog box will be displayed. Select the tool as required.

- Click on the **Geometry** tab of **2D CONTOUR** dialog box. The **Geometry** tab will be displayed.
- The **Nothing** button of **Pocket Selections** section is active by default. You need to select the slot; refer to Figure-26.

Figure-24. Slot Tool

Figure-25. SLOT dialog box

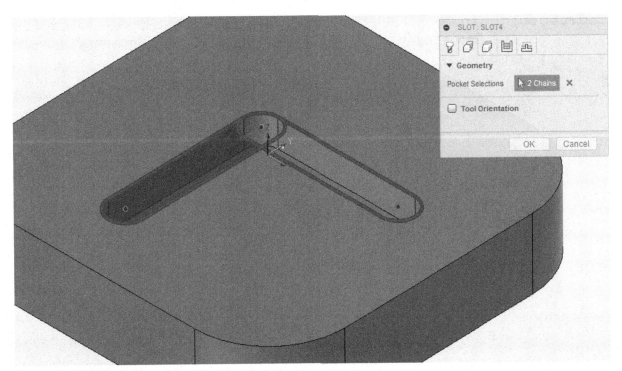

Figure-26. Selection for slot

- Select the desired option from the **Ramp Type** drop-down. In our case, we have applied **Plunge** option of **Ramp Type** drop-down from **Linking** tab for better generation of toolpath.

- The other tools of this dialog box have been discussed earlier.
- After specifying the parameters, click on the **OK** button from **SLOT** dialog box. The toolpath will be generated and displayed on the model; refer to Figure-27.

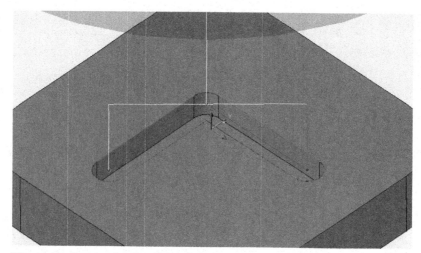

Figure-27. Generated toolpath for slot

Trace

The **Trace** tool is used to trace the selected path for machining. The procedure to use this tool is discussed next.

- Click on the **Trace** tool of **2D** drop-down from **Toolbar**; refer to Figure-28. The **TRACE** dialog box will be displayed; refer to Figure-29.
- Click on the **Select** button of **Tool** tab from **TRACE** dialog box. The **Select Tool** dialog box will be displayed. Select the tool as required.
- Click on the **Geometry** tab of **TRACE** dialog box. The **Geometry** tab will be displayed.

Figure-28. Trace tool

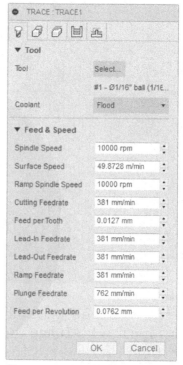

Figure-29. TRACE dialog box

- The **Nothing** button of **Curve Selections** section of **Geometry** tab is active by default. You need to select the geometry for trace; refer to Figure-30.

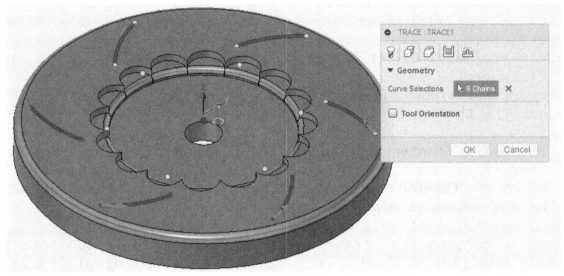

Figure-30. Selection of geometry for trace

- Click on the **Passes** tab of **TRACE** dialog box. The **Passes** tab will be displayed; refer to Figure-31.

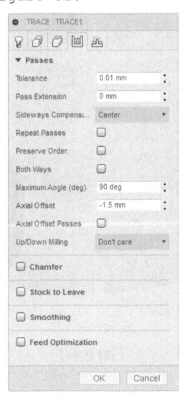

Figure-31. Passes tab of TRACE dialog box

- Click in the **Pass Extension** edit box and enter the value of distance to extend the toolpath along length.
- Select the **Repeat Passes** check box from **Passes** tab to perform an additional finishing pass with zero stock.
- Select the **Preserve Order** check box from **Passes** tab to specify that the geometry machined in the order in which they are selected.

- Select the **Both Ways** check box from **Passes** tab to use the Climb and Conventional machining to the open profile of the selected geometry.
- Click in the **Maximum Angle (deg)** edit box and specify the maximum plunge angle for machining.
- Click in the **Axial Offset** edit box and specify the value of axial offset for the toolpath of selected geometry.
- Select the **Axial Offset Passes** check box from **Passes** tab to enable multiple depths machining.
- Click on the **Up/Down Milling** drop down and select the required option.
- Select the **Chamfer** check box of **Passes** tab from **TRACE** dialog box to perform a chamfer operation on selected geometry.
- Click in the **Chamfer Width** edit box of **Chamfer** section from **Passes** tab and enter the value of width for machining of chamfer.
- Click in the **Chamfer Depth** box of **Chamfer** section from **Passes** tab and enter the value of depth of chamfer for machining.
- The other tools of this dialog box were discussed earlier.
- After specifying the parameters, click on the **OK** button from **TRACE** dialog box. The toolpath will be generated and displayed on the model; refer to Figure-32.

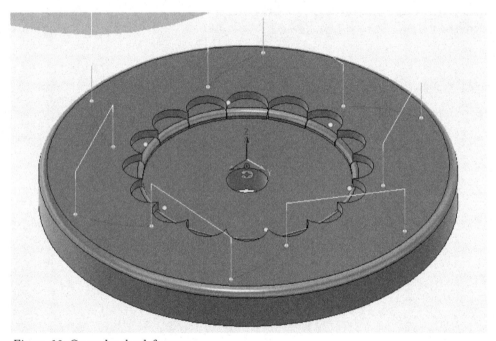

Figure-32. Created toolpath for trace

Thread

The **Thread** tool is used to internal or external threads on the selected geometry. The procedure to use this tool is discussed next.

- Click on the **Thread** tool of **2D** drop-down from **Toolbar**; refer to Figure-33. The **THREAD** dialog box will be displayed; refer to Figure-34.
- Click on the **Select** button of **Tool** tab from **THREAD** dialog box. The **Select Tool** dialog box will be displayed. Select the tool as required.
- Click on the **Geometry** tab of **THREAD** dialog box. The **Nothing** button of **Circular Face Selections** option from **Geometry** tab is active by default.

- You need to click on the circular face to select; refer to Figure-35

Figure-33. *Thread tool*

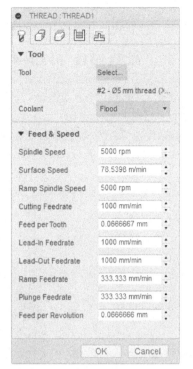

Figure-34. *THREAD dialog box1*

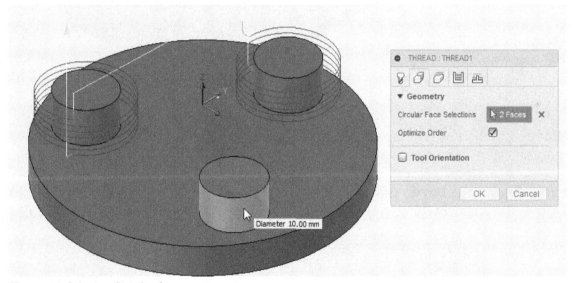

Figure-35. *Selecting Circular faces*

- Click on the **Passes** tab from **TREAD** dialog box. The **Passes** tab will be displayed; refer to Figure-36.
- Select the **Right handed** option of **Threading Hand** drop-down from **Passes** tab to create a right handed thread direction.
- Select the **Left handed** option of **Threading Hand** drop-down from **Passes** tab to create a left handed thread direction.
- Click in the **Thread Pitch** edit box of **Passes** tab to enter the value of distance between two thread.
- Click in the **Pitch Diameter Offset** edit box of **Passes** tab to enter the value of difference between major and minor thread diameter.

- Select the **Do Multiple Threads** check box of **Passes** tab to enter the number of threads and enter the value on **Number of Threads** edit box.
- Select the required option from **Compensation Type** drop-down to specify the compensation type.
- Select the **Multiple Passes** check box of **Passes** tab to enter a value for multiple depth cuts when Thread Milling.
- The other tools of this dialog box were discussed earlier.
- After specifying the parameters, click on the **OK** button from **THREAD** dialog box. The toolpath will be generated and displayed on the model; refer to Figure-37.

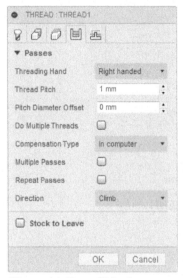

Figure-36. Passes Tab of
THREAD dialog box

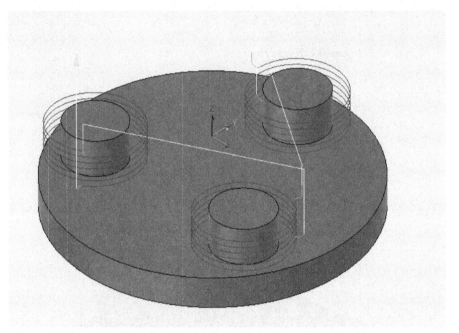

Figure-37. Generated Toolpath for thread

Bore

The **Bore** tool is used to increase the diameter of hole in tight tolerance. The procedure to use this tool is discussed next.

- Click on the **Bore** tool of **2D** drop-down from **Toolbar**; refer to Figure-38. The **BORE** dialog box will be displayed; refer to Figure-39.

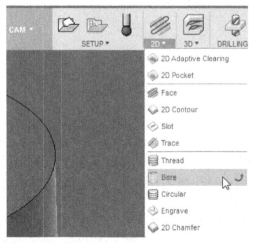

Figure-38. Bore tool

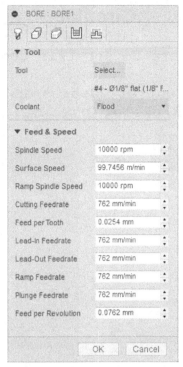

Figure-39. BORE dialog box

- Click on the **Select** button of **Tool** tab from **BORE** dialog box. The **Select Tool** dialog box will be displayed. Select the tool as required.
- Click on the **Geometry** tab of **THREAD** dialog box. The **Geometry** tab will be displayed.
- The **Nothing** button of **Circular Face Selections** option from **Geometry** tab is active by default. You need to click on the circular face of model to select; refer to Figure-40.

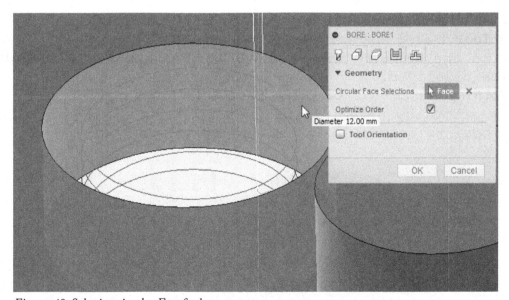

Figure-40. Selecting circular Face for bore

- The other options of the dialog box were discussed earlier.
- After specifying the parameter, click on the **OK** button from **BORE** dialog box. The toolpath will be generated and displayed on the model.

Circular

The **Circular** tool is used for milling the circular or round pocket/boss features. The procedure to use this tool is discussed next.

- Click on the **Circular** tool of **2D** drop-down from **Toolbar**; refer to Figure-41. The **CIRCULAR** dialog box will be displayed; refer to Figure-42.

Figure-41. Circular tool

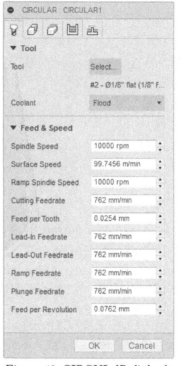

Figure-42. CIRCULAR dialog box

- Click on the **Select** button of **Tool** tab from **CIRCULAR** dialog box. The **Select Tool** dialog box will be displayed. Select the tool as required.
- Click on the **Geometry** tab of **CIRCULAR** dialog box. The **Geometry** tab will be displayed.
- The **Nothing** button of **Circular Face Selections** option from **Geometry** tab is active by default. You need to click on the circular face from model to select; refer to Figure-43.

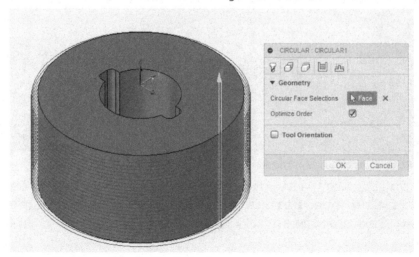

Figure-43. Selecting face for Circular tool

- The other tools of the dialog box are same as discussed earlier.
- After specifying the parameter, click on the **OK** button from **CIRCULAR** dialog box. The toolpath will be generated and displayed on the model.

Engrave

The **Engrave** tool is used to create artistic machining over the workpiece. These toolpath are also used to print text on the press dies. The procedure to use this tool is discussed next.

- Click on the **Engrave** tool of **2D** drop-down from **Toolbar**; refer to Figure-44. The **ENGRAVE** dialog box will be displayed; refer to Figure-45.

Figure-44. Engrave tool

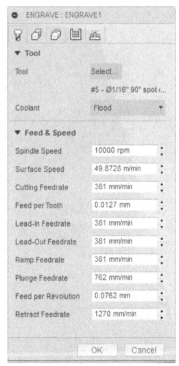

Figure-45. ENGRAVE dialog box

- Click on the **Select** button of **Tool** tab from **ENGRAVE** dialog box. The **Select Tool** dialog box will be displayed. Select the tool as required.
- Click on the **Geometry** tab of **ENGRAVE** dialog box. The **Geometry** tab will be displayed.
- The **Nothing** button of **Contour Selection** from **Geometry** tab is active by default. You need to select the contour for engraving; refer to Figure-46.
- The other tools of the dialog box were discussed earlier.
- After specifying the various parameters, click on the **OK** button from **ENGRAVE** dialog box to complete the process. The toolpath will be generated and displayed on the model; refer to Figure-47.

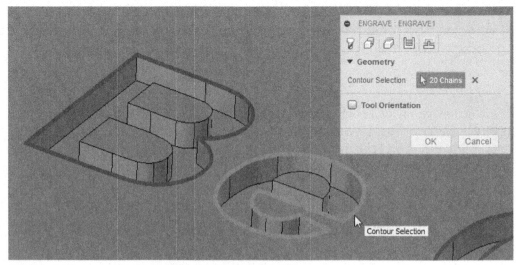

Figure-46. Contour selection for engraving

Figure-47. Generated toolpath for engrave

Chamfer

The **Chamfer** tool is used to create chamfer profile on the edge of model. The procedure to use this tool is discussed next.

- Click on the **Chamfer** tool of **2D** drop-down from **Toolbar**; refer to Figure-48. The **CHAMFER** dialog box will be displayed; refer to Figure-49.
- Click on the **Select** button of **Tool** tab from **2D CHAMFER** dialog box. The **Select Tool** dialog box will be displayed. Select the tool as required.
- Click on the **Geometry** tab of **2D CHAMFER** dialog box. The **Geometry** tab will be displayed.

- The **Nothing** button of **Contour Selection** section from **Geometry** tab is active by default. You need to select the edge for creating chamfer; refer to Figure-50.

Figure-48. 2D Chamfer tool

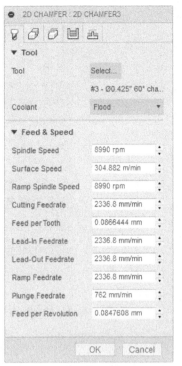

Figure-49. CHAMFER dialog box

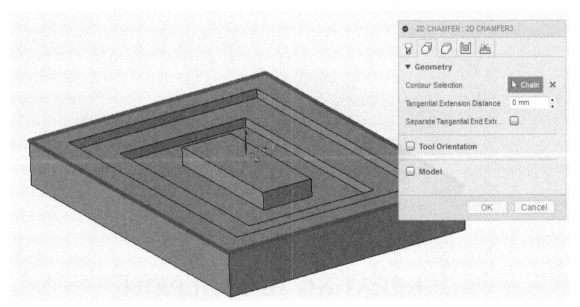

Figure-50. Selection of edge for chamfer

- Click in the **Passes** tab from **2D CHAMFER** dialog box. The options of **Passes** tab will be displayed; refer to Figure-51.
- Click in the **Chamfer Width** edit box of **Chamfer** section from **Passes** tab and enter the value of width of chamfer.
- Click in the **Chamfer Tip Offset** edit box of **Chamfer** section from **Passes** tab and enter the value of tip offset.
- Click in the **Chamfer Clearance** edit box of **Chamfer** section and enter the value of clearance.

- The other tools of the dialog box are same as discussed earlier.
- After specifying the parameters, click on the **OK** button from **2D CHAMFER** dialog box. The toolpath is generated and displayed on the model; refer to Figure-52.

Figure-51. Passes tab 2D CHAMFER dialog box

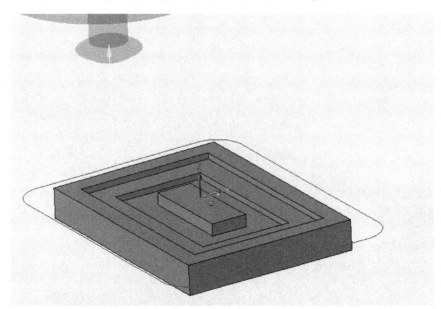

Figure-52. Toolpath generated for chamfer

GENERATING 3D TOOLPATH

Till now, we have discussed the procedure of generating the 2D Toolpaths. In this section, we will discuss the tool used to create 3D Toolpath. 3D toolpaths enable to the simultaneous movement of tool in horizontal and vertical direction.

Adaptive Clearing

The **Adaptive Clearing** tool is used to remove the material in bulk from workpiece. It is generally a roughing process. It uses an advanced strategy of milling motion because of which there is less load on the tool. The procedure to use this tool is discussed next.

- Click on the **Adaptive Clearing** tool of **3D** drop-down from **Toolbar**; refer to Figure-53. The **ADAPTIVE** dialog box will be displayed; refer to Figure-54.

Figure-53. Adaptive tool

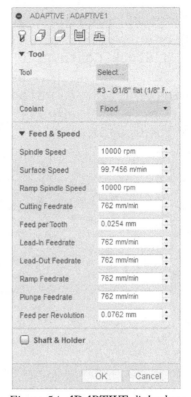

Figure-54. ADAPTIVE dialog box

- Click on the **Select** button of **Tool** tab from **ADAPTIVE** dialog box. The **Select Tool** dialog box will be displayed. Select the tool as required.
- Select the **Shaft & Holder** check box of **Tool** tab from **ADAPTIVE** dialog box to specify how the shaft and holder of the tool is used to avoid collisions with stock or workpiece.
- Click on the **Shaft and Holder Mode** drop-down of **Shaft and Holder** section from **Tool** tab and select the required option.
- Select the **Use Holder** check box of **Shaft and Holder Mode** to specify the software to use holder of the specified tool in the toolpath calculation to avoid collisions.
- Click in the **Holder Clearance** edit box of **Shaft and Holder Mode** and enter the value of tool clearance to keep away the holder with the part.

Geometry

- Click on the **Geometry** tab of **ADAPTIVE** dialog box. The **Geometry** tab will be displayed; refer to Figure-55.
- Select the **None** option of **Machining Boundary** drop-down from **Geometry** tab to machine all the stock without limitation. The **None** option is not available for all machining strategies.

- Select the **Bounding Box** option of **Machining Boundary** drop-down from **Geometry** tab to machine the toolpath within a specified box defined by the maximum extents of the part viewed from the WCS.

- Select the **Silhouette** option of **Machining Boundary** drop-down from **Geometry** tab to machine the toolpath within a defined boundary by the geometry of part or model viewed from WCS.

- Select the **Selection** option of **Machining Boundary** drop-down from **Geometry** tab to machine the toolpath within a specified region by a selected boundary.

- Select the **Stock Contours** check box from **Geometry** tab to select the geometry for machining.

- The **Nothing** button of **Stock Selection** section from **Stock Contours** check box is active by default. You need to select the geometry; refer to Figure-56.

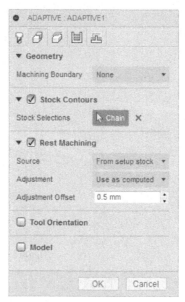

Figure-55. Geometry tab of ADAPTIVE dialog box

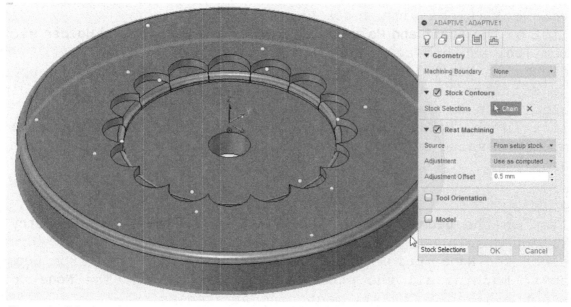

Figure-56. Selection of geometry

- Select the **Rest Machining** check box of **Geometry** tab to limit the operation to remove the material which is not removed by the previous tool.
- Click on the **Source** drop-down of **Rest Machining** check box and select the required option.
- Click on the **Adjustment** drop-down of **Rest Machining** check box and select the required option.
- Click in the **Adjustment Offset** edit box of **Rest Machining** check box and enter the value of stock to be removed, depending on the rest material adjustment setting.

Passes

- Click on the **Passes** tab of **ADAPTIVE** dialog box. The **Passes** tab will be displayed; refer to Figure-57.

Figure-57. Passes tab of ADAP-TIVE dialog box

- Click in the **Tolerance** edit box of **Passes** tab and enter the value of tolerance.
- Select the **Machine Shallow Areas** check box of **Passes** tab to remove the excessive cusps from the shallow areas of Z-level.
- Click in the **Optimal Load** edit box of **Passes** tab and enter the value of engagement which is followed by adaptive strategies.
- Click in the **Minimum Cutting Radius** edit box of **Passes** tab and enter the value of cutting radius.
- Select the **Machine Cavities** check box of **Passes** tab to machine the pockets of the model.

- Select the **Use Slot Clearing** check box of **Passes** tab to start pocket clearing with a slot along its middle, before continuing with a spiral motion towards the pocket walls.
- Select the **Fillets** check box of **Passes** tab to enter a value of fillet radius.
- The other tools of the dialog box were discussed earlier.
- After specifying the parameters, click on the **OK** button of from **ADAPTIVE** dialog box. The toolpath will be generated and displayed on the model; refer to Figure-58.

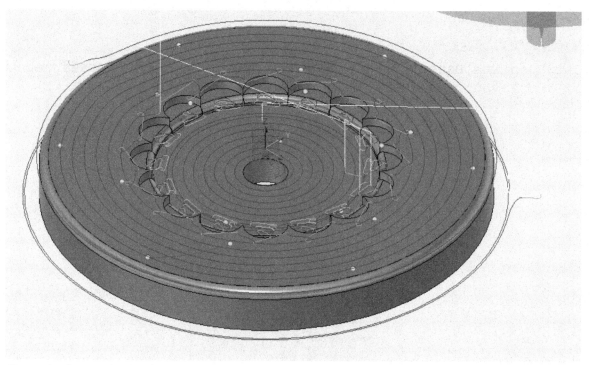

Figure-58. Generated toolpath of adaptive clearing

Pocket Clearing

The **Pocket Clearing** tool is a another tool mainly used for clearing large quantity of material. The procedure to use this tool is discussed next.

- Click on the **Pocket Clearing** tool of **3D** drop-down from **Toolbar**; refer to Figure-59. The **POCKET** dialog box will be displayed; refer to Figure-60
- Click on the **Select** button of **Tool** tab from **POCKET** dialog box. The **Select Tool** dialog box will be displayed. Select the tool as required.
- The options of this dialog box were discussed earlier.
- After specifying the parameters, click on the **OK** button from **POCKET** dialog box. The toolpath will be generated and displayed on the model; refer to Figure-61.

Figure-59. Pocket Clearing tool

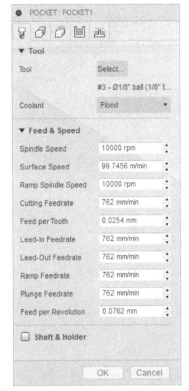

Figure-60. POCKET dialog Box

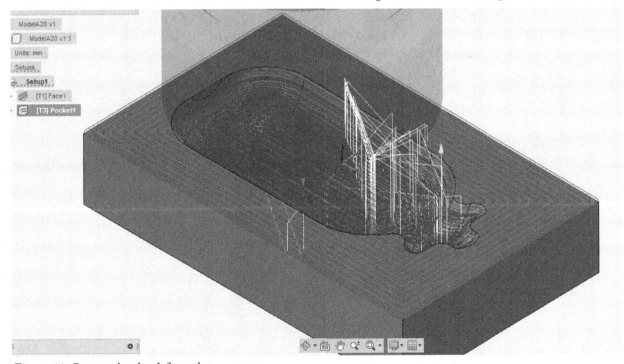

Figure-61. Generated toolpath for pocket

Parallel

The **Parallel** tool is used mainly for finishing strategies. This toolpath is used when parallel cutting passes can machine the workpiece. The procedure to use this tool is discussed next.

* Click on the **Parallel** tool of **3D** drop-down from **Toolbar**; refer to Figure-62. The **PARALLEL** dialog box will be displayed; refer to Figure-63.

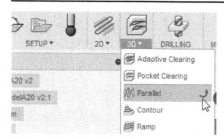

Figure-62. Parallel tool

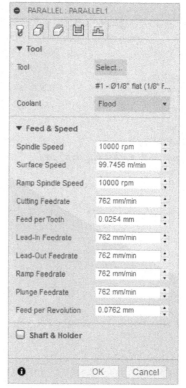

Figure-63. PARALLEL dialog box

- Click on the **Select** button of **Tool** tab from **POCKET** dialog box. The **Select Tool** dialog box will be displayed. Select the tool as required.
- Click on the **Geometry** tab of **PARALLEL** dialog box. The **Geometry** tab will be displayed; refer to Figure-64.

Figure-64. Geometry tab of PARALLEL dialog box

- Select the **Contact Point Boundary** check box of **Geometry** section to specify the boundary limits where the tool touches the surface of model rather than the tool center location.
- Select the **Contact Only** check box of **Geometry** section to not generate the toolpath on the model where the tool is not in contact with the machining surface.
- Select the **Slope** check box of **Geometry** tab to specify the range of slope within which the faces of part will be considered for machining.
- Click in the **From Slope Angle** edit box of **Slope** check box and enter the angular value greater than 0 degree.
- Click in the **To Slope Angle** edit box of **Slope** check box and enter the value less than 90 degree. Only area equal to or less then this values will be machined.
- Select the **Avoid/Touch Surface** check box of **Geometry** tab to avoid or touch the selected surface while machining with a specified distance.
- The **Nothing** button of **Avoid/Touch Surfaces** node is active by default. You need to click on the surface to select.
- Click in the **Avoid/Touch Surface Clearance** edit box of **Avoid/Touch Surface** check box and enter the required value.
- Select the **Touch Surfaces** check box of **Avoid/Touch Surfaces** check box to invert the meaning of **Avoid/Touch Check Surface** check box.
- The other options of the dialog box were discussed earlier.
- After specifying the parameters, click on the **OK** button from **PARALLEL** dialog box. The toolpath will be generated and displayed on the model; refer to Figure-65.

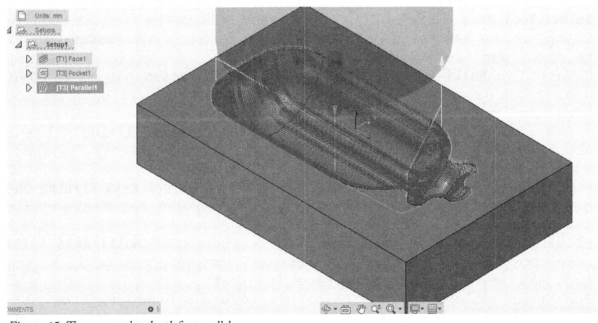

Figure-65. The generated toolpath for parallel

Contour

The **Contour** tool is generally used for finishing step walls. This tool can be used for machining vertical walls. The procedure to use this tool is discussed next.

- Click on the **Contour** tool of **3D** drop-down from **Toolbar**; refer to Figure-66. The **CONTOUR** dialog box will be displayed; refer to Figure-67.

Figure-66. Contour tool

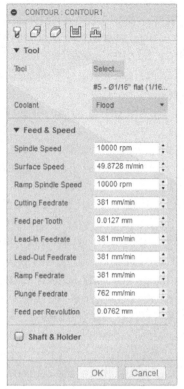

Figure-67. CONTOUR dialog Box

- Click on the **Select** button of **Tool** tab from **POCKET** dialog box. The **Select Tool** dialog box will be displayed. Select the tool as required.
- Click on the **Passes** tab of **CONTOUR** dialog box. The **Passes** tab will be displayed; refer to Figure-68
- Select the **Multi-Axis Tilting** check box of **Passes** tab from **CONTOUR** dialog box to enable multi axis tilt to avoid collision of model with tool holder when using short tools.
- Click in the **Maximum Tilt** edit box of **Multi-Axis Tilting** check box and specify the maximum allowed tilt from the selected operation tool axis.
- Click in the **Maximum Segment Length** edit box of **Multi-Axis Tilting** check box and specify the length of a single segment for the generated toolpath.
- Click in the **Maximum Tool Axis Sweep** edit box of **Multi-Axis Tilting** check box and specify the value of maximum angle change in a single tool axis sweep for the generated toolpath.
- The other options of the dialog box were discussed earlier.
- After specifying the parameters, click on the **OK** button from **CONTOUR** dialog box. The toolpath will be generated and displayed on the model; refer to Figure-69

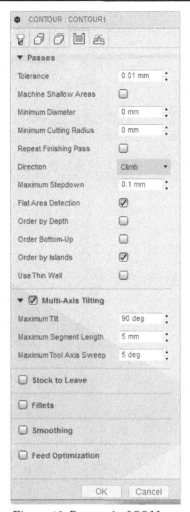

Figure-68. Passes tab of CON-
TOUR dialog box

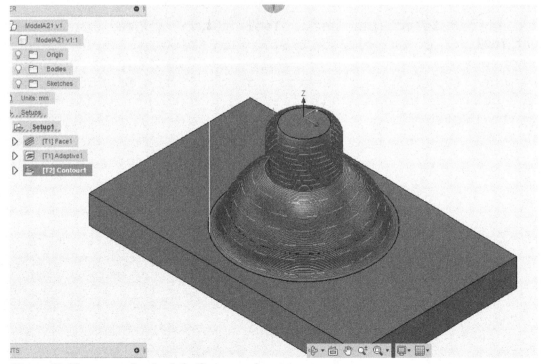

Figure-69. The generated toolpath for Contour tool

Ramp

The **Ramp** tool is used to create a finishing operations meant for step areas similar to Contour tool. The procedure to use this tool is discussed next.

- Click on the **Ramp** tool of **3D** drop-down from **Toolbar**; refer to Figure-70. The **RAMP** dialog box will be displayed; refer to Figure-71.

Figure-70. Ramp tool

Figure-71. RAMP dialog box

- Click on the **Select** button of **Tool** tab from **RAMP** dialog box. The **Select Tool** dialog box will be displayed. Select the tool as required.
- The options of the dialog box were discussed earlier.
- After specifying the parameters, click on the **OK** button from **RAMP** dialog box. The generated toolpath will be displayed on the model; refer to Figure-72.

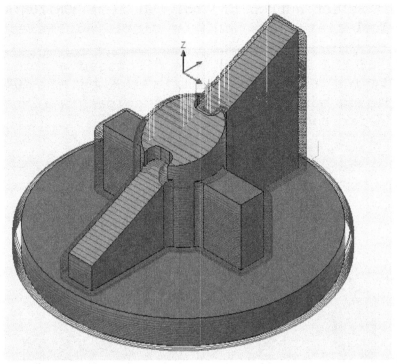

Figure-72. Generated toolpath for Ramp tool

Horizontal

The **Horizontal** tool is used for machining flat surfaces of model which are surrounded by other features. The procedure to use this tool is discussed next.

* Click on the **Horizontal** tool of **3D** drop-down from **Toolbar**; refer to Figure-73. The **HORIZONTAL** dialog box will be displayed; refer to Figure-74

Figure-73. Horizontal tool

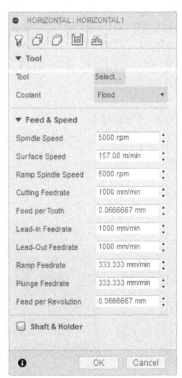

Figure-74. HORIZONTAL dialog box

- Click on the **Select** button of **Tool** tab from **HORIZONTAL** dialog box. The **Select Tool** dialog box will be displayed. Select the tool as required.
- The options of the dialog box were discussed earlier.
- After specifying the parameters, click on the **OK** button from **HORIZONTAL** dialog box. The generated toolpath will be displayed on the model; refer to Figure-75.

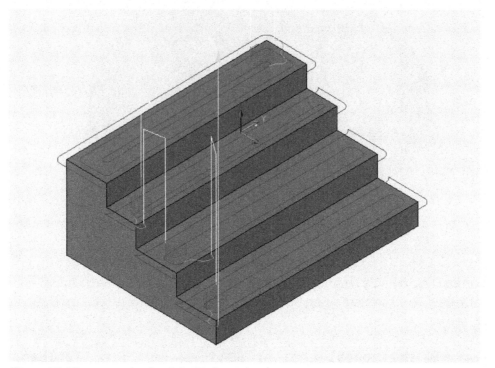

Figure-75. The generated toolpath for Horizontal tool

Pencil

The **Pencil** tool is used to create toolpaths along internal and sharp corners with small radii tool. This tool is generally used to remove material where no other toolpath can work. The procedure to use this tool is discussed next.

- Click on the **Pencil** tool of **3D** drop-down from **Toolbar**; refer to Figure-76. The **PENCIL** dialog box will be displayed; refer to Figure-77

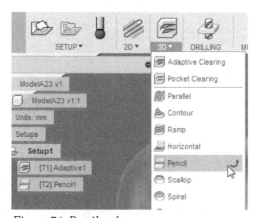

Figure-76. Pencil tool

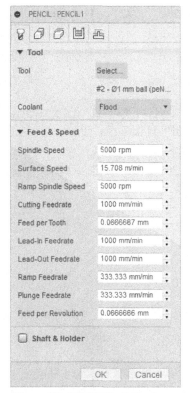

Figure-77. PENCIL dialog box

- Click on the **Select** button of **Tool** tab from **PENCIL** dialog box. The **Select Tool** dialog box will be displayed. Select the tool as required.
- The options of the dialog box were discussed earlier.
- After specifying the parameters, click on the **OK** button from **PENCIL** dialog box. The generated toolpath will be displayed on the model; refer to Figure-78.

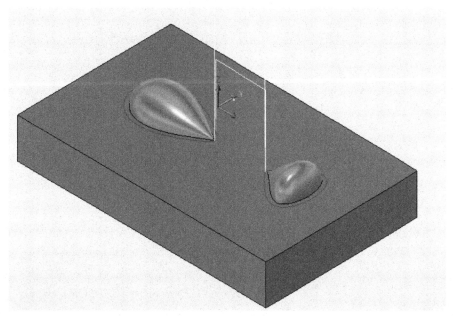

Figure-78. The generated toolpath for Pencil tool

Scallop

The **Scallop** tool is used to finish round faces of the model with slopes. This tool can also be used to machine fillets. The procedure to use this tool is discussed next.

- Click on the **Scallop** tool of **3D** drop-down from **Toolbar**; refer to Figure-79. The **SCALLOP** dialog box will be displayed; refer to Figure-80.

Figure-79. Scallop tool

- Click on the **Select** button of **Tool** tab from **SCALLOP** dialog box. The **Select Tool** dialog box will be displayed. Select the tool as required.
- Click on the **Passes** tab of **SCALLOP** dialog box. The **Passes** tab will be displayed; refer to Figure-81.

Figure-80. SCALLOP dialog box

Figure-81. Passes tab of SCALLOP dialog box

- Select the **Link from Inside to Outside** check box of **SCALLOP** dialog box to specify that linking should be done by ordering from inside passes to outside passes.

- Click on the **Inside/Outside Direction** drop-down of **Passes** tab and select the required option.
- Select the **Limit Number of Stepovers** check box of **Passes** tab to limit the number of steps of the tool so that the tool will not collapse with the surface of model.
- The options of the dialog box were discussed earlier.
- After specifying the parameters, click on the **OK** button from **SCALLOP** dialog box. The generated toolpath will be displayed on the model; refer Figure-82.

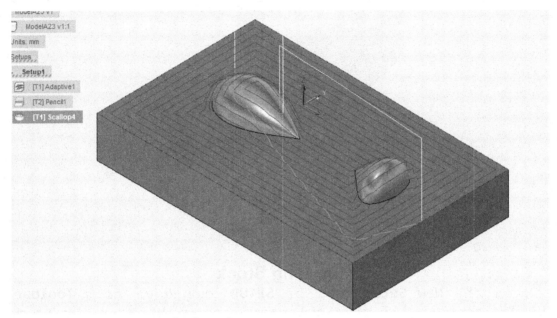

Figure-82. Generated toolpath of scallop tool

PRACTICAL

Generate the toolpath of the given model shown in Figure-83 using 2D tools.

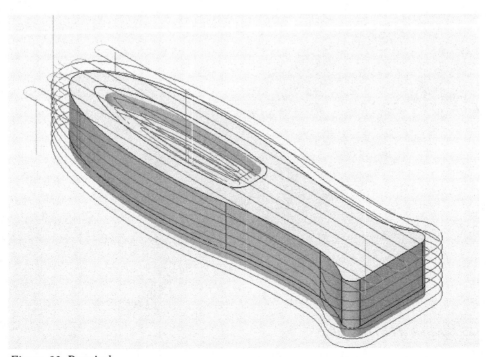

Figure-83. Practical

Adding Model to CAM

- Create and save the part in **MODEL Workspace**. The part file is available in the respective chapter folder of **Autodesk Fusion 360 Resources**.
- Click on the **CAM Workspace** from **Workspace** drop-down. The model will be displayed in the **CAM Workspace**; refer to Figure-84.

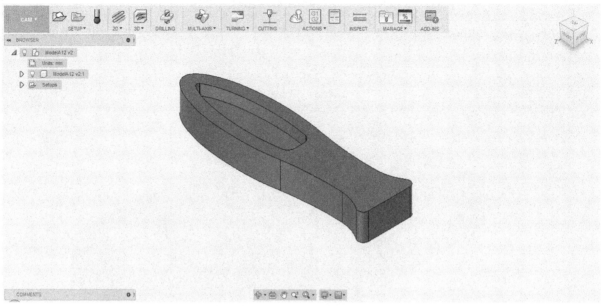

Figure-84. Practical1

Creating Stock

- Click on the **New Setup** tool of **SETUP** drop-down from **Toolbar**. The **SETUP** dialog box will be displayed along with the stock of model.
- Click on the **Setup** tab of **SETUP** dialog box and enter the parameters as displayed in Figure-85.

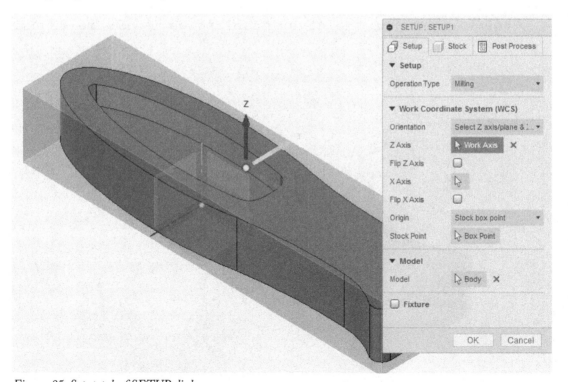

Figure-85. Setup tab of SETUP dialog

- Click on the **Stock** tab of **SETUP** dialog box and enter the parameters as displayed in Figure-86.

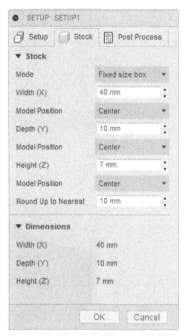

Figure-86. Stock tab of practical

- Click on the **Post Process** tab of **SETUP** dialog box and enter the parameters as displayed in Figure-87.

Figure-87. Post Process tab of SETUP dialog box

- After specifying the parameters, click on the OK button from SETUP dialog box. The stock will be created and displayed on the model.

Generating Face toolpath

- Click on the **Face** tool of **2D** drop-down from **Toolbar**. The **FACE** dialog box will be displayed.
- Click on the **Select** button of **Tool** option from **Tool** tab and select the tool from **Select Tool** dialog box as displayed; refer to Figure-88.
- After selecting the tool, click on the **OK** button from **Select Tool** dialog box. The tool will be selected for machining.
- Specify the parameters of **Tool** tab as displayed in Figure-89.

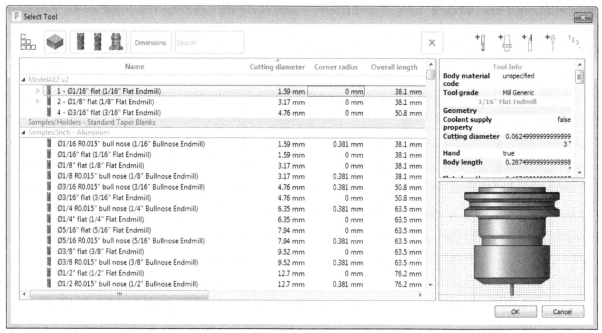

Figure-88. Selecting tool for facing

- Click in the Passes tab of **FACE** dialog box and specify the parameters as displayed in Figure-90.

Figure-89. Tool tab

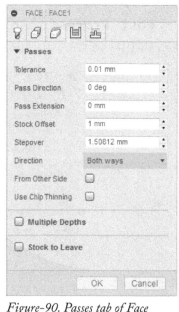

Figure-90. Passes tab of Face dialog box

- Click on the **Linking** tab of **FACE** dialog box and specify the parameters as displayed in Figure-91.

Figure-91. Linking Tab of FACE dialog box

- After specifying the parameters, click on the **OK** button from **FACE** dialog box. The toolpath will be generated and displayed on the model; refer to Figure-92.

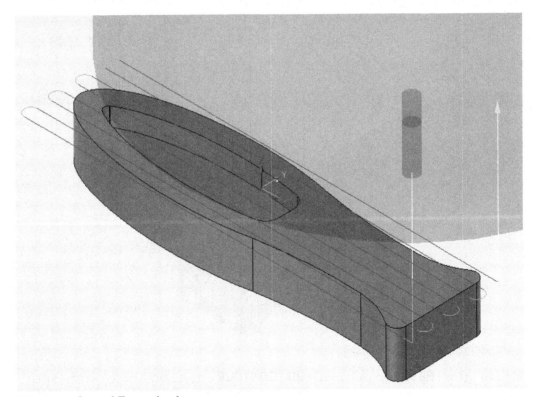

Figure-92. Created Face toolpath

Creating 2D Contour toolpath

- Click on the **2D Contour** tool of **2D** drop-down from **Toolbar**. The **2D CONTOUR** dialog box will be displayed.
- Click on the **Select** button of **Tool** option from **Tool** tab and select the tool as displayed from **Select Tool** dialog box; refer to Figure-93.

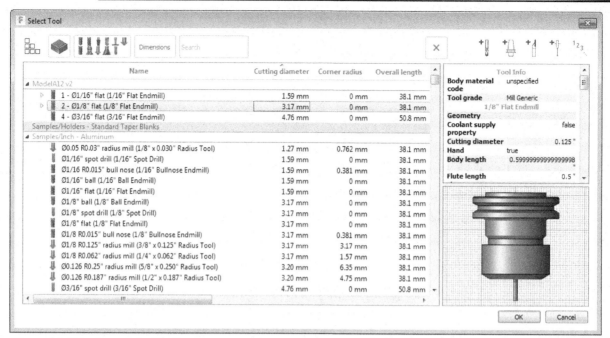

Figure-93. Selecting tool for 2D Contour

- After selecting, click on the **OK** button from **Select Tool** dialog box.
- Click on the **Geometry** tab of **2D CONTOUR** dialog box and specify the parameters as displayed in Figure-94.

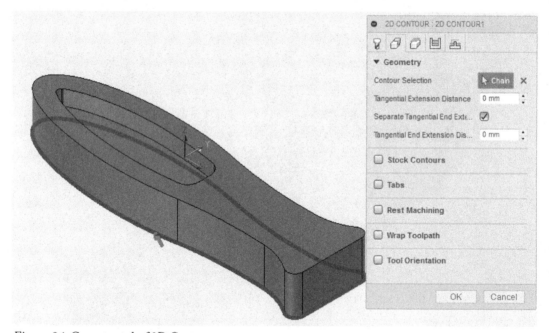

Figure-94. Geometry tab of 2D Contour

- Click in the **Passes** tab of **2D CONTOUR** dialog box and specify the parameters of **Passes** and **Multiple Depths** check box as displayed in Figure-95.
- Click on the **Linking** tab of **2D CONTOUR** dialog box and specify the parameters as displayed in Figure-96.

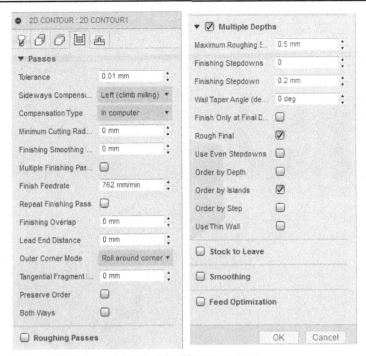

Figure-95. Passes tab of 2D CONTOUR dialog box

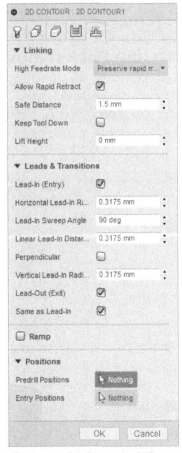

*Figure-96. Linking tab of 2D
CONTOUR dialog box*

- After specifying the parameters, click on the **OK** button from **2D CONTOUR** dialog box. The toolpath will be generated and displayed on the model; refer to Figure-97.

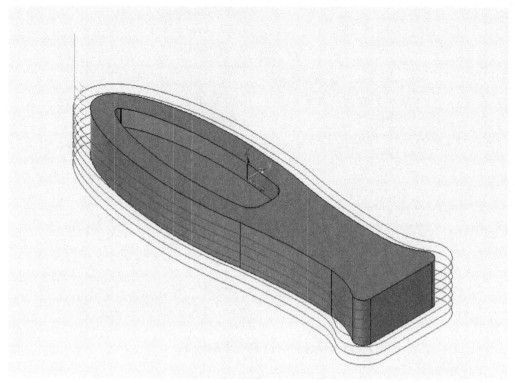

Figure-97. Generated toolpath for 2D Contour tool

Generating Pocket toolpath

- Click on the **2D Pocket** tool of **2D** drop-down from **Toolbar**. The **2D POCKET** dialog box will be displayed.
- Click on the **Select** button of **Tool** option from **Tool** tab and select the tool from **Select Tool** dialog box as displayed; refer to Figure-98.

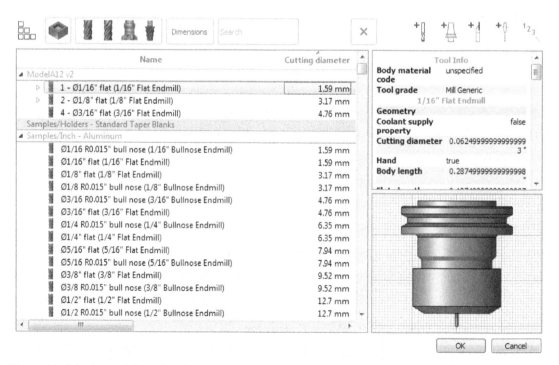

Figure-98. Selecting tool for pocket

- Specify the parameters of **Tools** tab as displayed in Figure-99.

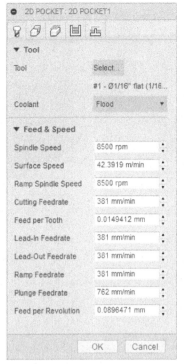

Figure-99. Tools tab of 2D POCKET dialog box

- Click on the **Geometry** tab of **2D POCKET** dialog box and specify the parameters as displayed in Figure-100.

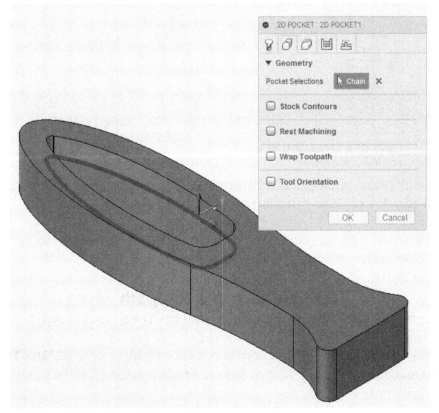

Figure-100. Geometry tab of 2D POCKET dialog box

- Click in the **Passes** tab of **2D POCKET** dialog box and specify the parameters as displayed in Figure-101.

- Click in the **Linking** tab of **2D POCKET** dialog box and specify the parameters as displayed in Figure-102.

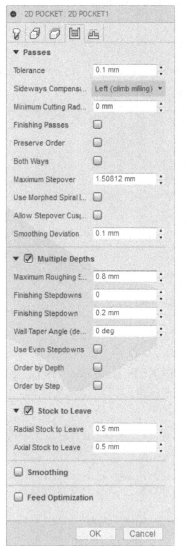

Figure-101. *Passes tab of 2D POCKET dialog box*

Figure-102. *Linking Tab of 2D POCKET dialog box*

- After specifying the parameters, click on the **OK** button from **2D POCKET** dialog box. The toolpath will be generated and displayed on the model; refer to Figure-103.
- After creating the required operations for a model, we need to simulate the generated toolpath.

Simulating the toolpath

- Click on the **Setup** option from **Browser Tree** and right-click on it. The marking menu will be displayed.
- Click on the **Simulate** tool from Marking Menu. The **SIMULATE** dialog box will be displayed along with simulation keys; refer to Figure-104.

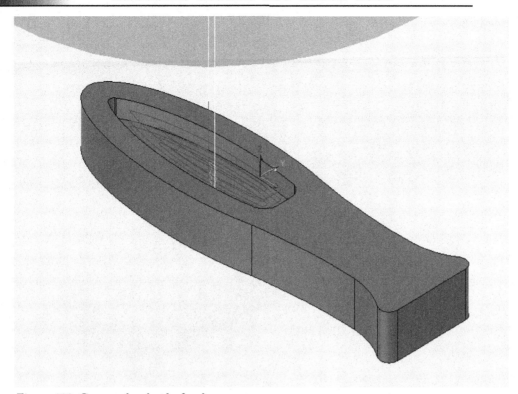

Figure-103. Generated toolpath of pocket

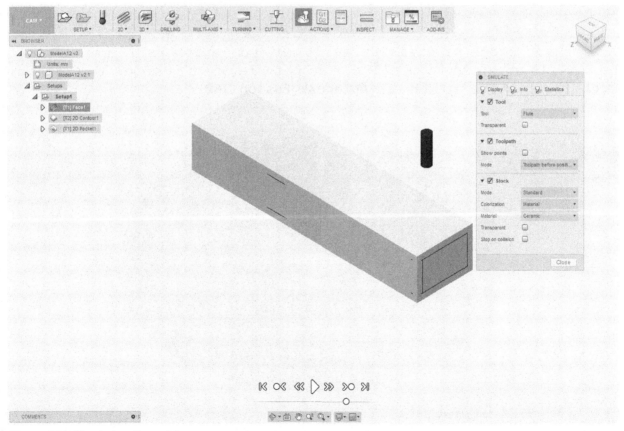

Figure-104. Simulating the model

- Click on the **Play** button to view the simulation process. One by one all generated toolpath will be applied to the model and at last the part will remain along with toolpaths on the model; refer to Figure-105.

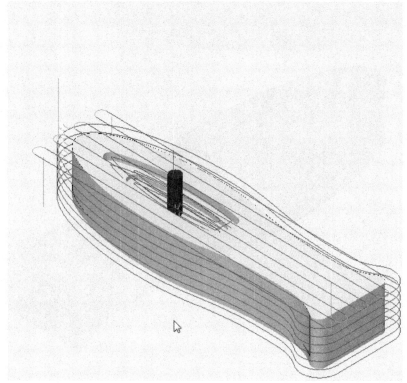

Figure-105. Running simulation process

PRACTICE 1

Machine the stock of Diameter as 55 mm and Length as 12 mm to create the part as shown in Figure-106. The pat file of this model is available in the respective folder of **Autodesk Fusion 360**.

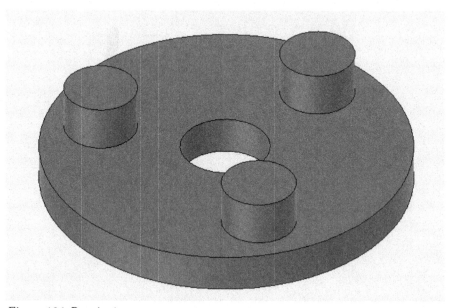

Figure-106. Practice 1

PRACTICE 2

Machine the stock of Stock Diameter as 55 mm and Length 8 mm to create the part as shown in Figure-107.

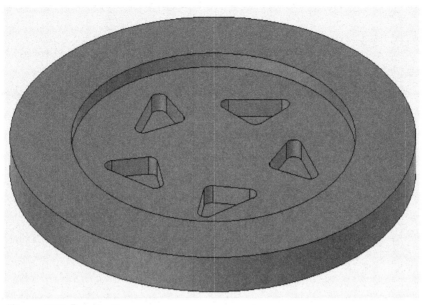

Figure-107. Practice 2

FOR STUDENT NOTES

FOR STUDENT NOTES

Chapter 17

Generating Milling Toolpaths - 2

Topics Covered

The major topics covered in this chapter are:

- *Radial Toolpath*
- *Morphed Spiral Toolpath*
- *Project Toolpath*
- *Swarf Toolpath*
- *Multi-Axis Contour Toolpath*
- *Drilling Toolpath*

3D TOOLPATHS

In previous chapter, we have worked on 2D toolpaths and some 3D toolpaths available in **Toolbar**. In this chapter, we will discuss rest of the 3D toolpaths. Later, we will also work on Multi-axis toolpaths and drilling operations.

Spiral

The Spiral toolpath is used to cut material in spiral fashion. This toolpath is useful when you need to finish rough objects like shown in Figure-1. The procedure to use this tool is discussed next.

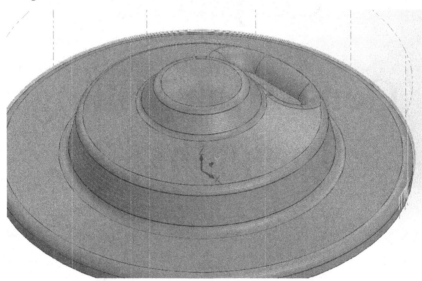

Figure-1. Spiral Toolpath

- Click on the **Spiral** tool of **3D** drop-down from **Toolbar**; refer to Figure-2. The **SPIRAL** dialog box will be displayed; refer to Figure-3.

Figure-2. Spiral tool

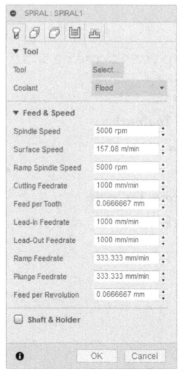

Figure-3. SPIRAL dialog box

- Click on the **Select** button of **Tool** section from **Tool** tab. The Select **Tool** dialog box will be displayed. Select the required tool.
- Click on the **Passes** tab of **SPIRAL** dialog box. The **Passes** tab will be displayed; refer to Figure-4.

Figure-4. Passes tab of SPIRAL dialog box

- Select the **Clockwise** check box of **Passes** tab to set the direction of spiral to clockwise.
- Select the **Spiral** option from **Spiral** drop-down to create a spiral toolpath which starts from the center and ends at the outermost boundary.
- Select the **Spiral with Circles** option from **Spiral** drop-down to create a circular toolpath at the minimum and maximum radius. This toolpath will only be created if the minimum radius is larger than zero and maximum radius is smaller than the radius of regulation boundary.
- Select the **Concentric Circles** option from **Spiral** drop-down to create a concentric circle toolpath.
- Click in the **Inner Limit** edit box of **Passes** tab to set the minimum inner radius.
- Click in the **Outer radius** edit box of **Passes** tab to set the maximum outer radius.
- Click in the **Stepover** edit box of **Passes** tab and enter the value.
- The other options of the dialog box are same as discussed earlier.
- After specifying the parameters, click on the **OK** button from **SPIRAL** dialog box. The toolpath will be generated and displayed on the model; refer to Figure-5

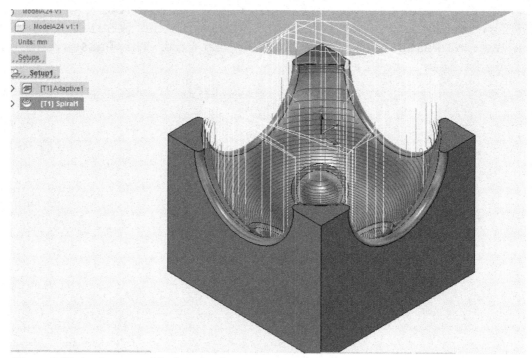

Figure-5. Toolpath generated for Spiral tool

Radial

The **Radial** tool is used to create toolpath along the radii of an arc. The toolpath created by radial is similar to the spokes of wheel which are then projected down on the surface. The procedure to use this tool is discussed next.

- Click on the **Radial** tool of **3D** drop-down from **Toolbar**; refer to Figure-6. The **RADIAL** dialog box will be displayed; refer to Figure-7.
- Click on the **Select** button of **Tool** section from **Tool** tab. The **Select Tool** dialog box will be displayed. Select the required tool.
- Click on the **Passes** tab of **RADIAL** dialog box. The **Passes** tab will be displayed; refer to Figure-8.
- Click in the **Angular Step** edit box of **Passes** tab and enter the angular value of step between radial passes.
- Click in the **Angle From** edit box of **Passes** tab and enter the value of radial starting angle measured from the X-axis.
- Click in the Angle To edit box of Passes tab and enter the value of radial ending value measured from the X-axis.
- The other options of the dialog box were discussed earlier.
- After specifying the parameters click on the **OK** button from **RADIAL** dialog box. The toolpath will be generated and displayed on the model; refer to Figure-9.

Figure-6. Radial tool

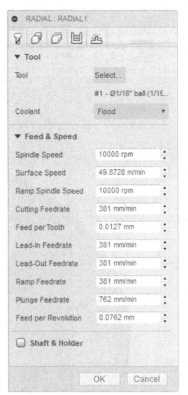

Figure-7. RADIAL dialog box

Figure-8. Passes tab of RA-
DIAL dialog box

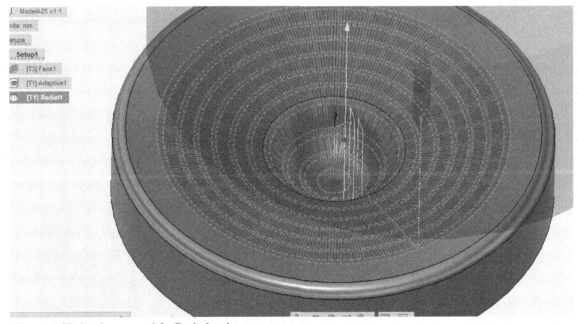

Figure-9. Toolpath generated for Radial tool

Morphed Spiral

The **Morphed Spiral** tool is mainly used to create toolpath of free-form surfaces. The procedure to use this tool is discussed next.

- Click on the **Morphed Spiral** tool of **3D** drop-down from **Toolbar**; refer to Figure-10. The **MORPHED SPIRAL** dialog box will be displayed; refer to Figure-11.

Figure-10. Morphed Spiral tool

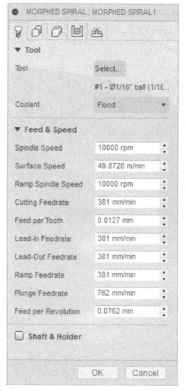

Figure-11. MORPHED SPIRAL
dialog box

- Click on the **Select** button of **Tool** section from **Tool** tab. The **Select Tool** dialog box will be displayed. Select the required tool.
- The options of the dialog box were discussed earlier.
- After specifying the parameters, click on the **OK** button from **MORPHED Spiral** dialog box. The toolpath will be created and displayed on the model; refer to Figure-12.

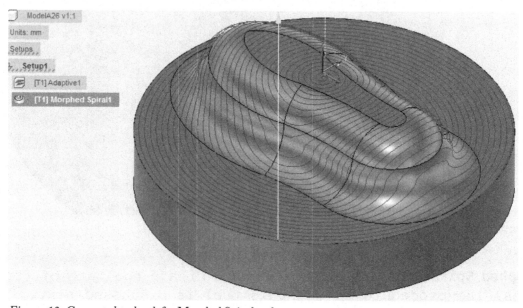

Figure-12. Generated toolpath for Morphed Spiral tool

Project

The **Project** tool is used to create a toolpath along the selected contour. The contour will be machined with the center of the tool. The procedure to use this tool is discussed next.

- Click on the **Project** tool of **3D** drop-down from **Toolbar**; refer to Figure-13. The **PROJECT** dialog box will be displayed; refer to Figure-14.

Figure-13. Project tool

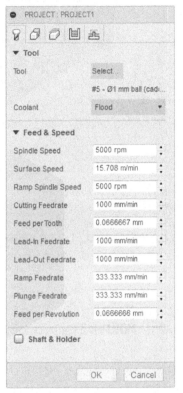

Figure-14. PROJECT dialog box

- Click on the **Select** button of **Tool** section from **Tool** tab. The **Select Tool** dialog box will be displayed. Select the required tool.
- Click on the required curve from model.
- The options of the dialog box were discussed earlier.
- After specifying the parameter, click on the **OK** button from **PROJECT** dialog box. The toolpath will be generated and displayed on the model; refer to Figure-15.

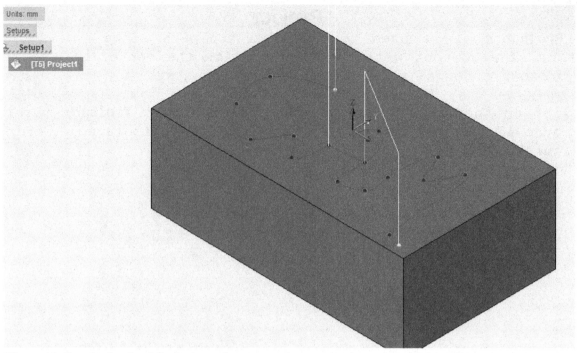

Figure-15. Generated toolpath for Project tool

Morph

The **Morph** tool is used to create the toolpath for machining shallow areas between selected contour. The procedure to use this tool is discussed next.

• Click on the **Morph** tool of **3D** drop-down from **Toolbar**; refer to Figure-16. The **MORPH** dialog box will be displayed; refer to Figure-17.

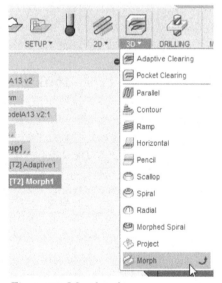

Figure-16. Morph tool

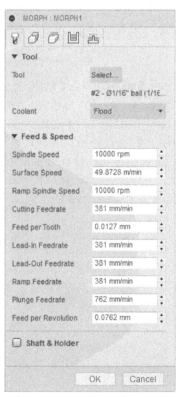

Figure-17. MORPH dialog box

- Click on the **Select** button of **Tool** section from **Tool** tab. The **Select Tool** dialog box will be displayed. Select the required tool.
- Select the curve as required from model; refer to Figure-18.

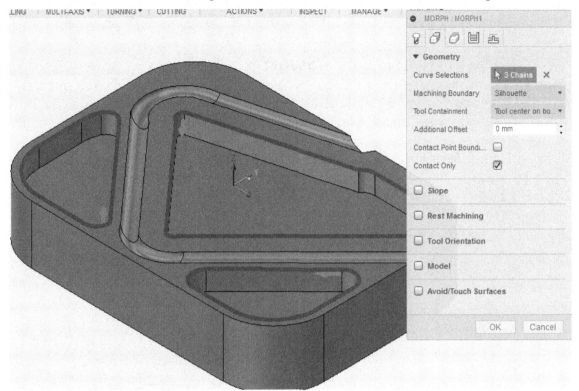

Figure-18. Selection of curve for Morph tool

- The options of the dialog box were discussed earlier.
- After specifying the various parameters, click on the **OK** button from **MORPH** dialog box. The toolpath will be generated and displayed on the model; refer to Figure-19.

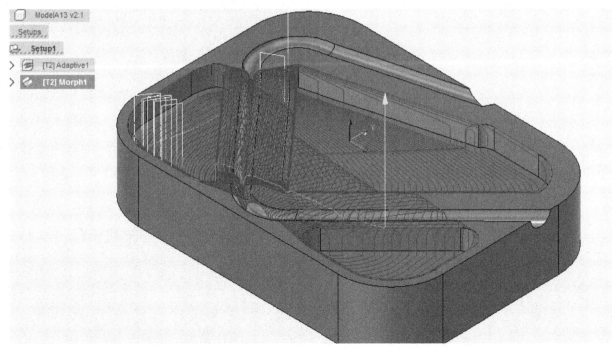

Figure-19. Toolpath generated for Morph tool

MULTI-AXIS TOOLPATH

Till now we have discussed the procedure of creating 2D and 3D toolpath. In this section we will discuss the procedure of creating the multi-axis toolpaths which are used to machine complex 3D shapes.

Swarf

The **Swarf** is a multi-axis tool which is used for side cutting of model. It is used for milling the beveled edges and tapered walls. The procedure to use this tool is discussed next.

- Click on the **Swarf** tool of **MULTI-AXIS** drop-down from **Toolbar**; refer to Figure-20. The **SWARF** dialog box will be displayed; refer to Figure-21.

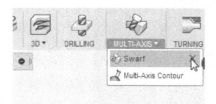

Figure-20. Swarf tool

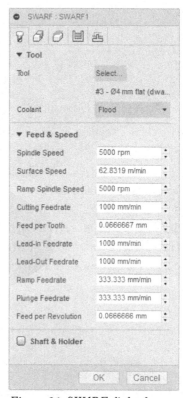

Figure-21. SWARF dialog box

- Click on the **Select** button of **Tool** section from **Tool** tab. The Select **Tool** dialog box will be displayed. Select the required tool.

Geometry

- Click on the **Geometry** tab of **SWARF** dialog box. The **Geometry** tab will be displayed; refer to Figure-22.

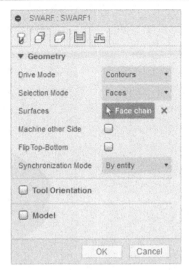

*Figure-22. Geometry tab of
SWARF dialog box*

- Select the **Contours** option from **Drive Mode** drop-down to select the contour from model.
- Select the **Surface** option from **Drive Mode** drop-down to select the faces from model for generation of toolpath.
- Select **Faces** option from **Selection Mode** drop-down to select continuous faces from model for selection.
- Select **Contour Pairs** option from **Selection Mode** drop-down to select the upper and lower edges chain from the tapered face which will drive the swarf toolpath.
- Select **Manual** option from **Selection Mode** drop-down to select the individual surfaces from the model.
- The **Nothing** button of **Surfaces** section is active by default. You need to select the face, edges, or contour.
- Select the **Machine other Side** check box of **Geometry** section to force the toolpath to machine to the other side of the selected contour.
- Select the **Flip Top Bottom** check box of **Geometry** section to change the direction of tool by switching the lower and upper contour. This option will only be applied when contours are selected.
- Click on the **Synchronization Mode** drop-down from **Geometry** tab and select the required option.

Passes

Click on the **Passes** tab of **SWARF** dialog box. The options of **Passes** tab will be displayed; refer to Figure-23.

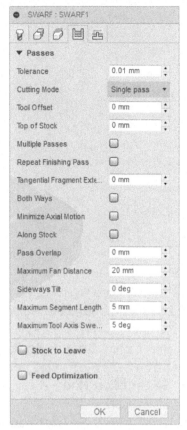

Figure-23. Passes tab of SWARF dialog box

- Click on the **Cutting Mode** drop-down of **Passes** tab and select the required option to specify how the tool will cut vertically.
- Click in the **Tool Offset** edit box of **Passes** tab and enter the value of tool offset along the tool axis relative to the bottom guide curve.
- Click in the **Top** of **Stock** edit box of **Passes** tab to specify the overall thickness of the stock.
- Select the **Repeat Finishing Passes** check box of **Passes** tab to perform an additional finishing pass with zero stock.
- Select the **Tangential Fragment Extension Distance** edit box of **Passes** tab to extends the cut tangentially at both ends of the pass.
- Select the **Minimize axial Motion** check box of **Passes** tab to minimize the motion of tool.
- Select the **Along Stock** check box of **Passes** tab to move the tool along stock for machining.
- Click in the **Pass Overlap** edit box of **Passes** tab and enter the value of distance to extend machining for a closed pass.
- Click on the **Maximum Fan Distance** edit box of **Passes** tab and specify the maximum distance over which to fan the tool axis.
- Click on the **Sideways tilt** edit box of **Passes** tab and specify the values in degrees the tool should be tilted sideways.
- Click on the **Maximum Segment Length** edit box of **Passes** tab and specify the value of maximum length for a single segment for the generated for toolpath.
- Click on the **Maximum Tool Axis Sweep** edit box of **Passes** tab and specify the maximum angle up to which tool can deviate about its axis while cutting.

- The other options of the dialog box were discussed earlier.
- After specifying the parameter, click on the **OK** button from **SWARF** dialog box. The toolpath will be generated and displayed on the model; refer to Figure-24.

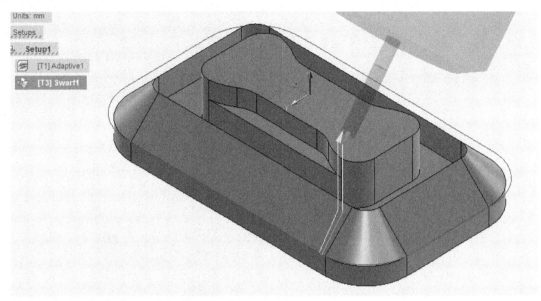

Figure-24. Toolpath generated for Swarf tool

Multi-Axis Contour

The **Multi-Axis Contour** tool is used for machining curves with the help of 5-Axis machining. These curves lie on the face of a model forming different 3D curvatures. The procedure to use this tool is discussed next.

- Click on the **Multi-Axis Contour** tool of **Multi-Axis** drop-down from **Toolbar**; refer to Figure-25. The **MULTI-AXIS CONTOUR** dialog box will be displayed; refer to Figure-26.

Figure-25. Multi-Axis Contour tool

- Click on the **Select** button of **Tool** section from **Tool** tab. The Select **Tool** dialog box will be displayed. Select the required tool.
- Click on the curve from the model to select for machining. You can also select more than 1 curves for machining.
- Click on the **Passes** tab of **MULTI-AXIS CONTOUR** dialog box. The **Passes** tab will be displayed; refer to Figure-27.

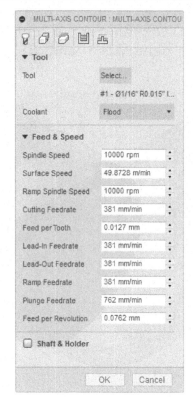

Figure-26. MULTI-AXIS CON-
TOUR dialog box

Figure-27. Passes tab of MULTI-
AXIS CPNTOUR dialog box

- Click on the **Cutting Mode** drop-down from **MULTI-AXIS CONTOUR** dialog box and select the required option to specify the method for machining along a specific contact path.
- Click on the **Sideways Compensation** drop-down and select the required option.
- Click on the **Forward Tilt** edit box of **MULTI-AXIS CONTOUR** dialog box and enter the angular value for tool should be tilted forward.
- Click in the **Minimum Tilt** edit box of **MULTI-AXIS CONTOUR** dialog box and specify the minimum value if allowed tilt from the selected operation tool axis.
- Click in the **Maximum Tilt** edit box of **MULTI-AXIS CONTOUR** dialog box and specify the minimum value if allowed tilt from the selected operation tool axis.
- The other options of the dialog box were discussed earlier.
- After specifying the parameter, click on the **OK** button from **MULTI-AXIS CONTOUR** dialog box. The toolpath will be generated and displayed on the model; refer to Figure-28.

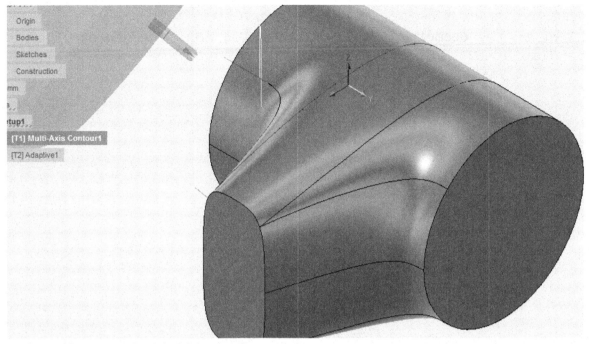

Figure-28. Generated toolpath for Multi-Axis Contour tool

CREATING DRILLING TOOLPATH

The **Drilling** tool is used to perform drilling at the specified locations. The procedure to use this tool is discussed next.

• Click on the **Drilling** tool from **Toolbar**; refer to Figure-29. The **DRILL** dialog box will be displayed; refer to Figure-30.

Figure-29. Drilling tool

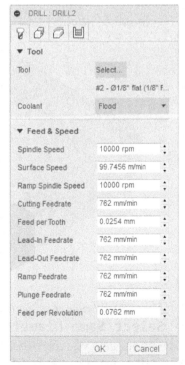

Figure-30. DRILL dialog box

• Click on the **Select** button of **Tool** section from **Tool** tab. The Select **Tool** dialog box will be displayed. Select the required tool for drilling.

Geometry

- Click on the **Geometry** tab of **DRILL** dialog box. The **Geometry** tab will be displayed; refer to Figure-31.

Figure-31. Geometry tab of DRILL dialog box

- Click on the **Selected Faces** option from **Hole Mode** drop-down to select the faces of hole from model for drilling.
- Click on the **Selected points** option from **Hole Mode** drop-down to select the points of holes for drilling.
- The **Nothing** button of **Hole Faces** option from **Geometry** tab is active by default. You need to click on the geometry from model; refer to Figure-32.

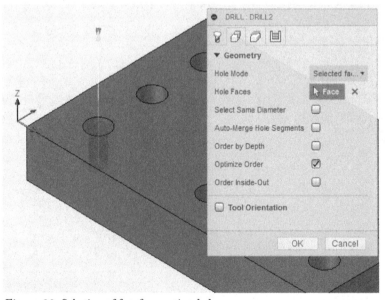

Figure-32. Selection of face for creating hole

- Select the **Select Same Diameter** check box of **Geometry** tab to select the hole of similar diameter.
- Select the **Auto-Merge Hole Segments** check box of **Geometry** tab to include the neighboring segments automatically while drilling a hole with multiple segments.

- Select the **Optimize Order** check box of **Geometry** tab to minimize the ordering of hole by ordering the holes.
- Select the **Order Inside-Out** check box of **Geometry** tab to rearranging the order of holes by machining the inner holes first and outer holes later.

Cycle

- Click on the **Cycle** tab of **DRILL** dialog box. The **Cycle** tab will be displayed; refer to Figure-33.

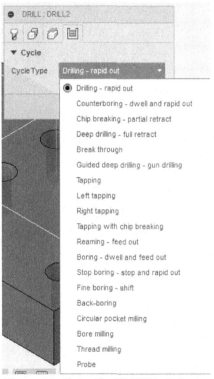

Figure-33. Cycle tab of DRILL dialog box

- Click on the **Cycle Type** drop-down of **Cycle** tab and select the required option.
- The other options of the dialog box were discussed earlier.
- After specifying the parameters, click on the **OK** button from **DRILL** dialog box. The toolpath will be generated and displayed on the model; refer to Figure-34.

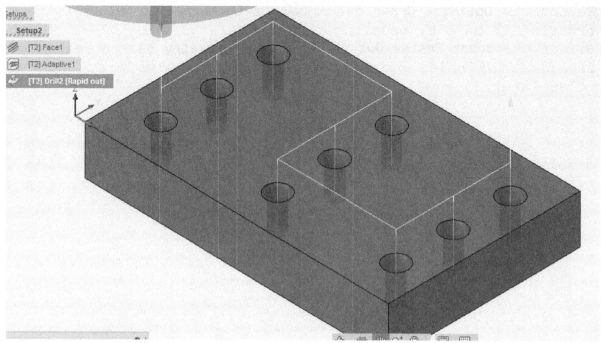

Figure-34. Toolpath generated for drill

PRACTICAL

Create the toolpath of the given model using 3D and 2D tools; refer to Figure-35.

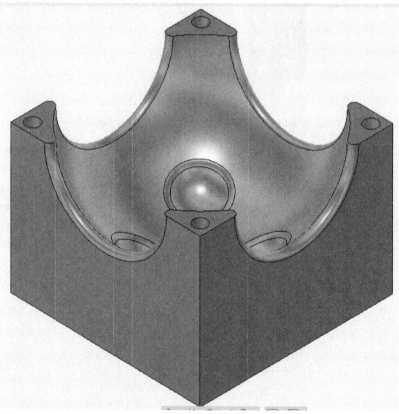

Figure-35. Practical 1

Adding Model to CAM

- Firstly, create and save the part in **MODEL Workspace**. The part file is available in the respective chapter folder of **Autodesk Fusion 360** folder.
- Click on the **CAM Workspace** from **Workspace** drop-down. The model will be displayed in the **CAM Workspace**.

Creating Stock

- Click on the **New Setup** tool of **SETUP** drop-down from **Toolbar**. The **SETUP** dialog box will be displayed along with the stock of model.
- Specify the parameters of **Setup** tab, **Stock** tab and **Post Process** tab of **SETUP** dialog box as shown in Figure-36.

Figure-36. Specifying parameters for stock

- After specifying the parameters, click on the **OK** button from **SETUP** dialog box. The stock will be created and displayed on the model.

Generating Adaptive Clearing Toolpath

- Click on the **Adaptive Clearing** tool of **3D** drop-down from **Toolbar**. The **ADAPTIVE** dialog box will be displayed.
- Click on the **Select** button of **Tool** tab and select the tool from the record list of **Select Tool** dialog box as shown in Figure-37.

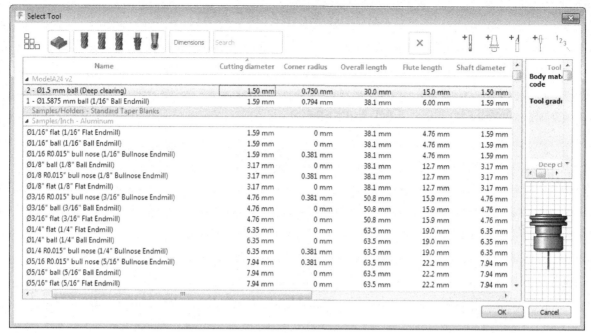

Figure-37. Selecting tool for adaptive clearing

- If the tool is not available in the list then create a new ball mill tool. After selecting the tool, click on the **OK** button from **Select Tool** dialog box. The tool will be added in the **ADAPTIVE** dialog box.
- Specify the parameters of **Geometry** tab, **Passes** tab, and **Linking** tab of **ADAPTIVE** dialog box; refer to Figure-38.

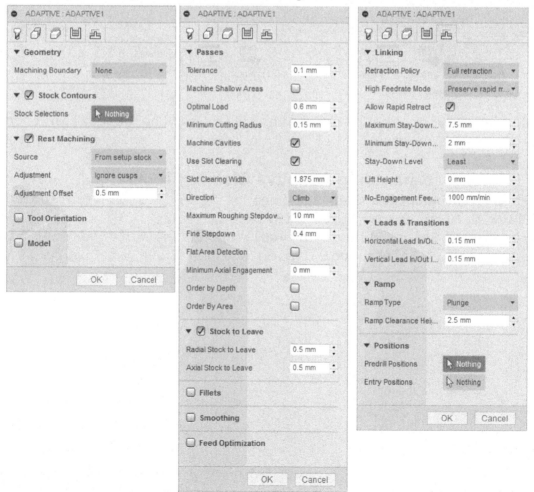

Figure-38. Parameters of Adaptive dialog box

- After specifying the parameters, click on the **OK** button from **ADAPTIVE** dialog box. The toolpath will be generated and displayed on the model; refer to Figure-39.

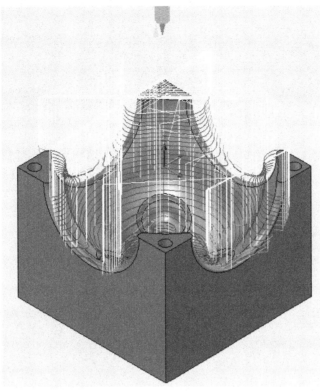

Figure-39. Toolpath of adaptive clearing

Generating Spiral Toolpath

- Click on the **Spiral** tool of **3D** drop-down from **Toolbar**. The **SPIRAL** dialog box will be displayed.
- Click on the **Select** button of **Tool** tab and select the tool from the record list of **Select Tool** dialog box as shown in Figure-40.

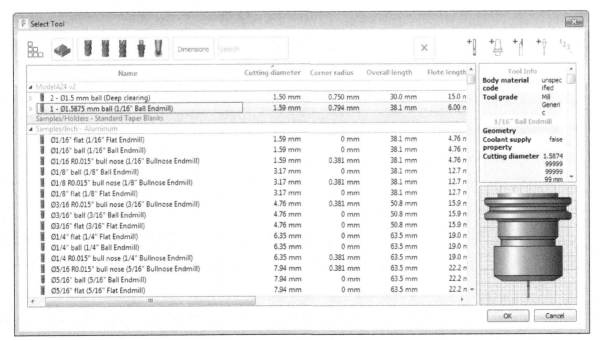

Figure-40. Selecting tool for spiral

- Specify the parameters of **Geometry** tab, **Passes** tab, and **Linking** tab of **SPIRAL** dialog box; refer to Figure-41.

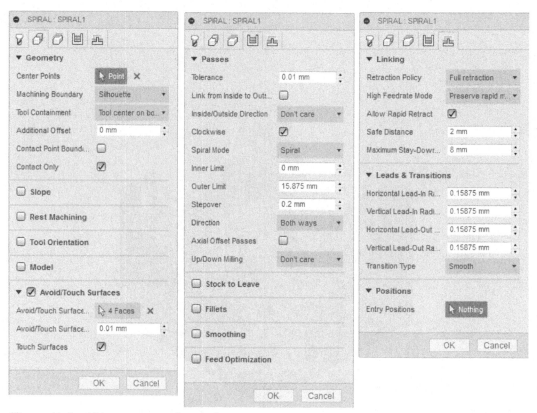

Figure-41. Specifying parameters for spiral

Generating Scallop Toolpath

- Click on the **Scallop** tool of **3D** drop-down from **Toolbar**. The **SCALLOP** dialog box will be displayed.
- Click on the **Select** button of **Tool** section and select the tool which is selected in last operation from **Select Tool** dialog box.
- Specify the parameters of **Tool** tab, **Geometry** tab, **Passes** tab, and **Linking** tab as shown in Figure-42.

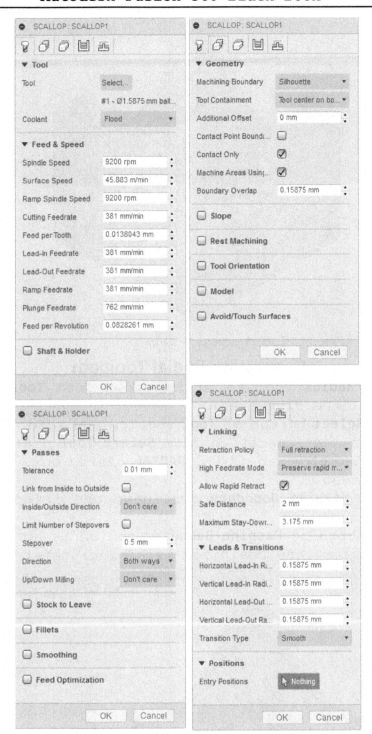

Figure-42. Parameters for Scallop tool

- After specifying the parameters, click on the **OK** button from **SCALLOP** dialog box. The toolpath will be created and displayed on the model; refer to Figure-43.

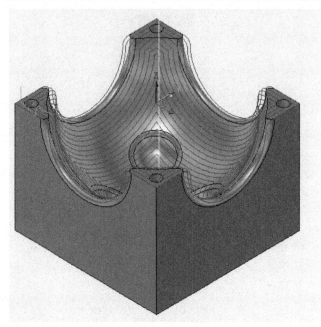

Figure-43. Toolpath generated for scallop

Generating Radial Toolpath

- Click on the **Radial** tool of **3D** drop-down from **Toolbar**. The **RADIAL** dialog box will be displayed.
- Click on the **Select** button of **Tool** section and select the tool which is selected in last operation from **Select Tool** dialog box.
- Click on the **Geometry** tab of the **RADIAL** dialog box. The **Geometry** tab will be displayed.
- Click on the **Machine Boundary** drop-down and select the **Selection** option.
- The **Nothing** button of **Machine Boundary** option is active by default. You need to select the chain for machining; refer to Figure-44.

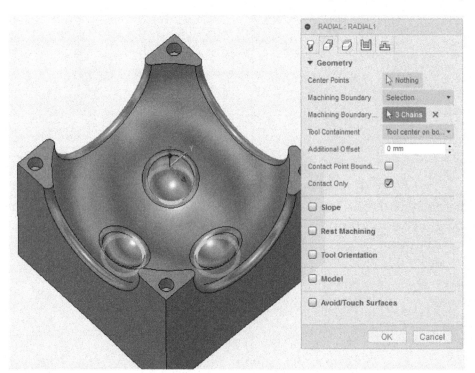

Figure-44. Selecting chains for machining

- Specify the parameters of **Passes** tab and **Geometry** tab as displayed in Figure-45.

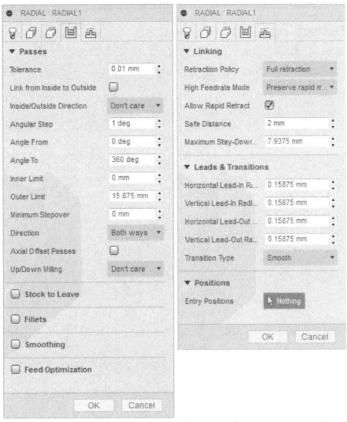

Figure-45. Parameters for Radial dialog box

- After specifying the parameters, click on the **OK** button from **RADIAL** dialog box. The toolpath will be created and displayed on the model; refer to Figure-46.

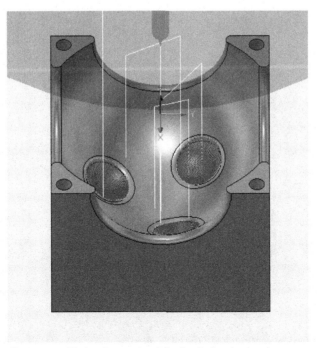

Figure-46. Toolpath generated for Radial tool

Generating Drill Toolpath

- Click on the **Drill** tool from **Toolbar**. The **DRILL** dialog box will be displayed.
- Click on the **Select** button of **Tool** section and select the tool from **Select Tool** dialog box as displayed in Figure-47.

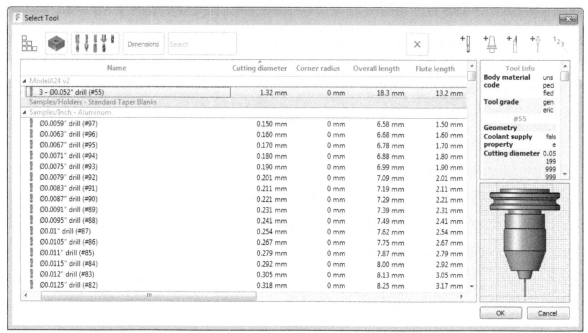

Figure-47. Selection of tool for drill

- After selecting the tool for drill, click on the **OK** button from **Select Tool** dialog box. The tool will we added in **DRILL** dialog box.
- Specify the parameters of Geometry tab from **DRILL** dialog box as displayed in Figure-48.

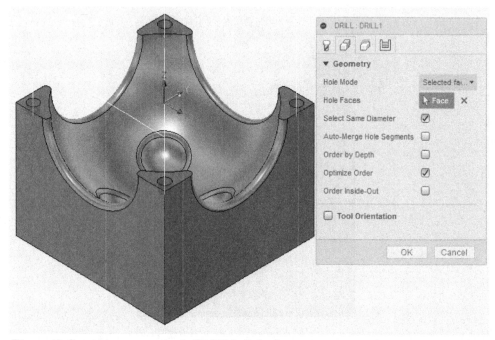

Figure-48. Sspecifying geometry tab of DRILL dialog box

- After specifying the parameters, click on the **OK** button from **DRILL** dialog box. The toolpath will be generated and displayed on the model; refer to Figure-49.

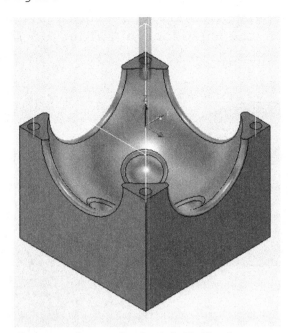

Figure-49. Generated toolpath for Drill tool

- Till now, we have applied the tools to create the required toolpath. Now, we will simulate the whole process to check whether the generated toolpath is working properly or the collision of tool with stock.
- Click on the **Simulate** button from Marking menu of created Setup. The **SETUP** dialog box will be displayed.
- Specify the parameters of **Simulate** dialog box as required and click on the **Play** button. The machining process will be started; refer to Figure-50.

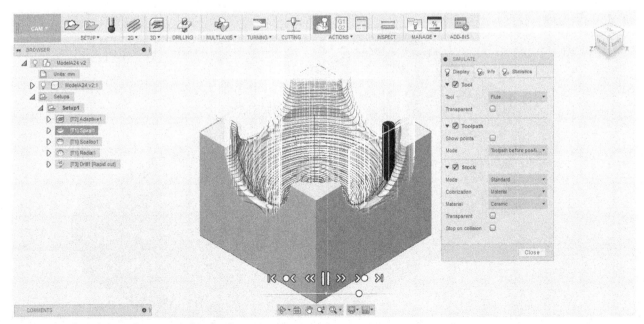

Figure-50. Simulating the generated toolpath

PRACTICE 1

Machine the stock of Width as 146.79 mm, Depth as 52 mm, Height as 21 mm to create the part as shown in Figure-51. The pat file of this model is available in the respective folder of **Autodesk Fusion 360**.

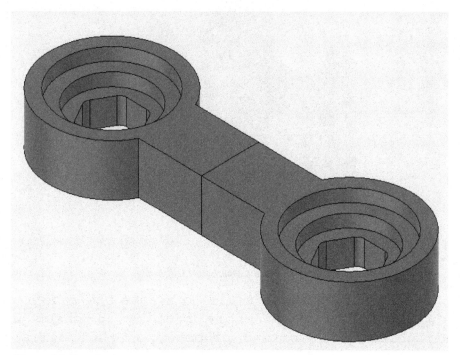

Figure-51. Practice 1

PRACTICE 2

Machine the stock of Width as 102 mm, Depth as 62 mm, Height as 23 mm to create the part as shown in Figure-52. The pat file of this model is available in the respective folder of **Autodesk Fusion 360**.

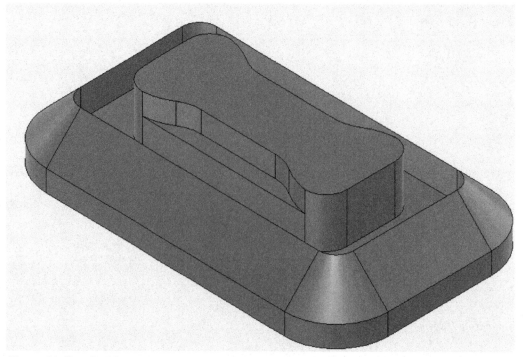

Figure-52. Practice 2

PRACTICE 3

Machine the stock of Stock Diameter as 80 mm, Length as 07 mm and stock align to the center to create the part as shown in Figure-53. The pat file of this model is available in the respective folder of **Autodesk Fusion 360**.

Figure-53. Practice 3

FOR STUDENT NOTES

FOR STUDENT NOTES

Chapter 18

Generating Turning and Cutting Toolpaths

Topics Covered

The major topics covered in this chapter are:

- *Turning Profile*
- *Turning Face*
- *Turning Chamfer*
- *Turning Thread*
- *Turning Groove*
- *Cutting Toolpaths*

GENERATING TURNING TOOLPATH

Till now we have discussed the procedure of creating milling toolpaths with the help of various tools. In this section we will discuss the tools used for creating the Turning toolpath.

Turning Profile

The **Turning Profile** tool is used for finishing and roughing the model with the use of various tools. The procedure to use this tool is discussed next.

- Click on the **Turning Profile** tool of **TURNING** drop-down from **Toolbar**; refer to Figure-1. The **PROFILE** dialog box will be displayed; refer to Figure-2.

Figure-1. Turning Profile tool

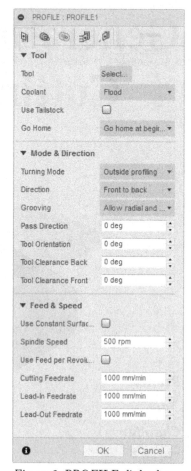

Figure-2. PROFILE dialog box

- Click on the **Select** button of **Tool** option from **Tool** tab. The **Select Tool** dialog box will be displayed. Select the required tool for profiling and click on the **OK** button. The tool will be selected and displayed on the **PROFILE** dialog box.

Tool

- Click on the **Coolant** drop-down of **Tool** tab and select the required option.

- Select the **Use Tailstock** check box of **Tool** tab to apply the support to the part which is to be machined.
- Click on the **Go Home** drop-down of **Tool** tab and select the required option to set the home position of tool.
- Click on the **Outside profiling** option of **Turning Mode** drop-down from **Mode & Direction** section to retract the tool outside of the stock and machine axially depending on the direction setting.
- Click on the **Face Profiling** option of **Turning Mode** drop-down from **Mode & Direction** section to approach the tool from front and machines radially depending on the direction setting.
- Click on the **Inside Profiling** option of **Turning Mode** drop-down from **Mode & Direction** section to approach or retract the tool from the center-line and machines radially depending on the direction setting.
- Click on the **Front to back** option of **Direction** drop-down from **Mode & Direction** section to cut from the front side of the stock towards the back side.
- Click on the **Back to Front** option of **Direction** drop-down from **Mode & Direction** section to cut the material of stock from back side towards the front side.
- Click on the **Both Ways** option of **Direction** drop-down from **Mode & Direction** section to cut the stock from both direction.
- Click on the **Grooving** drop-down from **Mode & Direction** section and select the required option for grooving.
- Click in the **Pass Direction** edit box of **Mode & Direction** section and enter the angular value to specify the direction of turning process.
- Click in the **Tool Orientation** edit box of **Mode & Direction** section and enter the angular value to specify the angle of tool for cutting process. This option is used for when your lathe turret has a programmable B-Axis. This option will help in machining process by cutting the stock corners of part.
- Click in the **Tool Clearance Back** edit box of **Mode & Direction** section and enter the angular value which is added to the tool angle to provide clearance behind the cutting edge.
- Click in the **Tool Clearance Back** edit box of **Mode & Direction** section and enter the angular value which is added to the tool angle to provide clearance front of the cutting edge.
- Select the **Use Constant Surface Speed** check box of **Feed & Speed** section from **Tool** tab to automatically adjust the spindle speed for maintaining a constant surface speed between tool and the workpiece.
- Click in the **Surface speed** edit box of **Feed & Speed** section to enter the value of spindle speed expressed as the speed of the tool on the surface.
- Click in the **Maximum Spindle Speed** edit box of **Feed & Speed** section and enter the value of maximum allowed spindle speed when using constant surface speed.
- Select the **Use Feed Per Revolution** check box of **Feed & Speed** section to automatically adjust the feed rate based on the RPM of the spindle to maintain a constant chip load.
- Click on the **Cutting Feed Per Revolution** edit box of **Feed & Speed** section and enter the value of adjusting the distance of tool cut

into the material for each 360 degree rotation of spindle when the tool is fully engaged.

- Click on the **Lead-In feed per Revolution** edit box of **Feed & Speed** section and enter the value of adjusting the distance of tool cut into the material for each 360 degree rotation of spindle, when the tool approaches and initially enters the material.

- Click on the **Lead-Out feed per Revolution** edit box of **Feed & Speed** section and enter the value of adjusting the distance of tool cut into the material for each 360 degree rotation of spindle when the tool exists the material.

Geometry

- Click on the **Geometry** tab of **PROFILE** dialog box. The **Geometry** tab will be displayed; refer to Figure-3.

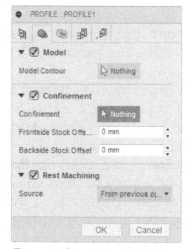

Figure-3. Geometry tab of PRO-FILE dialog box

- Select the **Model** check box of **Geometry** tab to select a model contour for machining.
- The **Nothing** button of **Model Contour** option is active by default. You need to click on the contour of model to select.
- Select the **Confinement** check box of **Geometry** tab to select a confinement region which can be defined with the combinations of edges, surfaces, or sketch point.
- The **Nothing** button of **Confinement** option from **Confinement** check box is active by default. You need to click on the surface, edge, or sketch point to select.
- Click in the **Frontside Stock Offset** edit box of **Geometry** tab and enter the value to specify the distance to machine beyond the frontside of the model.
- Click in the **Backside Stock Offset** edit box of **Geometry** tab and enter the value to specify the distance to machine beyond the backside of the model.
- Select the **Rest Machining** check box of **Geometry** tab to specify that only stock left after the previous operations should be machined.

- Click on the **Source** drop-down of **Rest Machining** check box and select the required option to specify the source from which the rest machining should be calculated.

Radii

Click on the **Radii** tab of **PROFILE** dialog box. The options of **Radii** tab will be displayed; refer to Figure-4.

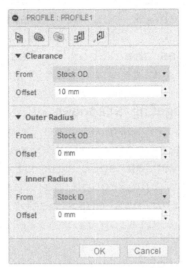

Figure-4. Radii tab of PROFILE dialog box

- Click on the **From** drop-down of **Clearance** section and select the required reference.
- Click in the **Offset** edit box of **Clearance** section and enter the desired value.
- Click on the **From** drop-down of **Outer Radius** section and select the required reference.
- Click in the **Offset** edit box of **Outer Radius** section and enter the desired value.
- Click on the **From** drop-down of **Inner Radius** section and select the required reference.
- Click in the **Offset** edit box of **Inner Radius** section and enter the desired value.

Passes

Click on the **Passes** tab of **PROFILE** dialog box. The **Passes** tab will be displayed; refer to Figure-5.

- Click in the **Tolerance** edit box of **Passes** tab and enter the value of tolerance.
- Click on the **Compensation Type** drop-down of **Passes** tab and select the required option.
- Select the **Make Sharp Corners** check box of **Passes** tab to specify that sharp corners must be forced.
- Select the **Finishing Passes** check box of **Passes** tab to specify the number of finishing passes.

- Click on the **Number of Stepovers** edit box of **Passes** tab and specify the number of finishing steps.
- Click on the **Stepover** edit box of **Passes** tab and enter the value of distance between two stepovers.
- Click on the **Finish Feedrate** edit box of **Passes** tab and enter the value of feedrate used for final finishing passes.
- Select the **Repeat Finishing Pass** check box of **Passes** tab and enter the value to perform an additional finishing pass with zero stock.
- Select the **No Dragging** check box of **Passes** tab to avoid the insert dragging along the stock.
- Select the **Roughing Passes** check box of **Passes** tab to enable roughing passes provides you with the ability to fine tune your grooving toolpath.
- Click on the **Maximum Roughing Stepdown** edit box of **PROFILE** dialog box to controls the aggressiveness of the roughing cuts. On increasing the value, the cutter removes more material with each roughing pass.
- Click in the **Roughing Overlap** edit box of **Passes** tab to specify the distance of the end of the roughing pass that how far the tool should move past the previous pass before it begins any lead-out.
- The other options of the dialog box were discussed earlier in this book.
- After specifying the parameter, click on the **OK** button from **PROFILE** dialog box. The toolpath will be generated and displayed on the part; refer to Figure-6.

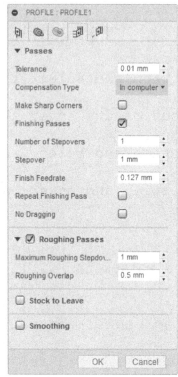

Figure-5. Passes tab of PROFILE dialog box

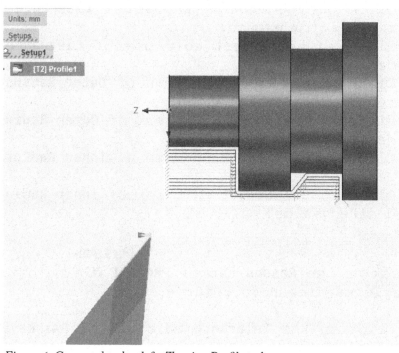

Figure-6. Generated toolpath for Turning Profile tool

Turning Groove

The **Turning Groove** tool is used for roughing and finishing strategy to create the groove in the model. The procedure to use this tool is discussed next.

- Click on the **Turning Groove** tool of **TURNING** drop-down from **Toolbar**; refer to Figure-7. The **GROOVE** dialog box will be displayed; refer to Figure-8.

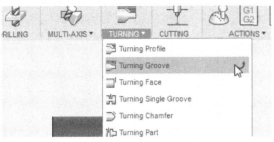

Figure-7. Turning Groove tool

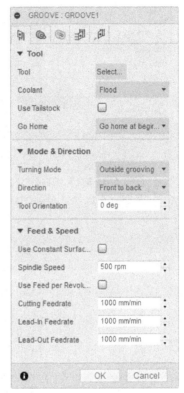

Figure-8. The GROOVE dialog box

- Click on the **Select** button of **Tool** section. The **Select Tool** dialog box will be displayed. Select the required tool and click on **OK** button. The tool will be added in the **GROOVE** dialog box.
- Click on the **Turning Mode** drop-down of **Mode & Direction** section from **TOOL** tab and select the required option according to your turning strategy.
- Click on the **Direction** drop-down of **Mode & Direction** section and select the required option according to the requirement.
- Select the **Confinement** check box of **Geometry** tab and select the faces which comes under groove of the part; refer to Figure-9.

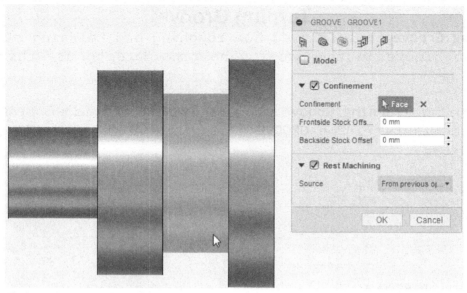

Figure-9. Selection of face for grooving

Passes tab

- Click on the **Passes** tab of **GROOVE** dialog box. The **Passes** tab will be displayed; refer to Figure-10.

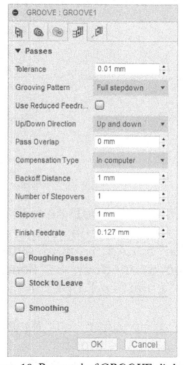

Figure-10. Passes tab of GROOVE dialog box

- Click on the **Tolerance** edit box of **Passes** tab and enter the value of tolerance value.
- Click on the **Full stepdown** option of **Grooving Pattern** drop-down from **Passes** tab to remove the stock material radially, before moving along the spindle axis to remove the next amount.
- Click on the **Partial stepdown** drop-down of **Grooving Pattern** drop-down to remove all the material along the spindle axis before starting to remove the material at the next depth which is specified by the Maximum Groove Stepdown value.

- Click on the **Sideways** with **Partial Stepdown** option of **Grooving Pattern** drop-down to remove the material from groove side by side. Due to this strategy the tool life is increased and it results in smaller chips formation.
- Select the **Use Pecking** check box of **Passes** tab to cut the material by a specified pecking depth using pecking force with the help of cutting tool. The gradual progressing down of tool to the cutting depth is required. The tool then retracts along its path by the specified pecking retract distance.
- Select the **Use Reduced Feedrate** check box of **Passes** tab to reduce the feedrate when grooving along the X-Axis.
- Click on the **Up/Down Direction** drop-down of **Passes** tab and select the required option to control the direction of finishing passes.
- Click in the **Pass Overlap** edit box of **Passes** tab to specify the overlap distance used for finishing pass split due to only up and only down machining directions.
- Click on the **Compensation Type** drop-down of **Passes** tab and select the required compensation type.
- Click on the **Backoff Distance** edit box of **Passes** tab and specify the distance to backoff from the stock before retracting.
- Select the **Finishing Passes** drop-down of **Passes** tab to remove the material from stock as a secondary finishing operation. This option is only available when **Roughing pass** option is enabled.
- Click in the **Number of Stepovers** check box of **Passes** tab and enter the value of number of finishing stepovers to apply.
- Click in the **Stepover** edit box of **Passes** tab and enter the value of distance which determine the amount of material initial toolpaths leave for the first finishing pass and for any subsequent finishing stepovers.
- Click in the **Finish Feedrate** edit box of **Passes** tab and enter the value of feedrates used for final finishing pass.
- Select the **Repeat Finishing Pass** check box of **Passes** tab to perform an additional finishing pass with zero stock.
- The other tools of the dialog box were discussed earlier.
- After specifying the parameters, click on the **OK** button from **GROOVE** dialog box. The toolpath will be generated and displayed on the model; refer to Figure-11.

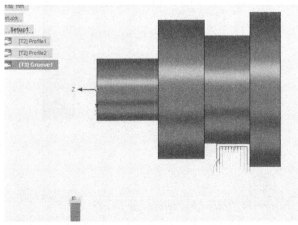

Figure-11. Generated toolpath Groove tool

Turning Face

The **Turning Face** tool is used for machining the front side of the part. The procedure to use this tool is discussed next.

- Click on the **Turning Face** tool of **TURNING** drop-down from **Toolbar**; refer to Figure-12. The **FACE** dialog box will be displayed; refer to Figure-13.

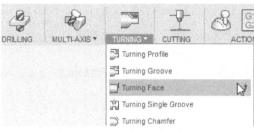

Figure-12. Turning Face tool

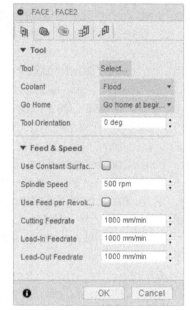

Figure-13. FACE dialog box

- Click on the **Select** selection button of **Tool** section. The **Select Tool** dialog box will be displayed. Select the required tool and click on **OK** button. The tool will be added in the **FACE** dialog box.
- Click on the **Tool Orientation** edit box of the **Tool** tab and enter the angular value to orient the tool as required.
- The options of the dialog box have already been discussed earlier.
- After specifying the parameters, click on the **OK** button from **FACE** dialog box. The toolpath will be generated and displayed on the model; refer to Figure-14.

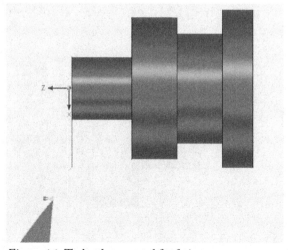

Figure-14. Toolpath generated for facing

Turning Single Groove

The **Turning Single Groove** tool is used for grooving at the selected position only. This tool will create a groove equal to the width of the tool. This is perfect for making a clearance groove behind the thread. The procedure to use this tool is discussed next.

- Click on the **Turning Single Groove** tool of **TURNING** drop-down from **Toolbar**; refer to Figure-15. The **SINGLE GROOVE** dialog box will be displayed; refer to Figure-16

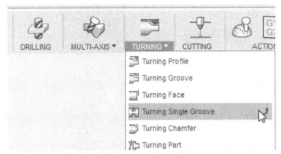

Figure-15. Turning Single Groove tool

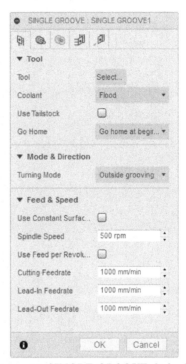

Figure-16. SINGLE GROOVE dialog box

- Click on the **Select button** of **Tool** section. The **Select Tool** dialog box will be displayed. Select the required tool and click on **OK** button. The tool will be added in the **SINGLE GROOVE** dialog box.

Geometry

- Click on the **Geometry** tab of **SINGLE GROOVE** dialog box. The **Geometry** tab will be displayed; refer to Figure-17.

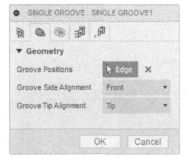

Figure-17. Geometry tab of
SINGLE GROOVE dialog box

- The **Nothing** button of **Groove Positions** option of **Geometry** tab is active by default. You need to click on the edge of groove to select.

- Click on the **Groove Side Alignment** drop-down of **Geometry** section and select the required option for tool alignment.
- Click on the **Groove Tip Alignment** drop-down of **Geometry** section and select the required option for tip alignment.
- The other options of the dialog box were discussed earlier.
- After specifying the various parameters, click on the **OK** button from **SINGLE GROOVE** dialog box. The toolpath will be generated and displayed on the model; refer to Figure-18.

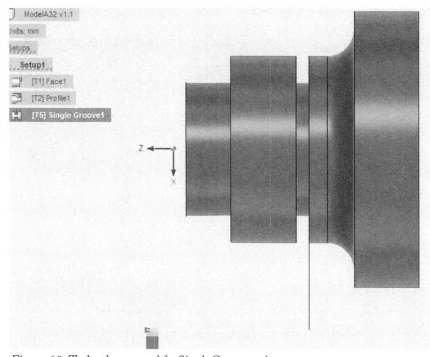

Figure-18. Toolpath generated for Single Groove tool

Turning Chamfer

The **Turning Chamfer** tool is used for chamfering the sharp corners that have not been chamfered in the design. The procedure to use this tool is discussed next.

- Click on the **Turning Chamfer** tool of **TURNING** drop-down from **Toolbar**; refer to Figure-19. The **CHAMFER** dialog box will be displayed; refer to Figure-20.

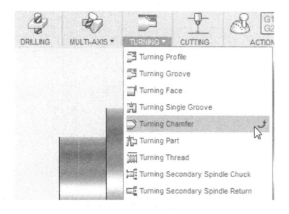

Figure-19. Turning Chamfer tool

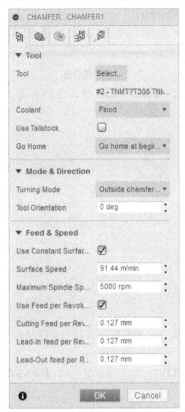

Figure-20. Tthe CHAMFER dialog box

- Click on the **Select button** of **Tool** section. The **Select Tool** dialog box will be displayed. Select the required tool and click on **OK** button. The tool will be added in the **CHAMFER** dialog box.
- Click on the **Turning Mode** drop-down of **Mode & Direction** section from **Tool** tab and select the required option according to the turning strategy.
- Click in the **Tool Orientation** edit box of **Mode & Direction** section from **Tool** tab and enter a specific value to orient the tool.

Geometry

Click on the **Geometry** tab of **CHAMFER** dialog box. The **Geometry** tab will be displayed; refer to Figure-21.

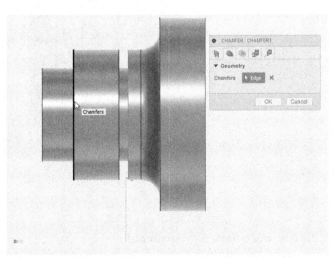

Figure-21. Geometry tab of CHAMFER dialog box

- The **Nothing** button of **Chamfers** option is active by default. You need to click on the edges from model to select.

Passes

Click on the **Passes** tab of **CHAMFER** dialog box. The **Passes** tab will be displayed; refer to Figure-22.

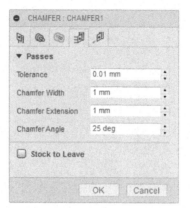

Figure-22. Passes tab of CHAM-FER dialog box

- Click in the **Chamfer Width** edit box of **Passes** tab and enter the value of width of chamfer.
- Click in the **Chamfer Extension** edit box of **Passes** tab and enter the value by which to extend the chamfer cutting pass.
- Click in the **Chamfer Angle** edit box of **Passes** tab and enter the value of angle of the chamfer measured from the Z-Axis.
- The other options of the dialog box were discussed earlier in this book.
- After specifying the parameters, click on the **OK** button from **CHAMFER** dialog box. The toolpath will be generated and displayed on the model; refer to Figure-23.

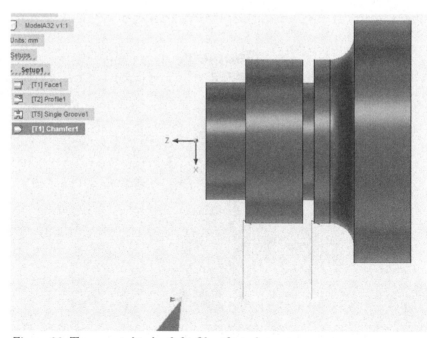

Figure-23. The generated toolpath for Chamfer tool

Turning Part

The **Turning Part** tool is used for cutting the part off with the bar. This tool is also known as cut off operation. The procedure to use this tool is discussed next.

- Click on the **Turning Part** tool of **TURNING** drop-down from **Toolbar**; refer to Figure-24. The **PART** dialog box will be displayed; refer to Figure-25.

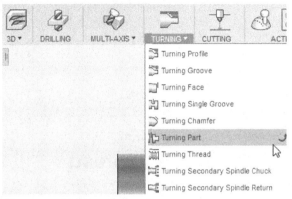

Figure-24. Turning Part tool

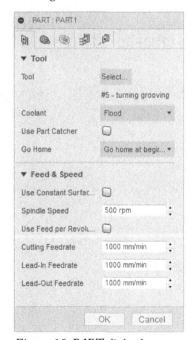

Figure-25. PART dialog box

- Click on the **Select button** of **Tool** section. The **Select Tool** dialog box will be displayed. Select the required tool and click on **OK** button. The tool will be added in the **PART** dialog box.
- The options of the dialog box were discussed earlier.
- After specifying the parameters, click on the **OK** button from **PART** dialog box. The toolpath will be generated and displayed on the model; refer to Figure-26.

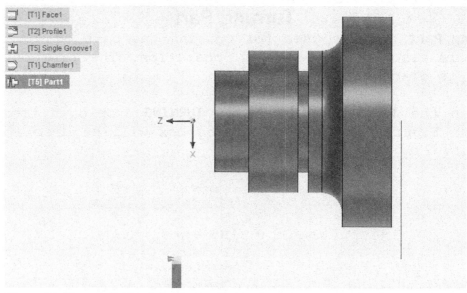

Figure-26. Generated toolpath for Turning Path tool

Turning Thread

The **Turning Thread** tool is used to create thread on cylindrical and conical surfaces. The procedure to use this tool is discussed next.

- Click on the **Turning Thread** tool of **TURNING** drop-down from **Toolbar**; refer to Figure-27. The **THREAD** dialog box will be displayed; refer to Figure-28.

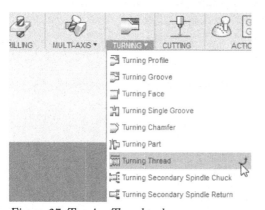

Figure-27. Turning Thread tool

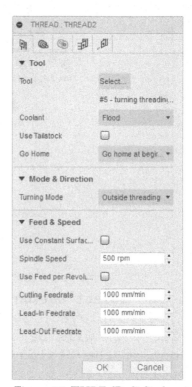

Figure-28. THREAD dialog box

- Click on the **Select** selection button of **Tool** section. The **Select Tool** dialog box will be displayed. Select the required tool and click on **OK** button. The tool will be added in the **THREAD** dialog box.

- Click on the **Nothing** button of **Thread Faces** option from **Geometry** tab and select the round faces where you want to machine the thread; refer to Figure-29.

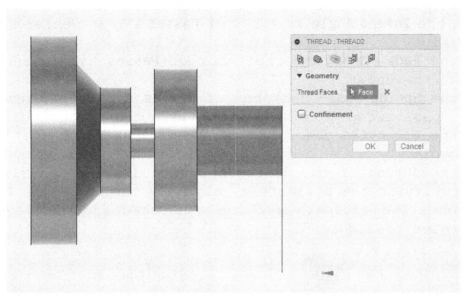

Figure-29. Selection of Face for thread

Passes

Click on the **Passes** tab of **THREAD** dialog box. The **Passes** tab will be displayed; refer to Figure-30.

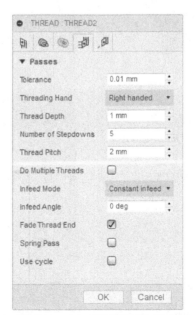

Figure-30. Passes tab of THREAD dialog box

- Click on the **Threading Hand** drop-down of **Passes** tab and select the required option.
- Click in the **Thread Depth** edit box of **Passes** tab and enter the desired value to specify the depth of thread.
- Click in the **Number of Stepdown** edit box of **Passes** tab and specify the desired number of step downs.
- Click in the **Thread Pitch** edit box of **Passes** tab and specify the value of thread pitch.

- Select the **Do Multiple Threads** check box of **Passes** tab to enter the number of threads.
- Click on the **Infeed Mode** drop-down of **Passes** tab and select the required option.
- Click in the **Infeed Angle** edit box of **Passes** tab and enter the angular infeed value.
- Select the **Fade Thread End** check box of **Passes** tab to fade out the thread at the end.
- Select the **Spring Pass** check box of **Passes** tab to perform the final finishing pass twice to remove stock left due to tool deflection.
- Select the **Use cycle** check box of **Passes** tab to request output as canned cycle.
- The other options of the dialog box were discussed earlier.
- After specifying the parameters, click on the **OK** button from **THREAD** dialog box. The toolpath will be generated and displayed on the model; refer to Figure-31.

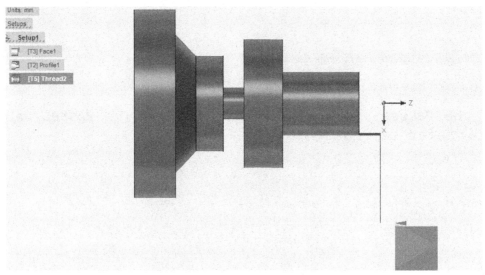

Figure-31. Toolpath generated for thread tool

- After simulating the Thread toolpath, the model will be displayed as shown in Figure-32.

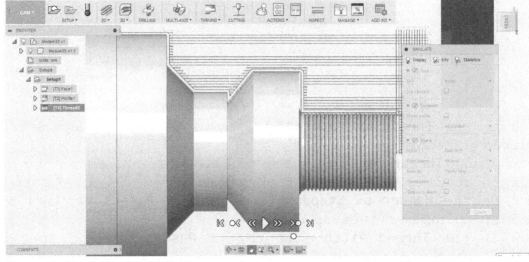

Figure-32. Simulating thread toolpath

Turning Secondary Spindle Chuck

The **Turning Secondary Spindle Chuck** tool is used to transfer stock from one chuck to another (Sub-spindle). This tool is useful when you want to machine back side of part after machining the front side. The procedure to use this tool is discussed next.

* Click on the **Turning Secondary Spindle Chuck** tool of **TURNING** drop-down from **Toolbar**; refer to Figure-33. The **SECONDARY SPINDLE CHUCK** dialog box will be displayed; refer to Figure-34

Figure-33. Turning Secondary Spindle Chuck tool

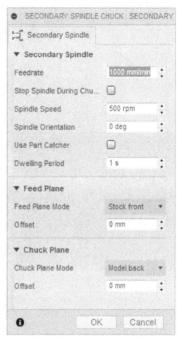

Figure-34. SECONDARY
SPINDLE CHUCK dialog box

* Click in the **Feedrate** edit box of **Secondary Spindle** section from **SECONDARY SPINDLE CHUCK** dialog box and enter the value of feed rate.
* Select the **Stop Spindle During Chuck** check box of **Secondary Spindle** section to keep spindle stopped during operations.
* Click in the **Spindle Orientation** edit box of **Secondary Spindle** section and enter the angular value between primary and secondary axis.
* Select the **Use Part Catcher** check box of **Secondary Spindle** section to activate part catcher when available.
* Click in the **Dwelling Period** edit box of **Secondary Spindle** section and enter a time for the operation to dwell.
* Click in the **Feed Plane Mode** drop-down of **Feed Plane** section and select the required option to specify the feed plane.
* Click in the **Offset** edit box of **Feed Plane** section and enter the offset value.
* Click in the **Chuck Plane Mode** drop-down of **Chuck Plane** section and select the required option to specify the chuck plane.
* Click in the **Offset** edit box of **Chuck Plane** section and enter the offset value.
* The operation will be created and displayed on the **Browser Tree**.

Turning Secondary Spindle Return

The **Turning Secondary Spindle Return** tool is used for automatic stock return from secondary chuck to main chuck. No toolpath is associated with the strategy. The procedure to use this tool is discussed next.

• Click on the **Turning Secondary Spindle Return** tool of TURNING drop-down from **Toolbar**; refer to Figure-35. The **SECONDARY SPINDLE RETURN** dialog box will be displayed; refer to Figure-36.

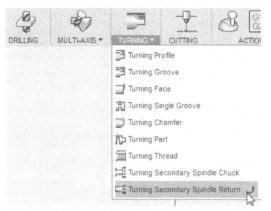

Figure-35. *Turning Secondary Spindle Return tool*

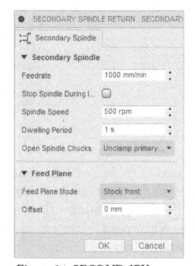

Figure-36. *SECONDARY SPINDLE RETURN dialog box*

• Click in the **Feedrate** edit box of **Secondary Spindle** section from **SECONDARY SPINDLE RETURN** dialog box and enter the value of feedrate.
• Select the **Stop Spindle During Return** check box of **Secondary Spindle** section to keep spindle stopped during operations.
• Click in the **Spindle Speed** edit box of **Secondary Spindle** section and enter the value of spindle speed in rpm for each machining operation.
• Click in the **Dwelling Period** edit box of **Secondary Spindle** section and enter a time for the operation to dwell.
• Click on the **Open Spindle Chucks** drop-down of **Secondary Spindle** section and select the required option to choose which spindle chucks to open before returning the secondary spindle.

- Click in the **Feed Plane Mode** drop-down of **Feed Plane** section and select the required option to specify the feed plane.
- Click in the **Offset** edit box of **Feed Plane** section and enter the offset value; refer to Figure-37.

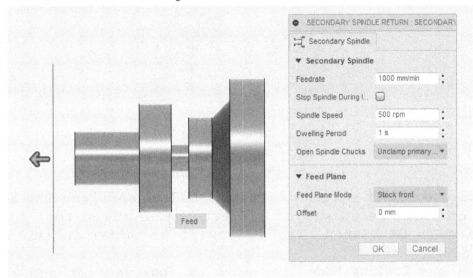

Figure-37. Entering offset value

- After specifying the parameters, click on the **OK** button from **SECONDARY SPINDLE RETURN** dialog box. The operation will be added in the **Browser Tree**.

GENERATING CUTTING TOOLPATHS

Till now, we have discussed the procedure to user turning toolpaths. In this section, we will discuss the procedure of generating toolpaths for cutting purpose.

Cutting

The **Cutting** tool is used to create toolpaths for machining with the help of Laser machine, Plasma cutters and Waterjet machine on a 2D Profile. The procedure to use this tool is discussed next.

- Click on the **CUTTING** tool from **Toolbar**; refer to Figure-38. The **2D PROFILE** dialog box will be displayed; refer to Figure-39.

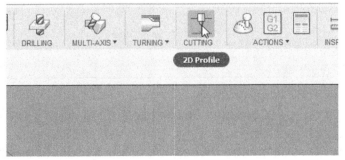

Figure-38. 2D Profile tool

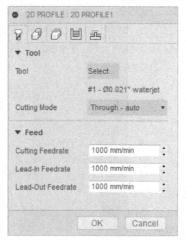

Figure-39. The 2D PROFILE dialog box

- Click on the **Select** selection button of **Tool** section. The **Select Tool** dialog box will be displayed. Select the required tool for cutting and click on **OK** button. The tool will be added in the **2D PROFILE** dialog box.
- Click on the **Cutting Mode** drop-down of **Tool** section and select the required option to set the appropriate cutting feeds for the material being cut. In some machines, there have internal quality tables to set the value.

Geometry

Click on the **Geometry** tab of **2D PROFILE** dialog box. The **Geometry** tab will be displayed; refer to Figure-40.

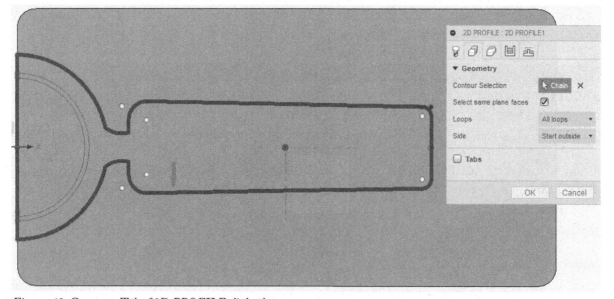

Figure-40. Geometry Tab of 2D PROFILE dialog box

- The **Nothing** button of **Contour Selection** option is active by default. You need to click on the closed contour from the model to select.
- Select the **Select same plane faces** check box of **Geometry** tab to select all the faces created on the same plane, from all parts within the stock boundary.
- Click on the **Loops** drop-down of **Geometry** tab and select the required option to filter outer or inner chain.

- Click on the **Side** drop-down of **Geometry** tab and select the required option to offset selected edges or sketches.

Linking

Click on the **Linking** tab of **2D PROFILE** dialog box. The **Linking** tab will be displayed; refer to Figure-41.

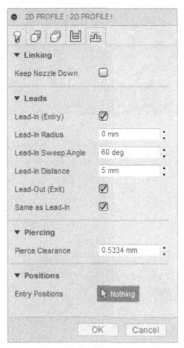

Figure-41. Linking Tab of 2D PROFILE dialog box

- Select the **Keep Nozzle Down** check box of **linking** tab to avoid retracts and to avoid previously cut areas.
- Select the **Lead-In (Entry)** check box of **Leads** section to enable a contour blend.
- Click in the **Lead-In Radius** edit box of **Leads** section and enter the value to specify the radius for lead in moves.
- Click in the **Lead-In Sweep Angle** edit box of **Leads** section and enter the value to specify the sweep angle of the lead-in arc.
- Click in the **Lead-In Distance** edit box of **Leads** section and enter the value to specify the length of the linear lead in the move.
- Select the **Lead-Out (Exit)** check box of **Linking** tab to enable a contour blend.
- Select the **Same as Lead-In** check box of **Linking** tab to set the lead-Out values be identical to the Lead-In values.
- Click in the **Pierce Clearance** edit box of **Piercing** section and enter the value to distance away from the part edge to start the cut.
- The **Nothing** button of **Entry Positions** option from **Positions** section is active by default. You need to click on the point from model to select.
- The other options of the dialog box have been discussed earlier.
- After specifying the parameters, click on the **OK** button from **2D PROFILE** dialog box. The toolpath will be generated and displayed on the tool; refer to Figure-42.

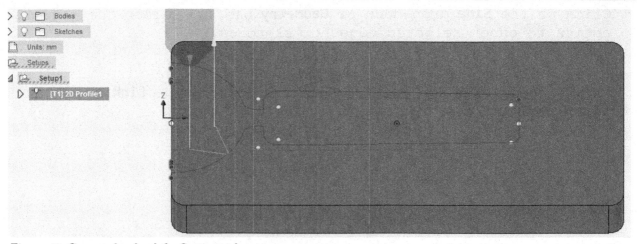

Figure-42. Generated toolpath for Cutting tool

PRACTICAL

Generate the toolpath of the given model shown in Figure-43 using Turning tools.

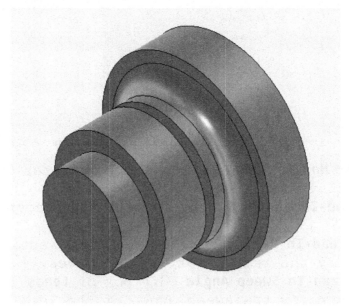

Figure-43. Practical

Adding Model to CAM

• Firstly, create and save the part in **MODEL Workspace**. The part file is available in the respective chapter folder of **Autodesk Fusion 360** folder.

• Click on the **CAM Workspace** from **Workspace** drop-down. The model will be displayed in the **CAM Workspace.**

Creating Stock

• Click on the **New Setup** tool of **SETUP** drop-down from **Toolbar**. The **SETUP** dialog box will be displayed along with the stock of model.

• Specify the parameters of **Setup** tab and **Stock** tab of **SETUP** dialog box to create the stock; refer to Figure-44.

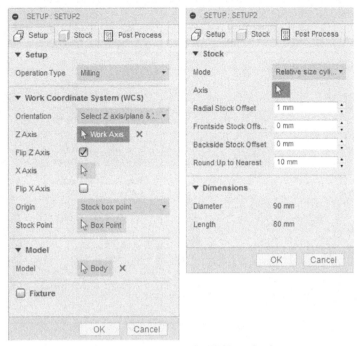

Figure-44. Specifying parameters for SETUP dialog box

- After specifying the parameters of **SETUP** dialog box, click on the **OK** button. The stock will be created and displayed on the model; refer to Figure-45.

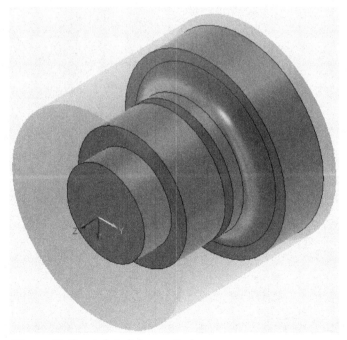

Figure-45. Created stock for practical

Generating Face Toolpath

- Click on the **Turning Face** tool of **TURNING** drop-down from **Toolbar**. The **FACE** dialog box will be displayed.
- Click on the **Select** button of **Tool** option and select the facing tool from **Select Tool** dialog box as displayed in Figure-46.

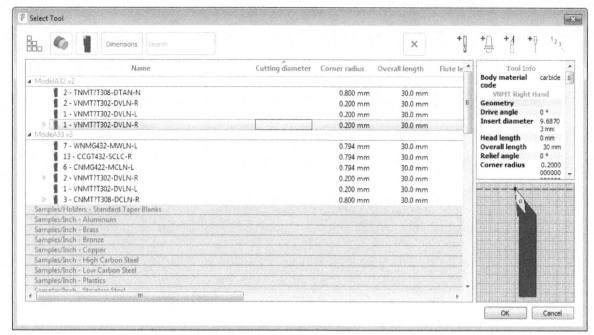

Figure-46. Selecting tool for turning

- After selecting the tool, click on the **OK** button from **Select Tool** dialog box. The tool will be added in the **Face** dialog box.
- Specify the parameters of **Tool** tab, **Geometry** tab, **Passes** tab and **Linking** tab of **FACE** dialog box as shown in Figure-47.

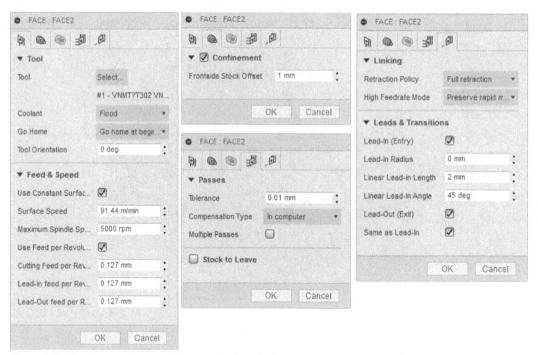

Figure-47. Specifying the parameters of FACE dialog box

- After specifying the parameters, click on the **OK** button from **FACE** dialog box to generate the toolpath. The toolpath will be generated and displayed on the model.

Generating Turning Profile toolpath

- Click on the **Turning Profile** tool of **TURNING** drop-down from **Toolbar**. The **PROFILE** dialog box will be displayed.
- Click on the **Select** button of **Tool** option and select the required tool from **Select Tool** dialog box.
- Specify the parameters of **Tool** tab, **Geometry** tab, **Passes** tab and **Linking** tab of **PROFILE** dialog box as shown in Figure-48.

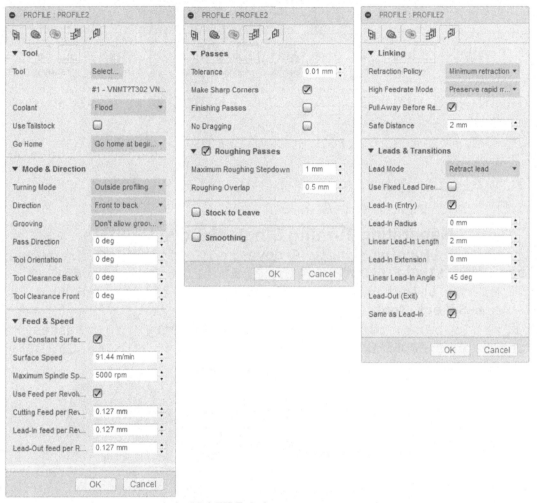

Figure-48. Specifying parameters for PROFILE dialog box

- After specifying the parameters, click on the **OK** button from **PROFILE** dialog box. The toolpath will be generated and displayed on the model; refer to Figure-49.

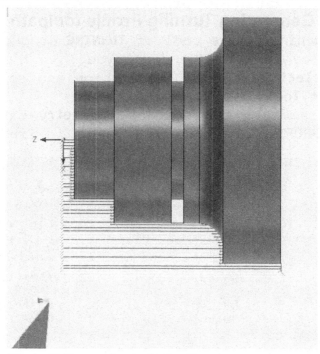

Figure-49. Generated toolpath of profile tool

Generating Turning Single Groove toolpath

- Click on the **Turning Single Groove** tool of **TURNING** drop-down from **Toolbar**. The **SINGLE GROOVE** dialog box will be displayed.
- Click on the **Select** button of **Tool** option and select the required tool from **Select Tool** dialog box.
- Specify the parameters of **Tool** tab, **Geometry** tab, **Passes** tab and **Linking** tab of **PROFILE** dialog box as shown in Figure-50.

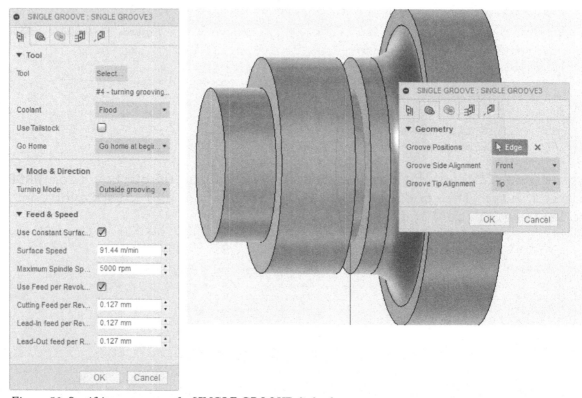

Figure-50. Specifying parameters for SINGLE GROOVE dialog box

- After specifying the parameters, click on the **OK** button from **SINGLE GROOVE** dialog box. The toolpath will be generated and displayed on the model; refer to Figure-51

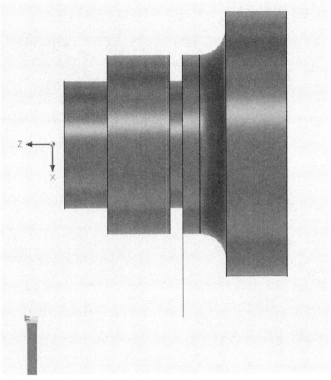

Figure-51. Generated toolpath for single groove

Generating Turning Part toolpath

- Click on the **Turning Part** tool **TURNING** drop-down from **Toolbar**. The **PART** dialog box will be displayed.
- Click on the **Select** button of **Tool** option from **PART** dialog box and select the grooving tool of parameters shown in Figure-52.

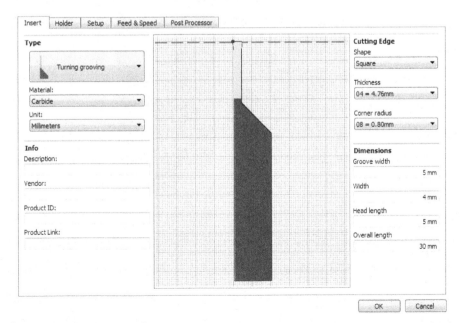

Figure-52. Selecting tool for part

- After selecting the tool of parameters displayed above, click on the **OK** button from **Select Tool** dialog box. The tool will be added in the **PART** dialog box.
- Specify the parameter of **Tool** tab, **Passes** tab, and **Linking** tab of **PART** dialog box as displayed in Figure-53.

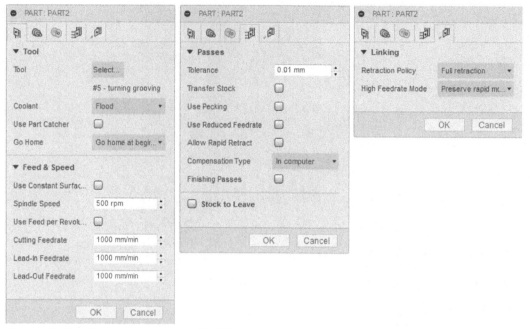

Figure-53. Specifying the parameters of PART dialog box

- After specifying the parameters, click on the **OK** button from **PART** dialog box. The toolpath will be generated and displayed on the model; refer to Figure-54.

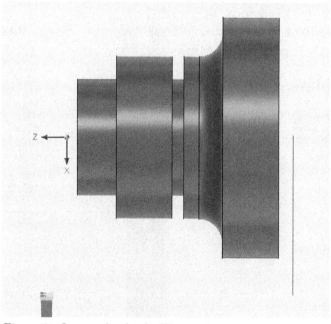

Figure-54. Generated toolpath of Part tool

Simulating the Toopath

- Right-click on the **Setup** from **Browser Tree** and click on the **Simulate** tool from the Marking menu. The **SETUP** dialog box will be displayed along with the model.
- Use the simulation keys to play the machining animation to check the errors occurred while machining process; refer to Figure-55.

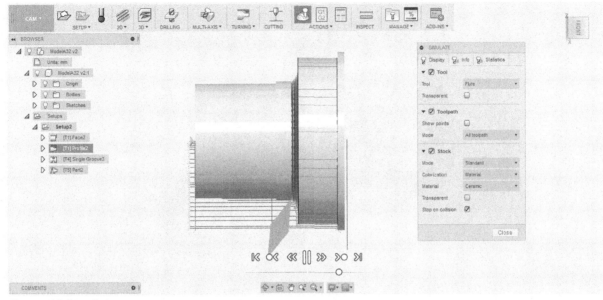

Figure-55. Simulation process

PRACTICE 1

Machine the stock of Diameter as 50 mm and Length as 62 mm to create the part as shown in Figure-56. The pat file of this model is available in the respective folder of **Autodesk Fusion 360.**

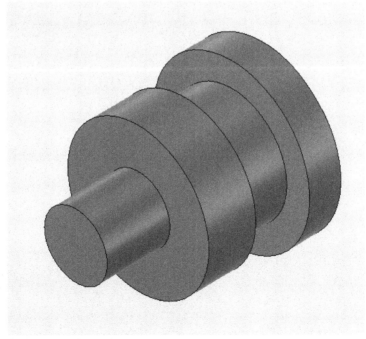

Figure-56. Practice 1

PRACTICE 2

Machine the stock of Diameter as 80 mm and Length as 130 mm to create the part as shown in Figure-57. The pat file of this model is available in the respective folder of **Autodesk Fusion 360.**

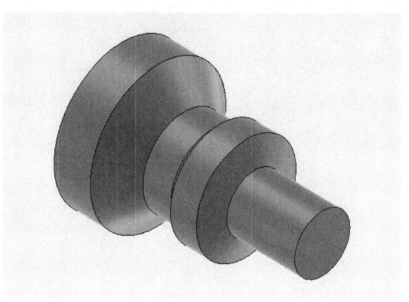

Figure-57. Practice

Chapter 19

Miscellaneous CAM Tools

Topics Covered

The major topics covered in this chapter are:

- *Linear Pattern*
- *Mirror Pattern*
- *Duplicate Pattern*
- *Manual NC*
- *Post Process*
- *Setup Sheet*

INTRODUCTION

Till now, we have learned the procedure of generating Milling and Turning toolpath. In this chapter, we will discuss some of the other tools used to organize and manipulate the toolpaths.

NEW FOLDER

The **New Folder** tool is used for creating a folder to combine similar group operation. The procedure to use this tool is discussed next.

- Select the operations from **Browser Tree** by holding **CTRL** key and right-click on any of the selected operations. A shortcut menu will be displayed; refer to Figure-1.

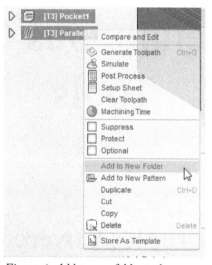

Figure-1. Add to new folder tool

- Click on the **Add to New Folder** button from the menu. The selected operations will be added in a new folder. The folder will be displayed in the **Browser Tree** with the name as **Folder**; refer to Figure-2.

Figure-2. Moved operations

- If you want to rename the newly created folder then double-click on the folder with a pause between the click and enter the desired name.

There is an another method for creating the folder which is discussed next.

- Click on the **New Folder** button of **SETUP** drop-down from **Toolbar**; refer to Figure-3. The folder will be created in the **Browser Tree**.

Figure-3. New Folder tool

- Select the operations by holding **CTRL** key from **Browser Tree** and drag the operations into newly created folder. The operation will be added in the created folder.

NEW PATTERN

The **New Pattern** tool is used for duplicating a generated toolpath on the same model in Linear, Circular, Mirror, and Duplicate pattern. The use of **New Pattern** tool can speed up your entire programming process since all changes to a pattern take effect immediately and no toolpath has to be updated. The procedure to use this tool is discussed next.

- Right-click on the specific operation which you want to create the linear pattern. The shortcut menu will be displayed; refer to Figure-4. This is applicable to all the pattern options of **FOLDER:PATTERN** dialog box.

Figure-4. The Add to New Pattern tool

- Click on the **Add to New Pattern** tool from the displayed menu. The **FOLDER : PATTERN** dialog box will be displayed; refer to Figure-5.

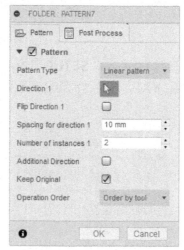

Figure-5. FOLDER PATTERN tool

Linear Pattern

In this section, we will discuss the procedure of creating linear pattern.

- Click on the **Liner Pattern** option of **Pattern Type** drop-down from **FOLDER : PATTERN** dialog box for creating linear pattern.
- The selection button of **Direction 1** section of **Pattern** tab is active by default. You need to click on the edge or face from model to select the direction.
- Select the **Flip Direction 1** check box of **Pattern** tab to flip the direction of pattern along selected edge or face.
- Click in the **Spacing for Direction 1** edit box of **Pattern** tab and enter the value of distance between two consecutive instances of the pattern.
- Click on the **Number of instances 1** edit box of **Pattern** tab and enter the number of instances required in the pattern along first direction.
- Select the **Additional Direction** check box of **Pattern** tab to use an additional direction for creating the pattern.
- Select the **Keep original** check box of **Pattern** tab to keep the original pattern while applying a new pattern.
- Select the **Preserve Order** option of **Operation Order** drop-down from **Pattern** tab to machine all operations in each instance of the pattern before moving to the next instance.
- Select **Order by Operation** option of **Operation Order** drop-down from **Pattern** tab to machine all occurrences of each operation in the pattern before moving to the next operation.
- Select **Order by tool** option of **Operation Order** drop-down from **Pattern** tab to machines all operations in the pattern that use the current tool before changing tools.
- The other tools of the drop-down were discussed earlier in this book.
- After specifying the parameters, click on the **OK** button from **FOLDER : PATTERN** dialog box; refer to Figure-6. The pattern will be created and displayed in the **Browser Tree**.

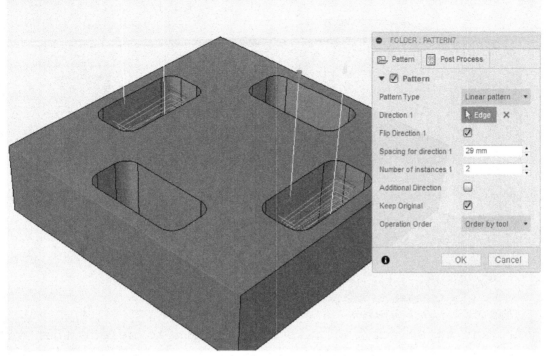

Figure-6. Applied linear pattern

Circular Pattern

In this section, we will discuss the procedure of creating the circular patter.

* Click on the **Circular Pattern** option of **Pattern Type** drop-down from **FOLDER : PATTERN** dialog box for creating circular pattern; refer to Figure-7.

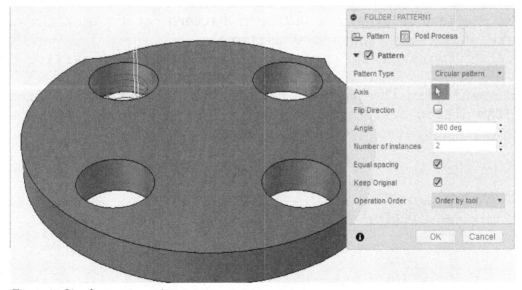

Figure-7. Circular pattern option

* The selection button of **Axis** section from **Pattern** tab is active by default. You need to click on the axis or edge from model to select; refer to Figure-8.

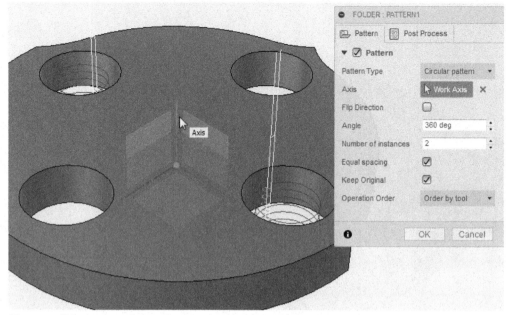

Figure-8. Selecting axis for pattern

- If you want to select the axis from origin to select then visible the origin from **Browser Tree** and select the required axis.
- Select the **Flip Direction** check box of **Pattern** tab to flip the direction of selected edge or axis.
- Click in the **Angle** edit box of **Pattern** tab and enter the angular value to place the pattern.
- Click on the **Number of instances** edit box of **Pattern** tab and enter the value to specify number of times the operation should be machined in the selected direction.
- Select the **Equal Spacing** check box of **Pattern** tab to equally distribute the selected patterns at a specified angle.
- Select the **Keep original** check box of **Pattern** tab to keep the original pattern while applying a new pattern.
- The other tools of the drop-down were discussed earlier in this book.
- After specifying the parameters, click on the **OK** button from **FOLDER : PATTERN** dialog box; refer to Figure-9. The preview of pattern will be created and displayed on the model.

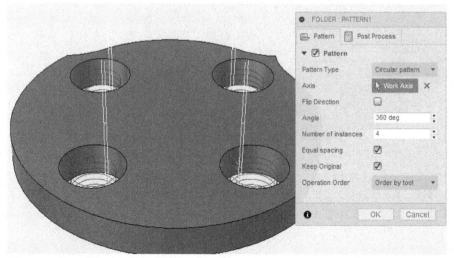

Figure-9. Preview of circular pattern

Mirror Pattern

In this section, we will discuss the procedure of creating the mirror pattern.

- Click on the **Mirror Pattern** option of **Pattern Type** drop-down from **FOLDER : PATTERN** dialog box for creating mirror pattern; refer to Figure-10.

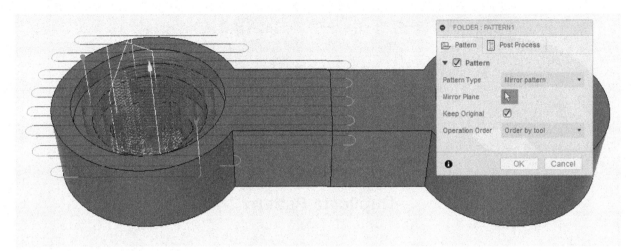

Figure-10. Mirror pattern option

- The selection button of **Mirror Plane** section from **Pattern** tab is active by default. You need to click on the plane or face from model to select; refer to Figure-11.

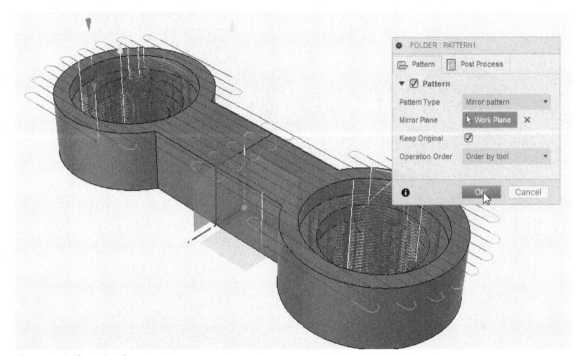

Figure-11. Completed mirror pattern

- The other options of the dialog box were discussed earlier.
- After specifying the parameters, click on the **OK** button from **FOLDER : PATTERN** dialog box; refer to Figure-12. The preview of pattern will be created and displayed on the model.

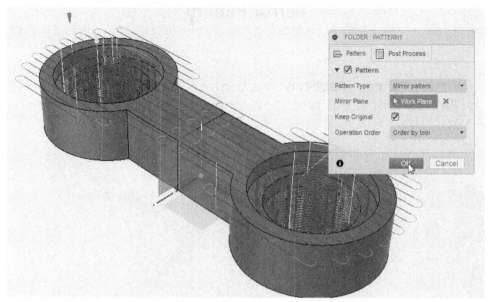

Figure-12. Completed mirror pattern

Duplicate Pattern

In this section, we will discuss the procedure of creating the duplicate pattern.

- Click on the **Duplicate Pattern** option of **Pattern Type** drop-down from **FOLDER : PATTERN** dialog box for creating duplicate pattern; refer to Figure-13.

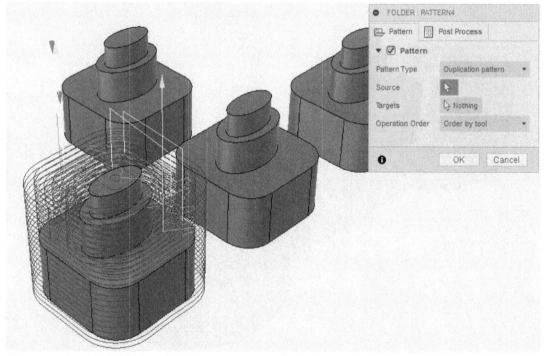

Figure-13. Duplication pattern option

- The selection button of **Source** section from **Pattern** tab is active by default. You need to click on the point from source model to select.

- The selection button of **Targets** section of **Pattern** tab is active by default. You need to click on the target points on the model on which you want to duplicate the toolpath.
- The other options of the dialog box have been discussed earlier.
- After specifying the parameters, click on the **OK** button from **FOLDER : PATTERN** dialog box; refer to Figure-14. The preview of duplicate pattern will be created and displayed on the model.

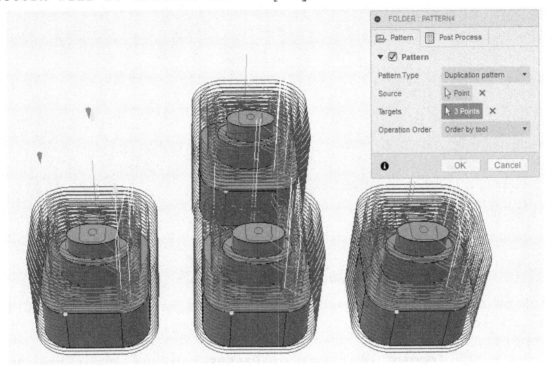

Figure-14. Selecting target body

Manual NC

The **Manual NC** tool is used to insert special manual NC entries in the CAM browser. The procedure to use this tool is discussed next.

- Click on the **Manual NC** tool of **SETUP** drop-down from **Toolbar**; refer to Figure-15. The **MANUAL NC** dialog box will be displayed; refer to Figure-16.

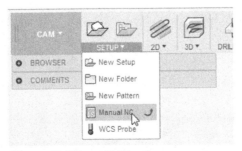

Figure-15. Manual NC tool

Figure-16. MANUAL NC dialog box

- Click on the **Manual Type** drop-down of **Passes** tab from **MANUAL NC** dialog box and select the required option to specify the type of manual NC operation.
- Click in the **Comment** edit box of **Passes** tab and enter the desired text to be output during post processing.

WCS Probe

The **WCS Probe** tool is used to output the NC code for the probing cycles of your CMM. The procedure to use this tool is discussed next.

- Click on the **WCS Probe** tool of **SETUP** drop-down from **Toolbar**; refer to Figure-17. The **WCS PROBE** dialog box will be displayed; refer to Figure-18.

Figure-17. WCS Probe tool

Figure-18. WCS PROBE dialog box

- Click on the **Select** button of **Tool** section from **Tool** tab. The **Select Tool** dialog box will be displayed; refer to Figure-19.

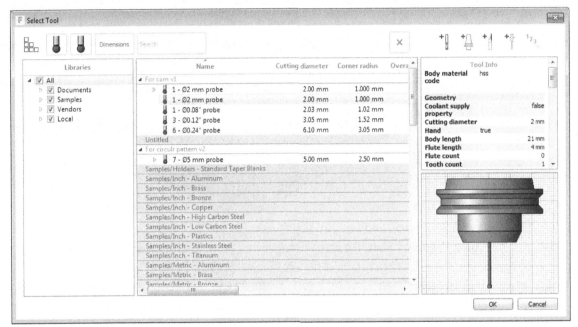

Figure-19. Select Tool dialog box for selecting probe tool

- Select the required probing tool from the table and click on the **OK** button from **Select Tool** dialog box.

Geometry

Click on the **Geometry** tab of **WCS PROBE** dialog box. The **Geometry** tab will be displayed; refer to Figure-20.

Figure-20. Geometry tab of WCS PROBE dialog box

- Click in the **Probe Mode** drop-down of **Geometry** tab from **WCS PROBE** dialog box and select the required option.
- The **Nothing** button of **Probe Surface(s)** section of **Geometry** tab is active by default. You need to click on the face of model to checked by probe; refer to Figure-21.

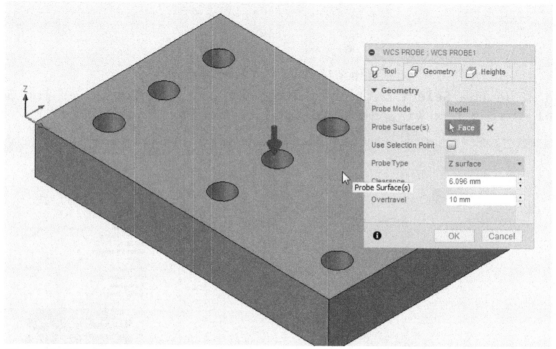

Figure-21. Selected face for probing

- Select the **Use Selection Point** check box of **Geometry** tab to set the position of probe based on the clicked position of selected face.
- Click in the **Clearance** edit box of **Geometry** tab and enter the distance value of probe to stay clear of the feature before the actual probing.
- Click in the **Overtravel** edit box of **Geometry** tab and enter the distance of the probe which is allowed to move beyond the expected strike point.
- The other options of the dialog box have been discussed earlier in this book.
- After specifying the parameters, click on the **OK** button from **WCS PROBE** dialog box. The operation will be created and displaced in the **Browser Tree**.

Simulate

The **Simulate** tool is used to review and simulate the created toolpaths. The procedure to use this tool is discussed next.

- Click on the **Simulate** tool of **ACTIONS** drop-down from **Toolbar**; refer to Figure-22. The **SIMULATE** dialog box will be displayed along with model and simulation keys; refer to Figure-23.

Figure-22. Simulate tool

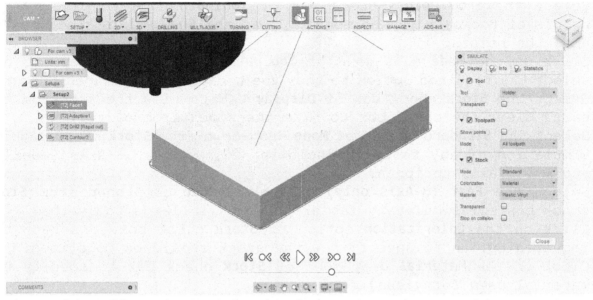

Figure-23. SIMULATE dialog box along with simulation keys

- Click on the Setup or desired operation from **Browser Tree** and then click on the **Simulate** tool from **Toolbar**. If a toolpath created is not valid then it shows the notification of No Valid Toolpath.
- There is also another way to select **Simulate** tool. Right-click on the specific operation or setup from **Browser Tree**. A shortcut menu will be displayed; refer to Figure-24. Click on the **Simulate** tool from the menu. The **SIMULATE** dialog box will be displayed.

Figure-24. Simulate tool from shortcut menu

Display tab

- Select the **Tool** check box of **Display** tab from **SIMULATE** dialog box to view the tool and holder while machining animation process.

- Click on the **Tool** drop-down of **Display** tab and select the required option to tell software that which segment of the tool to show while machining.
- Select the **Transparent** check box of **Display** tab to make the tool transparent.
- Select the **Toolpath** check box of **Display** tab to display the toolpath on the model. Clear the **Toolpath** check box to hide the toolpath.
- Select the **Show Points** check box of **Toolpath** node to display the points of toolpath on the model. Clear the **Show Points** check box to hide the points.
- Click on the **Mode** drop-down of **Toolpath** node from **Display** tab and select the required option to show the toolpath of the specific part.
- Select the **Stock** check box of **Display** tab to view the stock on the part. Clear the check box to hide the stock from part.
- Select the **Standard** option of **Mode** drop-down from **Stock** check box for simulation of any toolpath including 2D, 3-axis, 3-Axis indexing and multi axis toolpath.
- Select the **Fast (3-Axis only)** option of **Mode** drop-down from **Stock** check box for simulating faster toolpath in large toolpaths.
- Click on the **Colorization** option of **Stock** check box to select the required option to specify how the stock should be colorized.
- Click on the **Material** drop-down of **Stock** check box to specify the material used for visualization.
- Select the **Transparent** check box of **Stock** tab to transparent the stock of model.
- Select the **Stop on collision** option of **Stock** tab to stop the tool on collision with stock of model. When the tool or holder collides with stock then it shows red sign in the animation bar; refer to Figure-25

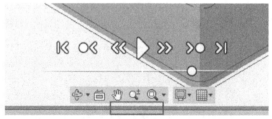

Figure-25. Holder collision with stock

Info tab

Click on the **Info** tab of **SIMULATE** dialog box. The **Info** tab will be displayed; refer to Figure-26.

Figure-26. Info tab of SIMUL-TAE dialog box

- In this **Info** tab, the information related to stock, operation, spindle speed, volume, cursor position, etc. will be displayed.

Statistics

Click on the **Statistics** tab of **SIMULATE** dialog box. The **Statistics** tab will be displayed; refer to Figure-27.

Figure-27. Statistics tab

- In the **Statistic** tab, the information like Machining time, Machining distance, Operations, and Tool change will be displayed.

Simulation Keys

The simulation keys are used to watch the animation of stock removal; refer to Figure-28.

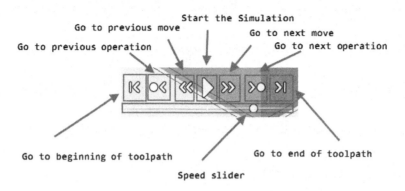

Figure-28. Simulation keys

Post Process

The **Post Process** tool is used to convert the machine-independent cutter location data into machine-specific NC code that can be run directly on CNC machines. The procedure to use this tool is discussed next.

- Click on the **Post Process** tab of **ACTIONS** tab from **Toolbar**; refer to Figure-29. The **Post Process** dialog box will be displayed; refer to Figure-30.

Figure-29. Post Process tool

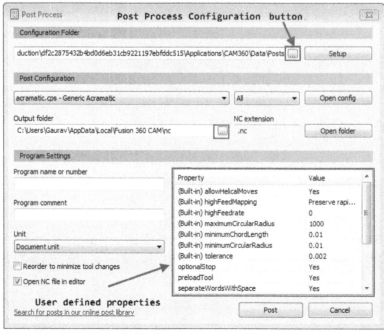

Figure-30. Post Process dialog box

- Click on the **Post Process Configuration** button from **Post Process** dialog box. The **Select Post Process Configuration folder** dialog box will be displayed; refer to Figure-44.

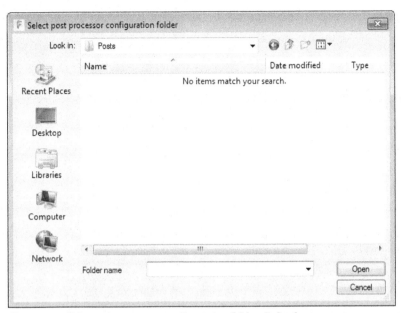

Figure-31. Select post processor configuration folder dialog box

- Create a new folder or select from the existing one to set the post processing configuration folder and click on the **Open** button.
- Click on the **Setup** button of **Post Process** dialog box and select the required option to reset the default post processor configuration.
- Click in the **Post Processing** drop-down of **Post Process** dialog box and select the required option to generate the NC codes.
- Click in the **All** drop-down of **Post Process** tab and select the required option to select the posts to show.
- Click on the **Open Config** button of **Post Process** dialog box to open the post configuration in the editor.
- Click in the **Select Output Folder** button to select the output folder for saving the NC file.
- Click on the **Open Folder** button of **Post Process** dialog box to open the output folder in windows explorer.
- Click in the **Program name or number** edit box of **Post Process** tab and enter the desired name or number of the output file.
- Click in the **Program comment** edit box of **Post Process** dialog box and enter the desired text related to the output file.
- Click in the **Unit** drop-down of **Post Process** tab and select the desired unit to specify the output unit. When unit of the output file is set to **Document unit** option then either inch or millimeter will be used.
- Select the **Record to minimize tool changes** check box of **Post Process** tab to record the operation between jobs to minimise the number of tool changes.
- Select the **Open NC file in editor** check box of **Post Process** dialog box to open the output NC file in editor.
- If you want to reset a single property or all the properties for output file then right-click on any property from **User defined property** box and select the required option. The property will be reset.

- After specifying the parameters, click on the **Post** button from **Post Process** dialog box. The **Post Process** output location folder will be displayed; refer to Figure-32.

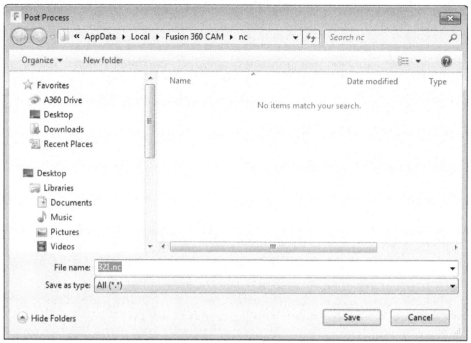

Figure-32. Post Process output location folder

- Click on the **Save** button to save the file. The output file will be save in specified folder.
- If you have selected the option to view the file in editor then editor will also be displayed; refer to Figure-33.

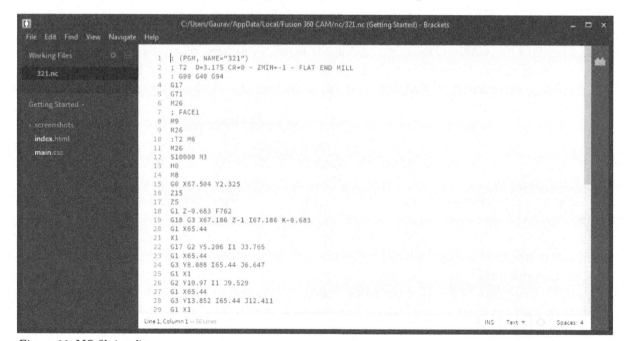

Figure-33. NC file in editor

- You can edit the codes as required by using the options of the editor.

Setup Sheet

The **Setup Sheet** tool is used to generate an overview for the NC program operator. Setup sheet provides the data related to stock, tool data, work piece position, and machine statistics. Before creating the setup sheet, you need to be sure that the proper setup is selected.

If you are on a network and the operator has access to a PC on the shop floor, you can save this to a folder that the operator can access. This will save paper and ensure the operator always has access to the most current setup information. The default **Setup Sheet** can be viewed in any standard web browser. The procedure to use this tool is discussed next.

- Click on the **Setup Sheet** tool of **ACTIONS** panel from **Toolbar**; refer to Figure-32. The **Select Setup Sheet Output Folder** dialog box will be displayed; refer to Figure-33.

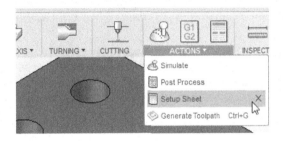

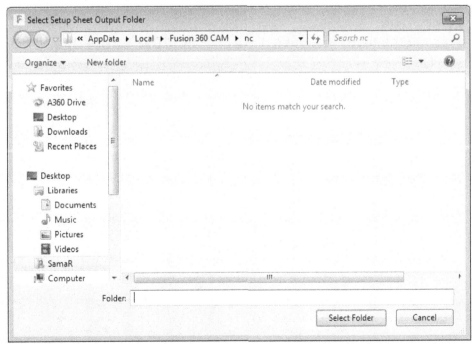

- Select the location folder as required and click on the **Select Folder** button from **Select Setup Sheet Output Folder** dialog box. The setup sheet will be created and open in the default internet browser; refer to Figure-34.

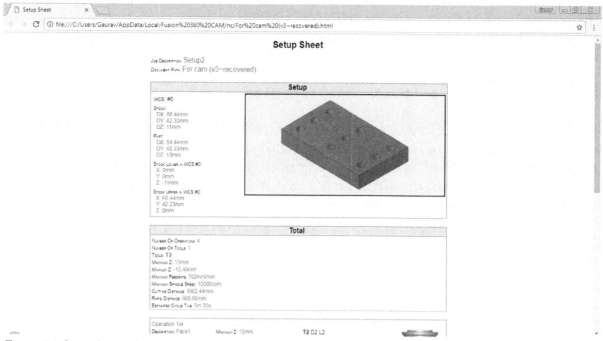

Figure-34. Created setup sheet

Generate Toolpath

The **Generate Toolpath** tool is used to regenerate the toolpath of the selected operation whenever a feature of model is modified on which the operation depends. The procedure to use this tool is discussed next.

- Click on the operation for which you want to regenerate the toolpath from **Browser Tree**.
- Click on the **Generate Toolpath** tool of **ACTIONS** drop-down from **Toolbar**; refer to Figure-35.
- You can also select the **Generate Toolpath** tool from shortcut menu; refer to Figure-36.

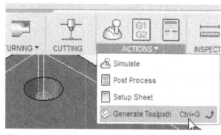

Figure-35. Generate Toolpath tool

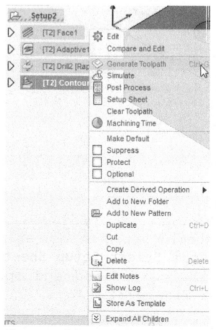

Figure-36. Selecting Generate Toolpath tool from Shortcut menu

- The toolpath will be regenerated in **Browser Tree**; refer to Figure-37.

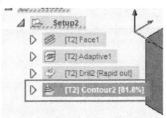

Figure-37. Regenerating toolpath

Clear Toolpath

The **Clear toolpath** tool is used to clear the toolpath of the selected operation. The procedure to use this tool is discussed next.

- Right-click on the required operation from **Browser Tree** and select the **Clear Toolpath** tool; refer to Figure-38.

Figure-38. Clear Toolpath tool

- The toolpath of selected operation will be deleted.

Machining Time

The **Machining Time** tool is used to measure and calculate the machining time for a selected operation with a high accuracy. The procedure to use this tool is discussed next.

- Right-click on the required operation from **Browser Tree** and select the **Machining Time** tool; refer to Figure-39.

Figure-39. Machining Time tool

Figure-40. MACHINING TIME dialog box

- Click in the **Feed Scale(%)** edit box of **MACHINING TIME** dialog box and enter the desired value.
- Click in the **Rapid Feed(mm/min)** edit box of **MACHINING TIME** dialog box and enter the required value.
- Click in the **Tool change Time(s)** edit box of **MACHINING TIME** dialog box and enter the desired value.
- In the **MACHINING TIME** dialog box other information related to tool will be displayed.

Tool Library

The **Tool Library** tool displays the **Tool Library** dialog box where you manage all the tools like milling, lathe, and cutting tools for your individual documents and operations, as well as libraries of predefined tools.

- Click on the **Tool Library** tool of **MANAGE** drop-down from **Toolbar**; refer to Figure-40. The **CAM Tool Library** dialog box will be displayed; refer to Figure-41.

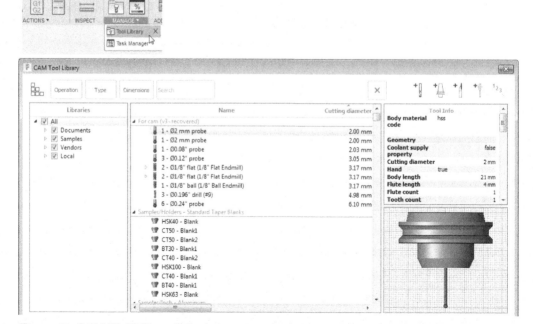

Figure-41. CAM Tool Library dialog box

- View or modify the tool as required from the **CAM Tool Library** dialog box.

Task Manager

The **Task Manager** tool is used for controlling toolpath generation. CAM allows you to continue working inside Fusion 360 while generating toolpaths in the background. The main interface for controlling toolpath generation is the **CAM Task Manager**.

- Click on the **Task Manager** tool of **MANAGE** drop-down from **Toolbar**; refer to Figure-42. The **CAM Task Manager** dialog box will be displayed; refer to Figure-43.

Figure-42. Task Manager tool

Figure-43. CAM Task Manager dialog box

- Now, generate a toolpath of any operation and you can see the progress on the **CAM Task Manager**; refer to Figure-44.

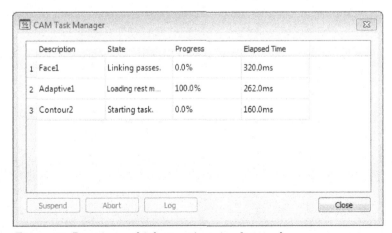

Figure-44. Running multiple operations simultaneously

FOR STUDENTS NOTES

Chapter 20

Introduction to
Simulation in Fusion 360

Topics Covered

The major topics covered in this chapter are:

- *Introduction.*
- *Types of Analyses performed in Fusion 360.*
- *FEA*
- *User Interface of Fusion 360 Simulation.*

INTRODUCTION

Simulation is the study of effects caused on an object due to real-world loading conditions. Computer Simulation is a type of simulation which uses CAD models to represent real objects and it applies various load conditions on the model to study the real-world effects. Fusion 360 is a CAD-CAM-CAE software package. In Fusion 360 Simulation, we apply loads on a constrained model under predefined environmental conditions and check the result (visually and/or in the form of tabular data). The types of analyses that can be performed in SolidWorks are given next.

TYPES OF ANALYSES PERFORMED IN FUSION 360 SIMULATION

Fusion 360 Simulation performs almost all the analyses that are generally performed in Industries. These analyses and their use are given next.

Static Analysis

This is the most common type of analysis we perform. In this analysis, loads are applied to a body due to which the body deforms and the effects of the loads are transmitted throughout the body. To absorb the effect of loads, the body generates internal forces and reactions at the supports to balance the applied external loads. These internal forces and reactions cause stress and strain in the body. Static analysis refers to the calculation of displacements, strains, and stresses under the effect of external loads, based on some assumptions. The assumptions are as follows.

- All loads are applied slowly and gradually until they reach their full magnitudes. After reaching their full magnitudes, load will remain constant (i.e. load will not vary against time).
- Linearity assumption: The relationship between loads and resulting responses is linear. For example, if you double the magnitude of loads, the response of the model (displacements, strains and stresses) will also double. You can make linearity assumption if:

1. All materials in the model comply with Hooke's Law that is stress is directly proportional to strain.
2. The induced displacements are small enough to ignore the change is stiffness caused by loading.
3. Boundary conditions do not vary during the application of loads. Loads must be constant in magnitude, direction, and distribution. They should not change while the model is deforming.

If the above assumptions are valid for your analysis, then you can perform **Linear Static Analysis**. For example, a cantilever beam fixed at one end and force applied on other end; refer to Figure-1.

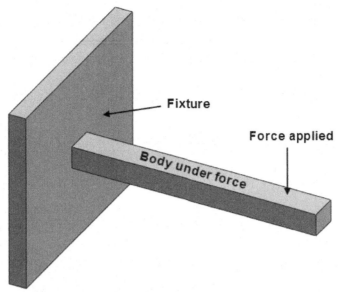

Figure-1. Linear static analysis example

If the above assumptions are not valid, then you need to perform the **Non-Linear Static analysis**. For example, force applied on an object attached with a spring; refer to Figure-2.

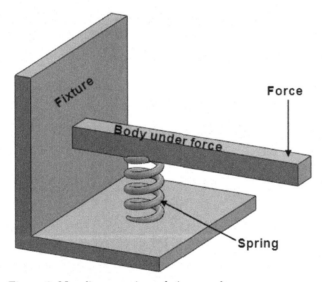

Figure-2. Non-linear static analysis example

Modal Analysis (Vibration Analysis)

By its very nature, vibration involves repetitive motion. Each occurrence of a complete motion sequence is called a "cycle." Frequency is defined as so many cycles in a given time period. "Cycles per seconds" or "Hertz". Individual parts have what engineers call "natural" frequencies. For example, a violin string at a certain tension will vibrate only at a set number of frequencies, that's why you can produce specific musical tones. There is a base frequency in which the entire string is going back and forth in a simple bow shape.

Harmonics and overtones occur because individual sections of the string can vibrate independently within the larger vibration. These various shapes are called "modes". The base frequency is said to vibrate in the first mode, and so on up the ladder. Each mode shape will have an associated frequency. Higher mode shapes have higher frequencies. The most disastrous kinds of consequences occur when a power-driven device such as a motor, produces a frequency at which an attached structure naturally vibrates. This event is called "resonance." If sufficient power is applied, the attached structure will be destroyed. Note that armies, which normally marched "in step," were taken out of step when crossing bridges. Should the beat of the marching feet align with a natural frequency of the bridge, it could fall down. Engineers must design in such a way that resonance does not occur during regular operation of machines. This is a major purpose of Modal Analysis. Ideally, the first mode has a frequency higher than any potential driving frequency. Frequently, resonance cannot be avoided, especially for short periods of time. For example, when a motor comes up to speed it produces a variety of frequencies. So, it may pass through a resonant frequency.

Thermal analysis

There are three mechanisms of heat transfer. These mechanisms are Conduction, Convection, and Radiation. Thermal analysis calculates the temperature distribution in a body due to some or all of these mechanisms. In all three mechanisms, heat flows from a higher-temperature medium to a lower temperature one. Heat transfer by conduction and convection requires the presence of an intervening medium while heat transfer by radiation does not.

There are two modes of heat transfer analysis.

Steady State Thermal Analysis

In this type of analysis, we are only interested in the thermal conditions of the body when it reaches thermal equilibrium, but we are not interested in the time it takes to reach this status. The temperature of each point in the model will remain unchanged until a change occurs in the system. At equilibrium, the thermal energy entering the system is equal to the thermal energy leaving it. Generally, the only material property that is needed for steady state analysis is the thermal conductivity. This type of analysis is available in Fusion 360.

Transient Thermal Analysis

In this type of analysis, we are interested in knowing the thermal status of the model at different instances of time. A thermos designer, for example, knows that the temperature of the fluid inside will eventually be equal to the room temperature(steady state), but designer is interested in finding out the temperature of the fluid as a function of time. In addition to the thermal conductivity, we also need to specify density, specific heat, initial temperature profile, and the period of time for which solutions are desired. Till the time of writing this book, the transient thermal analysis was not available in Fusion 360.

Thermal Stress Analysis

The Thermal Stress Analysis is performed to check the stresses induced in part when thermal and structural loads act on the part simultaneously. Thermal Stress Analysis is important in cases where material expands or contracts due to heating or cooling of the part to certain temperature in irregular way. One example where thermal stress analysis finds its importance is two material bonded strip working in a high temperature environment.

Event Simulation

The Event Simulation analysis is used to study the effect of object velocity, initial velocity, acceleration, time dependent loads, and constraints in the design. The results of this analysis include displacements, stresses, strains, and other measurements throughout a specified time period. You can perform this analysis when you need to check the effect of throwing a phone from some height or similar cases where motion is involved.

Shape Optimization

The Shape Optimization in Fusion 360 is not an analysis but a study to find the shape of part which utilizes minimum material but sustains the applied load up to required factor of safety.

Till this point, you have become familiar with the analyses that can be performed by using Fusion 360. But, do you know how the software analyze the problems. The answer is FEA.

FEA

FEA, Finite Element Analysis, is a mathematical system used to solve real-world engineering problems by simplifying them. In FEA by Fusion 360, the model is broken into small elements and nodes. Then, distributed forces are applied on each element and node. The cumulative result of forces is calculated and displayed in results. Note that Fusion 360 uses **Linear Tetrahedron** element with 4 nodes to mesh 3D solids and **Line** element for bolt connectors (only available in Fusion 360 Ultimate). 2D and Planar (shell) elements are not currently supported in Autodesk Fusion 360 till the time we are writing this book.

As we are ready with some basic information about simulation in Fusion 360. Let us get started with initiating the simulation environment of Fusion 360.

STARTING SIMULATION IN FUSION 360

In Fusion 360, every workspace is available in a seamless manner. To start simulation in Fusion 360, click on the **Workspace** drop-down and select the **Simulation** option; refer to Figure-3. The simulation workspace will become active; refer to Figure-4. Also, the **New Study** dialog box will be displayed asking you to select the analysis type you want to perform.

Figure-3. Simulation option

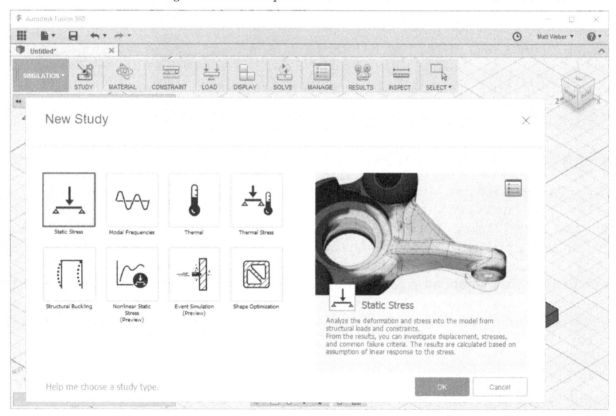

Figure-4. Simulation workspace of Fusion 360

PERFORMING AN ANALYSIS

The static stress analysis is performed when the load is stable and the object deforms according to Hooke's Law. The procedure to start static stress analysis is given next. You can apply the same procedure for starting other analyses too.

- Click on the **Static Stress** button from the **New Study** dialog box and click on the **OK** button. The tools required to perform static stress analysis will be displayed in the **Toolbar**; refer to Figure-5.

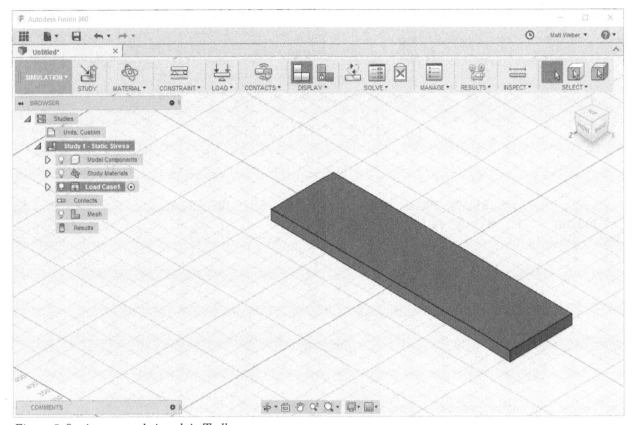

Figure-5. Static stress analysis tools in Toolbar

Various tools and options of **Toolbar** in Simulation environment are discussed next.

STARTING NEW SIMULATION STUDY

While you are working on an analysis if you need to start another analysis then you can do so by using the **New Simulation Study** button. The procedure is given next.

- Click on the **Study** tool from the **Toolbar**. The **New Study** dialog box will be displayed as discussed earlier.
- Double-click on the desired button to perform respective analysis.

STUDY MATERIAL

Study material is the material applied to model with all the physical properties so that you can check the effect of load on actual material conditions. The tools related to study material are available in the **MATERIAL** drop-down of the **Toolbar**; refer to Figure-6. These tools are discussed next.

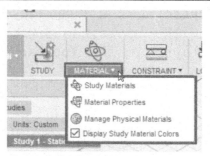

Figure-6. MATERIAL drop-down

Applying Study Material

Study material is applied to assign physical properties of material to the model. The procedure to apply study material is given next.

- Click on the **Study Material** tool from the **MATERIAL** drop-down in the **Toolbar**. The **Study Materials** dialog box will be displayed; refer to Figure-7.

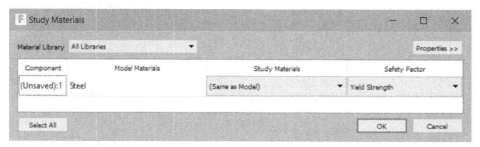

Figure-7. Study Materials dialog box

- Click in the drop-down of **Study Materials** column in the dialog box for the current component and select the desired material; refer to Figure-8.

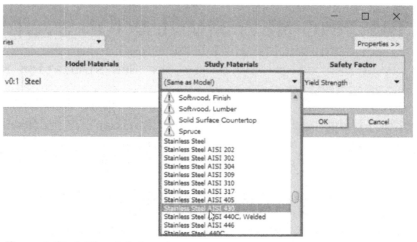

Figure-8. Study Materials drop-down

- Select the desired safety factor criteria from the drop-down in **Safety Factor** column of the dialog box. There are two options in this drop-down; **Yield Strength** and **Ultimate Tensile Strength**. Yield Strength is the point where metal starts to permanently deform. Ultimate Tensile Strength is the point after which the metal becomes so weak that it can break.

* After specifying all the desired parameters, click on the **OK** button.
  The material will be applied.

Displaying Material Properties

All the properties of different materials in material library can
be checked by using the **Material Properties** button. The procedure is
discussed next.

* Click on the **Material Properties** tool from the **Material** drop-down in
  the **Toolbar**. The **Material Properties** dialog box will be displayed;
  refer to Figure-9.

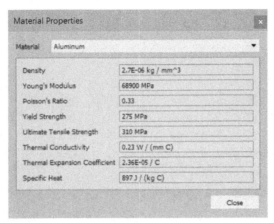

Figure-9. Material Properties dialog box

* Select the material from the **Material** drop-down at the top in the
  dialog box to check the material properties.
* Click on the **Close** button to exit the dialog box.

Managing Physical Material

The **Manage Physical Materials** tool is used to manage physical properties
of material. If you want to edit any parameter of material before using
it in analysis then you can do so by using this tool. The procedure
is given next.

* Click on the **Manage Physical Materials** tool from the **Material** drop-down
  in the **Toolbar**. The **Material Browser** dialog box will be displayed;
  refer to Figure-10.

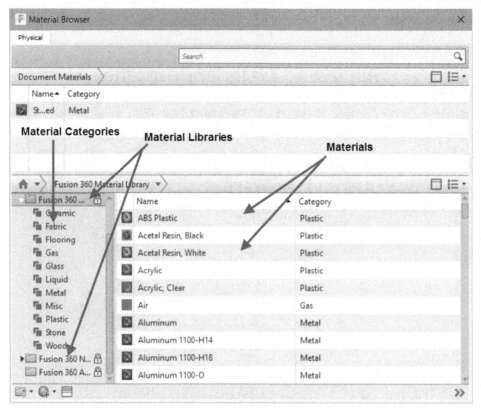

Figure-10. Material Browser dialog box

- Select the desired material library and category from the left area of the **Material Browser** dialog box.
- Hover the cursor on the material you want to edit from the right area and click on the **Adds material to favorites and displays in editor** button; refer to Figure-11. The editing options will be displayed at the right in dialog box; refer to Figure-12.

Figure-11. Editing material

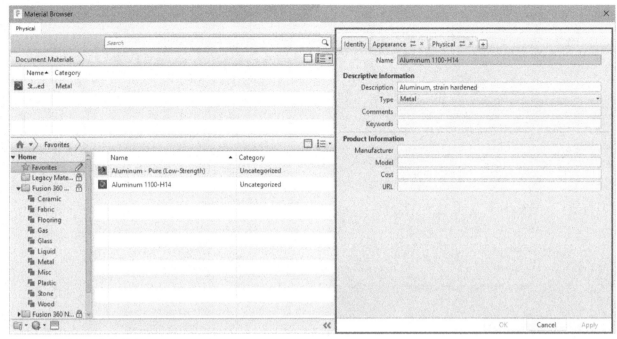

Figure-12. Material editing area

- Click at the desired tab in the right area of the dialog box and specify the parameters related to material.
- Click on the **OK** button from the right area to apply the changes.

Creating New Material

If you want to create a new material then follow the procedure given next.

- Click on the **Create New Library** tool from the **Creates, opens, and edits user-defined libraries** drop-down at the bottom left corner of the **Material Browser** dialog box; refer to Figure-13. The **Create Library** dialog box will be displayed; refer to Figure-14.

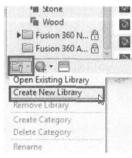

Figure-13. Create New Library option

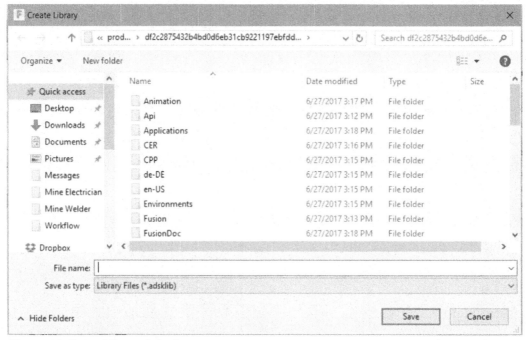

Figure-14. Create Library dialog box

- Specify the desired name of the library in **File name** edit box and save the library file at desired location. A new library will be added.

- Select the newly added library and click on the **Create Category** tool from the **Creates, opens, and edits user-defined libraries** drop-down at the bottom left corner of the **Material Browser** dialog box. A new category will be added to the library. Right-click on the category and select **Rename** if you want to rename it as desired.

- Click on the **Create New Material** option from the **Creates and duplicates materials** drop-down at the bottom in the dialog box as shown in Figure-15. The **Select Material Browser** dialog box will be displayed along with editing options in the **Material Browser** dialog box.

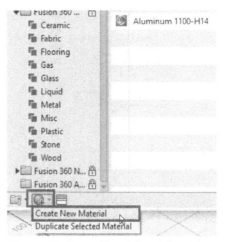

Figure-15. Create New Material option

- Close the **Select Material Browser** dialog box and specify the desired parameters of material in the right area.

- To apply physical or appearance properties to material, click on the + sign next to **Identity** tab in the editing area of the dialog box. A drop-down will be displayed; refer to Figure-16. Select the desired option from the drop-down (like, we have selected the **Physical** option). The **Asset Browser** dialog box will be displayed; refer to Figure-17.

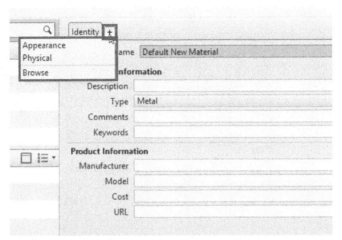

Figure-16. Adding properties of material

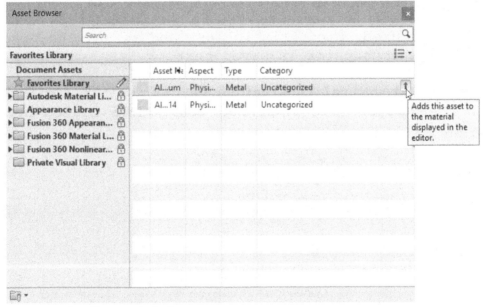

Figure-17. Asset Browser dialog box

- Click on the **Adds this asset to the material displayed in the editor** button as shown in Figure-17 to copy the physical properties of material.
- Close the **Asset Browser** dialog box. The physical properties have been assigned to the new material. Modify the values as required.
- Click on the **Apply** button. The material will be added in the library.
- Add more materials as required and then close the dialog box.

Displaying Study Material Colors

By default, appearance assigned to the part in the **Model** workspace is displayed in the **Simulation** workspace. If you want to display appearance of study material then select the **Display Study Material Colors** check box from the **MATERIAL** drop-down.

APPLYING CONSTRAINTS

Constraint are used to restrict motion of part when load is applied to form equilibrium. The tools to apply constraints are available in **CONSTRAINT** drop-down of the **Toolbar**; refer to Figure-18.

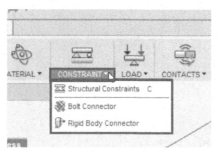

Figure-18. CONSTRAINT drop-down

The procedures to apply different type of constraints are discussed next.

Applying Structural Constraints

The structural constraints are used to apply different type of structural constraints like fixed, pin, frictionless and so on. The procedure to apply structural constraint is given next.

*   Click on the **Structural Constraints** tool from the **CONSTRAINT** drop-down in the **Toolbar**. The **STRUCTURAL CONSTRAINTS** dialog box will be displayed; refer to Figure-19.

Figure-19. STRUCTURAL CONSTRAINTS dialog box

Fixed Constraint

*   Select the **Fixed** option from the **Type** drop-down if you want to fix selected faces/edges/vertices of the part.
*   Select the face/edge/vertex that you want to be fixed.
*   Select the desired axis button from the **Axis** section. Like, select the **Ux** button if you want to restrict movement along X axis. By default, all the three buttons are selected and hence the movement along all the three axes is restricted. Refer to Figure-20.

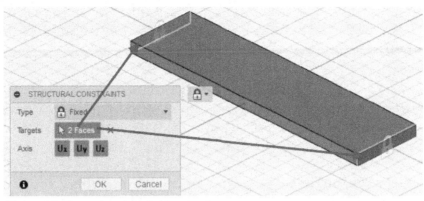

Figure-20. Faces selected for fixed constraint

- Click on the **OK** button from the dialog box to fix the selected geometries.

Pin Constraint

The Pin constraint is used to restrict radial, axial, and tangential movement of a cylindrical part. The procedure to use this constraint is given next.

- Select the **Pin** option from the **Type** drop-down in the **STRUCTURAL CONSTRAINTS** dialog box. The options in the dialog box will be displayed as shown in Figure-21.

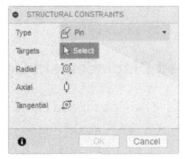

Figure-21. STRUCTURAL CONSTRAINTS dialog box with Pin option selected

- Select the cylindrical face on which you want to apply pin constraint; refer to Figure-22.

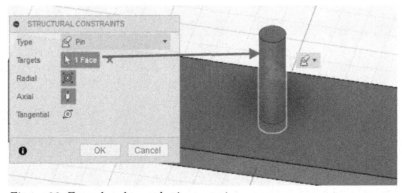

Figure-22. Face selected to apply pin constraint

- Select the desired buttons to restrict the respective motion. For

example, select the **Radial** button to restrict radial motion.
* Similarly, specify the other parameters as required and click on the **OK** button.

Frictionless Constraint

The Frictionless constraint is used to restrict the movement of object perpendicular to selected face. However, the object is free to move in the plane. The procedure to apply this constraint is given next.

* Select the **Frictionless** option from the **Type** drop-down in the **STRUCTURAL CONSTRAINTS** dialog box. The options in the dialog box will be displayed as shown in Figure-23.

Figure-23. STRUCTURAL CONSTRAINTS dialog box with Frictionless option selected

* Select the desired face on which you want to apply the frictionless constraint.
* Click on the **OK** button to exit.

Prescribed Displacement Constraint

The **Prescribed displacement** constraint is used to apply fixed constraint at specified displacement. The procedure to use this constraint is given next.

* Select the **Prescribed Displacement** option from **Type** drop-down in the **STRUCTURAL CONSTRAINTS** dialog box. You will be asked to select the face/edge/vertex.
* Select the desired geometry. The options in the dialog box will be modified; refer to Figure-24.

Figure-24. STRUCTURAL CONSTRAINTS dialog box with Prescribed Displacement option selected

* Specify the desired parameters and click on the **OK** button.

Applying Bolt Connector Constraint

Bolt connector is used to apply connection similar to bolt fastener connection in assemblies. Note that bolt connector represents the nut-bolt connection or threaded nut connection mathematically. The procedure to apply bolt connector is given next.

* Click on the **Bolt Connector** tool from the **CONSTRAINT** drop-down in the **Toolbar**. The **BOLT CONNECTOR** dialog box will be displayed; refer to Figure-25.

Figure-25. BOLT CONNEC-TOR dialog box

* Select the round edge of the part where you want the bolt head to be placed. The **BOLT CONNECTOR** dialog box will be displayed as shown in Figure-26.

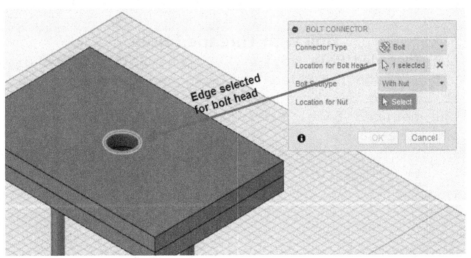

Figure-26. BOLT CONNECTOR dialog box after selecting bolt head location

Bolt Fastener with Nut

* Select the **With Nut** option from the **Bolt Subtype** drop-down if you want to create a bolt-nut fastener constraint. You will be asked to select the round edge where nut will be placed.
* Select the desired edge. The **BOLT CONNECTOR** dialog box will be displayed with preview of bolt-nut fastener; refer to Figure-27.

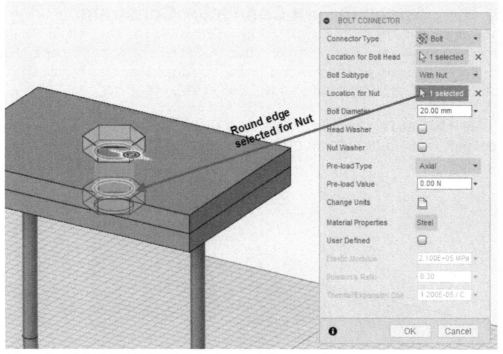

Figure-27. BOLT CONNECTOR dialog box after selecting nut location

- Specify the parameters like bolt diameter, bolt washer, nut washer, pre-load etc. and click on the **OK** button to create the constraint.

Bolt with Threaded Hole

- Select the **Threaded Hole** option from the **Bolt Subtype** drop-down if you want to create threaded bolted connection. You will be asked to select face to be threaded.
- Select the desired face. Preview of the bolt will be displayed; refer to Figure-28.

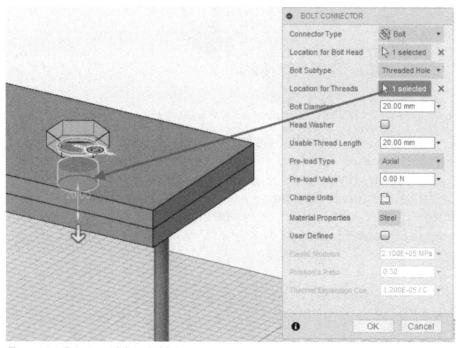

Figure-28. Preview of threaded bolt

- Specify the parameters as required and click on the **OK** button.

Rigid Body Connector Constraint

The Rigid Body Connector constraint is used where a vertex of one component is to be rigidly connected with face, edge, or vertex of other body. The procedure to use this constraint is given next.

- Click on the **Rigid Body Connector** tool from the **CONSTRAINT** drop-down in the **Toolbar**. The **Rigid Body Connector** dialog box will be displayed; refer to Figure-29.

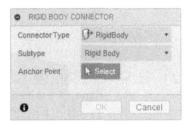

Figure-29. RIGID BODY CON-
NECTOR dialog box

- Select the desired vertex that you want to be anchor point for connection on the first body/component. You will be asked to select the dependent entities.
- Select the points/faces/edges that are dependent on the anchor point for movement; refer to Figure-30.

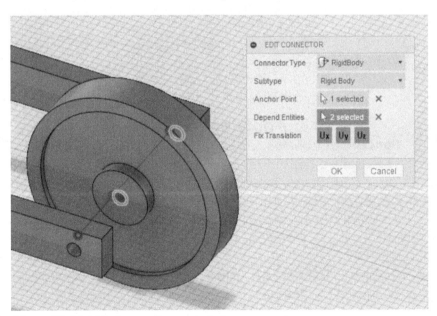

Figure-30. Rigid body connector constraint

- Similarly, you can use the **Interpolation** option from the **Subtype** drop-down to create rigid connection with translational and rotational constraining.

APPLYING LOADS

Loads in Fusion 360 are the representation of forces and loads applied on the part in real application. The tools to apply loads are available in the **LOAD** drop-down of the **Toolbar**; refer to Figure-31. Various tools in this drop-down are discussed next.

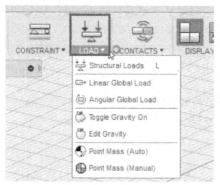

Figure-31. LOAD drop-down

Applying Structural Loads

There are various structural loads that can be applied on the object like force, pressure, moment, remote force, bearing load, and hydrostatic pressure. The procedures to apply different loads are discussed next.

* Click on the **Structural Loads** tool from the **LOAD** drop-down in the **Toolbar**. The **STRUCTURAL LOADS** dialog box will be displayed; refer to Figure-32.

Figure-32. STRUCTURAL LOADS dialog box

Applying Force

* By default, **Force** option is selected in the **STRUCTURAL LOADS** dialog box. If not selected then select the **Force** option from the **Type** drop-down. You will be asked to select face/edge/point on which force is to be applied.
* Select the desired face/edge/point. The options in the **STRUCTURAL LOADS** dialog box will be modified according to geometry selected; refer to Figure-33.
* Select the desired **Direction Type** button from the dialog box. If you have selected the **Normal** button 🔳 then force will be applied perpendicular to the selected face. You can use the **Flip** 🔳 button below it to reverse direction of force; refer to Figure-34.

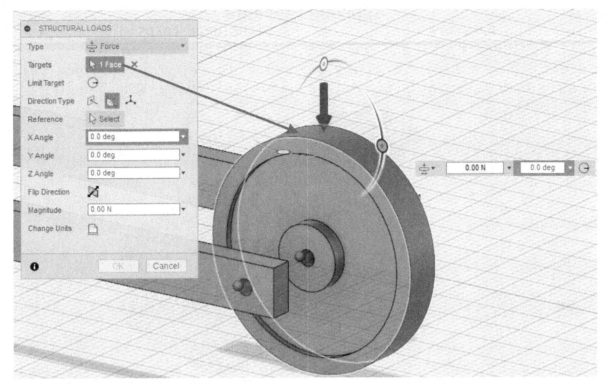

Figure-33. Face selected to apply force

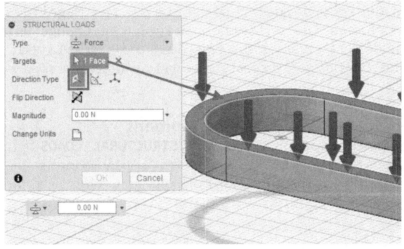

Figure-34. Force applied with Normal button selected

- Select the **Angle (delta)** button ⬚ from the **Direction Type** section of the dialog box if you want to apply force at some angle; refer to Figure-33. Specify the desired angle values in the **X Angle, Y Angle,** and **Z Angle** edit boxes. Select the **Flip** button to reverse the direction if required. Note that the **Limit Target** button is also available in the dialog box. Select this button and specify the radius range in which the force will be applied.
- Select the **Vector** button from the **Direction Type** section if you want to specify force value along each vector direction.
- To change the unit of load, click on the **Change Units** button if you want to change the unit for load.
- Click on the **OK** button from the dialog box to apply the load.

Applying Pressure

- Select the **Pressure** option from the **STRUCTURAL LOADS** dialog box. You will be asked to select the faces to apply pressure.
- Select the face(s) to apply pressure force. The options in the dialog box will be displayed as shown in Figure-35.

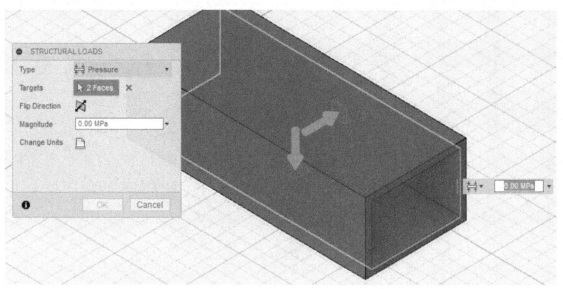

Figure-35. STRUCTURAL LOADS dialog box with pressure option

- Specify the desired value of pressure in the **Magnitude** edit box. If you want to change the unit then select the **Change Units** button and specify the desired value in the edit box displayed.
- Click on the **OK** button from the dialog box to apply the pressure load.

Applying Moment

- Select the **Moment** option from the **STRUCTURAL LOADS** dialog box. You will be asked to select the faces to apply moment.
- Select the desired face. The options in the dialog box will be modified; refer to Figure-36.

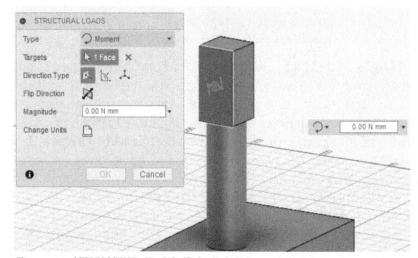

Figure-36. STRUCTURAL_LOADS_dialog_box_with_moment_option.png

- Set the desired value of moment in the **Magnitude** edit box.

- Set the other options as discussed earlier and click on the **OK** button.

Applying Remote Force

The remote force is used to represent effect of load applied at different location on the selected location; refer to Figure-37. The procedure to apply load is given next.

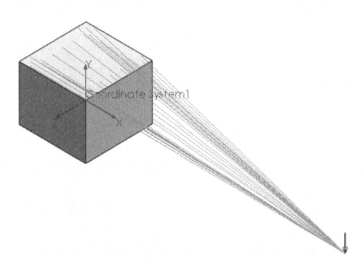

Figure-37. Remote load applied

- Select the **Remote Force** option from the **Type** drop-down in the **STRUCTURAL LOADS** dialog box. You will be asked to select a location to apply force.
- Select the desired face/edge/point. The options in the dialog box will be displayed as shown in Figure-38.

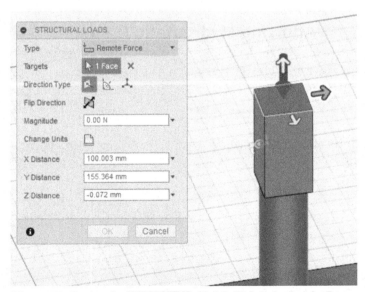

Figure-38. Options in STRUCTURAL LOADS dialog box for Remote Force

- Set the X, Y, and Z distances of the load location in the **X Distance**, **Y Distance**, and **Z Distance** edit boxes of the dialog box, respectively.
- Specify the other parameters as discussed earlier.
- Click on the **OK** button to apply force.

Applying Bearing Load

Bearing load is the force exerted by bearing on round face of the part. The procedure to apply bearing load is given next.

- Select the **Bearing Load** option from the **Type** drop-down in the **STRUCTURAL LOADS** dialog box. You will be asked to select the round face on which bearing load is to be applied.
- Select the desired face. The options in the dialog box will be displayed as shown in Figure-39.

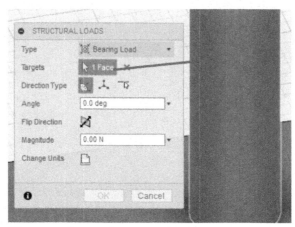

Figure-39. Options for Bearing Load

- Specify the desired parameters as discussed earlier. Note that bearing load is a directional force and applicable on only half of the full 360 cylindrical face.
- After specifying the parameters, click on the **OK** button to apply load.

Applying Hydrostatic Pressure

Hydrostatic pressure is a linearly varying pressure exerted by fluid on the surface of part. This force is applicable when the part is in contact with high volume of fluid. The procedure to apply this load is given next.

- Select the **Hydrostatic Pressure** option from the **Type** drop-down in the **STRUCTURAL LOADS** dialog box. You will be asked to select the face(s) to apply hydrostatic pressure.
- Select the desired face(s). The options in the dialog box will be displayed as shown in Figure-40. Note that if you are applying hydrostatic pressure for the first time then a message box will be displayed prompting you to activate gravity. Activate the gravity by clicking on the **Yes** button.
- Click on the **Select** button for **Select Surface Point** section in the dialog box and select the desired point up to which the fluid is filled in the system.

*Figure-40. Options in STRUC-
TURAL LOADS dialog box for
Hydrostatic Pressure*

- Specify the desired offset value for fluid surface point in the **Offset Distance** edit box.
- Select the desired fluid type from the **Fluid Type** drop-down. If you have a different fluid that the options available then select the **Custom** option and specify the density of fluid in the **Density** edit box.
- Click on the **OK** button after specifying the desired values to apply load.

Applying Linear Global Load (Acceleration)

Linear Global Load is the applied when the whole system is under acceleration like an object placed in an accelerating car. The procedure to apply linear global load is given next.

- Click on the **Linear Global Load** tool from the **LOAD** drop-down in the **Toolbar**. The **LINEAR GLOBAL LOAD** dialog box will be displayed as shown in Figure-41 and you will be asked to select a reference for acceleration direction.

*Figure-41. LINEAR GLOBAL
LOAD dialog box*

- Select the desired face/edge to specify the direction of acceleration.
- Specify the desired value of acceleration in the **Magnitude** edit box.
- Similarly, specify the desired angle value for X, Y, and Z in the **X Angle**, **Y Angle**, and **Z Angle** edit boxes.
- Specify the other parameters as required and click on the **OK** button.

Applying Angular Global Load

The Angular Global Load is applied to give angular velocity or angular acceleration to the system. The procedure to apply angular global load is given next.

- Click on the **Angular Global Load** tool from the **LOAD** drop-down in the **Toolbar**. The **ANGULAR GLOBAL LOAD** dialog box will be displayed; refer to Figure-42.

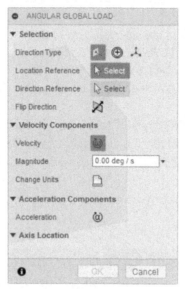

Figure-42. ANGULAR GLOB-AL LOAD dialog box

- Select the desired location for the applying velocity or acceleration. The input boxes will be displayed to apply angular velocity and specify the location of the exerting point along X direction.
- Click on the **Select** button from the **Direction Reference** section and select the face to define axis for angular velocity/acceleration.
- Specify the other parameters as discussed earlier.
- Click on the **Acceleration** button if you want to specify the acceleration also from the **Acceleration Components** rollout in the dialog box. Specify the related parameters as discussed earlier.
- Click on the **OK** button to apply angular velocity/acceleration.

Toggling Gravity On/Off

Anyone who has passed high school should be knowing what is gravity!! The procedure to activate and de-activate gravity is given next.

- Click on the **Toggle Gravity On** button from the **LOAD** drop-down in the **Toolbar** if the gravity is off and you want to activate it.
- Click on the **Toggle Gravity Off** button from the **LOAD** drop-down in the **Toolbar** if the gravity is on and you want to de-activate it.

Editing Gravity

The **Edit Gravity** tool is used to edit the value and direction of gravity acting the system. The procedure is given next.

- Click on the **Edit Gravity** tool from the **LOAD** drop-down in the **Toolbar**. The **EDIT GRAVITY** dialog box will be displayed; refer to Figure-43.

Figure-43. EDIT GRAVITY dialog box

- Select the desired direction reference (face/edge) to specify the direction of gravity.
- Specify the desired value of gravity in the **Magnitude** edit box.
- Click on the **OK** button to change the value.

Apply Point Mass (Auto)

The **Point Mass (Auto)** tool is used to replace the real component with a point mass. This phenomena is used to simplify simulation calculations. The procedure to use this tool is given next.

- Click on the **Point Mass (Auto)** tool from the **LOAD** drop-down in the **Toolbar**. The **POINT MASS (AUTO)** dialog box will be displayed; refer to Figure-44.

Figure-44. POINT MASS (AUTO) dialog box

- Select the object that you want to be replaced by point mass.
- Click on the **Select** button for **Geometries** section and select the face on which you want to place the mass.
- Specify the desired mass value in the **Mass** edit box of the dialog box; refer to Figure-45.

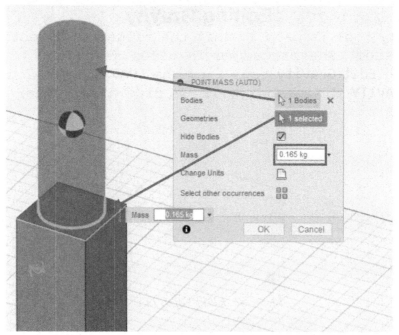

Figure-45. Specifying mass value of point mass

- Click on the **OK** button to apply point mass.

Apply Point Mass (Manual)

The **Point Mass (Manual)** tool works in the same way as **Point Mass (Auto)**. The only difference between the two is that in case of **Point Mass (Manual)** tool, you do not need to select a body to be replaced by mass.

APPLYING CONTACTS

Contacts are applied when there are two or more bodies/components in contact with each other and load is transferred between them during simulation. The tools to apply contact are available in the **CONTACTS** drop-down; refer to Figure-46. These tools are discussed next.

Figure-46. CONTACTS drop-down

Applying Automatic Contacts

If you are performing analysis on an assembly with multiple components then applying automatic contact is a very important steps. Without applying automatic contact, you can not perform analysis of assembly in Fusion 360. The procedure to apply automatic contact is given next.

- Click on the **Automatic Contacts** tool from the **CONTACTS** drop-down in the **Toolbar**. The **AUTOMATIC CONTACTS** dialog box will be displayed; refer to Figure-47.

Figure-47. AUTOMATIC CONTACTS dialog box

- Specify the desired value of tolerance in the edit box of the dialog box. The tolerance specified here is the maximum gap up to which the software will apply contacts. If the gap between two components is more than the specified value then Fusion will not apply any contact automatically.
- Click on the **Generate** button. The contacts will be generated automatically.

Modifying Contacts

Automatic Contacts tool applies the same contact to the components in the assembly. The procedure to check and modify automatically applied contacts is given next.

- To check the automatically applied contacts, click on the **Edit** button displayed on hovering the cursor over **Contacts** node of the **BROWSER**; refer to Figure-48. The **CONTACTS MANAGER** will be displayed; refer to Figure-49. Here, you can check the contacts automatically applied.

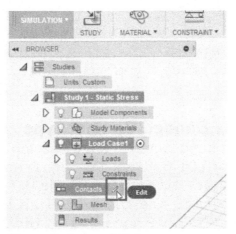

Figure-48. Contacts Edit button

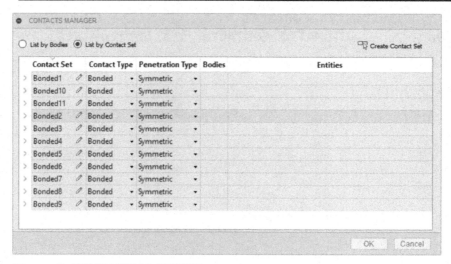

Figure-49. CONTACTS MANAGER

- Select the contact that you want to edit from the **Contact Set** column in the **CONTACTS MANAGER.** The contact will be highlighted in the model.
- To modify the contact, click in the **Contact Type** column for the selected contact. List of different available contacts will be displayed; refer to Figure-50.

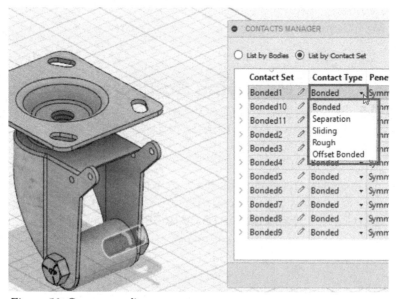

Figure-50. Contact type list

- Select the desired contact type to change. Various contact types are discussed next.

Bonded Contact Type

The bonded contact is used when there is no relative displacement between two connected solid bodies is required. This type of contact is used to glue together different solids of an assembly. The two surfaces that are in contact are classified as master and slave. Every node in the slave surface(slave nodes) is tied to a node in the master surface(master node) by a constraint. You will learn about master surface and slave surface in the next topic.

Separation Contact Type

The separation contact is applied when separation between parts is allowed but prohibits part penetration.

Sliding Contact Type

The sliding contact is a type of contact which allows displacement tangential to the contacting surface but no relative movement along the normal direction. This type of contact constraint is used to simulate sliding movement in the assembly. The two surfaces that are in contact are classified as master and slave. Every node in slave surface(slave nodes) is tied to a node in the master surface(master node) by this constraint.

Rough Contact Type

The rough contact is used when two parts cannot slide over each other as friction between them is very high. Note that the parts cannot penetrate in each other if this contact type is selected.

Offset Bonded Type

The offset bonded contact is used when two parts are at a distance in assembly but you want them to be bonded as bonded contact type.

Applying Manual Contacts

Applying automatic contacts is the first step for performing analysis on the assembly but **Automatic Contacts** apply the same contact to all the assembly joints which can be changed by using **CONTACTS MANAGER.** But what to do if automatic contacts are not generated for required faces. The **Manual Contacts** tool is used to apply these contacts. The procedure to use this tool is given next.

- Click on the **Manual Contacts** tool from the **CONTACTS** drop-down in the **Toolbar**. The **MANUAL CONTACTS** dialog box will be displayed as shown in Figure-51. Also, you will be asked to select the master body.

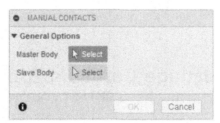

Figure-51. MANUAL CONTACTS
dialog box

- Select the first body. You will be asked to select the slave body.
- Select the second body. You will be asked to select the face/edge on the first body.
- Select the desired face/edge at which the body is in contact with other body.

- Click on the **Select** button for **Selection Set 2** section in the dialog box and select the contacting face/edge on the second body. The options in the dialog box will be displayed as shown in Figure-52.

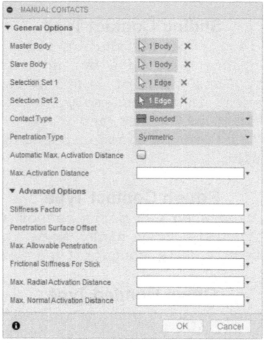

Figure-52. Options in MANUAL CONTACTS dialog box

- Select the desired contact type from the **Contact Type** drop-down in the dialog box.
- Select the desired option from the **Penetration Type** drop-down. If you have selected the **Symmetric** option then both master component and slave component cannot penetrate into each other. If the **Unsymmetric** option is selected then the master component can penetrate the slave component.
- Specify the desired maximum activation distance in the **Max. Activation Distance** edit box. This parameter is useful when parts are not coincident and a small gap is present between them. Choose a small value to prevent conflicting contact interactions.
- Similarly, specify the other parameters as required.
- Click on the **OK** button to create the contact.

Manage Contacts Tool

The **Manage Contacts** tool in the **CONTACTS** drop-down is used to edit contacts earlier applied. On clicking this tool, the **CONTACTS MANAGER** will be displayed. The options in the **CONTACTS MANAGER** have already been discussed.

SOLVING ANALYSIS

Once you have applied all the information required to perform analysis, you need to perform a check whether you have specified the required information or not. Once the system says, it has the required information then you are good to go for analysis. The tools to perform pre-check and analysis are available in the **SOLVE** drop-down of the **Toolbar**; refer to Figure-53. These tools are discussed next.

Figure-53. SOLVE drop-down

Performing Pre-check

Pre-checking is an important step before performing analysis. Although the tool will not tell you that you have specified load at wrong place or other design faults but the tool will warn you that you have not specified load, constraint, contact like parameters which need to be specified before performing analysis. The procedure to perform pre-check is given next.

• Click on the **Pre-check** tool from the **SOLVE** drop-down in the **Toolbar**. If there is any parameter left to be specified then **Cannot Solve** dialog box will be displayed; refer to Figure-54. Apply the parameters which are not specified. If all the parameters are specified then **Ready to Solve** dialog box will be displayed. Click on the **OK** button and perform meshing.

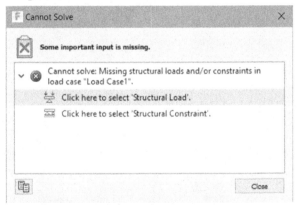

Figure-54. Cannot Solve dialog box

Meshing

Meshing is the base of FEM. Meshing divides the solid/shell models into elements of finite size and shape. These elements are joined at some common points called nodes. These nodes define the load transfer from one element to other element. Meshing is a very crucial step in design analysis. The automatic mesher in the software generates a mesh based on a global element size, tolerance, and local mesh control specifications. Mesh control lets you specify different sizes of elements for components, faces, edges, and vertices.

The software estimates a global element size for the model taking into consideration its volume, surface area, and other geometric details. The size of the generated mesh (number of nodes and elements) depends on the geometry and dimensions of the model, element size, mesh tolerance, mesh control, and contact specifications. In the early stages of design analysis where approximate results may suffice, you can specify a larger element size for a faster solution. For a more accurate solution, a smaller element size may be required.

Meshing generates 3D tetrahedral solid elements and 1D beam elements. A mesh consists of one type of elements unless the mixed mesh type is specified. Solid elements are naturally suitable for bulky models. Shell elements are naturally suitable for modeling thin parts (sheet metals), and beams and trusses are suitable for modeling structural members.

The procedure to create the mesh of the solid is given next.

- Click on the **Generate Mesh** tool from the **SOLVE** drop-down in the **Toolbar**. The system will start creating mesh and progress bar will be displayed. Once the operation is complete. The mesh will be displayed; refer to Figure-55.

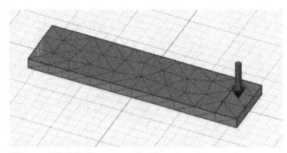

Figure-55. Mesh automatically created

- If you want to change the automatic mesh settings then click on the **Edit** button displayed on hovering the cursor over **Mesh** in the **BROWSER**; refer to Figure-56. The **Mesh Settings** dialog box will be displayed; refer to Figure-57.

Figure-56. Edit button for mesh

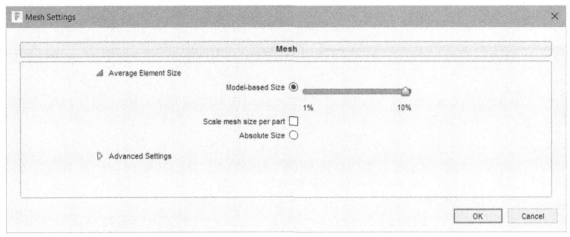

Figure-57. Mesh Settings dialog box

- Move the slider towards left to decrease the average size of mesh. Note that decreasing the size will increase the analysis solution time.
- Select the **Scale mesh size per part** check box if you want to system to scale mesh size of each individual part in assembly based on its size. In other words, if there are 10 parts in assembly with different sizes then mesh elements of each part will have different size.
- If you want to specify a value for all mesh element sizes then select the **Absolute Size** radio button and specify the desired value for element size in the edit box next to it.
- Select the desired element order from the **Element Order** drop-down in the **Advanced Settings** node of the dialog box. Select the **Parabolic** option element order for complex parts which require higher degree of elements. For simple parts, select the **Linear** option from the drop-down.
- Select the **Create Curved Mesh Elements** check box if you want the mesh elements to follow curvature of round/curved faces of the part.
- Similarly, specify the other parameters in the **Advanced Settings** node and click on the **OK** button.

Applying Local Mesh Control

Local mesh control is used when you need to increase or decrease the size of elements in a finite area of the part. The procedure to apply local mesh control is given next.

- Click on the **Local Mesh Control** tool from the **MANAGE** drop-down in the **Toolbar**. The **LOCAL MESH CONTROL** dialog box will be displayed; refer to Figure-58. You will also be asked to select a face/edge.
- Select the face/edge(s) for which you want to increase/decrease the element size. If you are working on an assembly then you can select the body after clicking on the **Select** button from **Body Selection** section of the dialog box.
- After selecting the desired geometries, move the slider towards coarse or fine to change the mesh size.

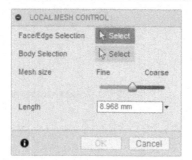

Figure-58. LOCAL MESH CONTROL dialog box

- Click on the **OK** button to apply the change.

Note that changes in mesh will not be reflected automatically. To update mesh, right-click on the **Mesh** in the **Browser** and select the **Generate Mesh** option from the shortcut menu; refer to Figure-59.

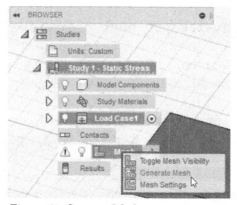

Figure-59. Generate Mesh option

Adaptive Mesh Refinement

Adaptive mesh refinement is used when dynamic refinement of mesh is required at the stress-strain locations to increase accuracy . The procedure to apply adaptive mesh refinement is given next.

- Click on the **Adaptive Mesh Refinement** tool from the **MANAGE** drop-down in the **Toolbar**. The **Adaptive Mesh Refinement** dialog box will be displayed; refer to Figure-60.

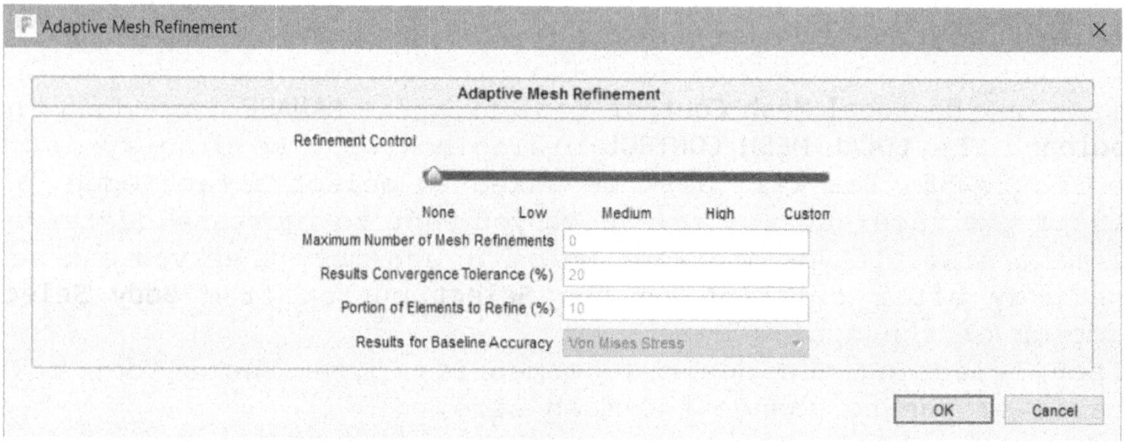

Figure-60. Adaptive Mesh Refinement dialog box

- Move the **Refinement Control** slider towards right to increase refinement level of meshing at stress/strain areas.
- Click on the **OK** button to apply refinement.

Solving Analysis

Once you have specified all the parameters then it is time to solve the analysis. The procedure to do so is given next.

- Click on the **Solve** button from the **SOLVE** drop-down in the **Toolbar**. The **Solve** dialog box will be displayed; refer to Figure-61.

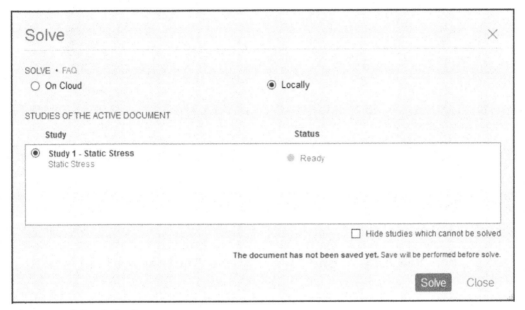

Figure-61. Solve dialog box

- Make sure **Ready** is displayed in the **Status** column of the study to be performed. Click on the **Solve** button. The results will be displayed; refer to Figure-62.

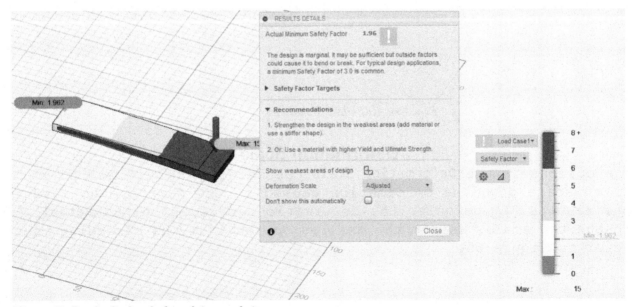

Figure-62. Results displayed after solving analysis

To check the weakest areas of design, click on the **Show weakest areas of design** button from the **RESULTS DETAILS** dialog box. The weakest areas of design will be displayed; refer to Figure-63.

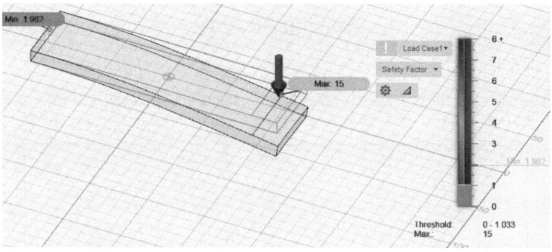

Figure-63. Weakest areas result

Move the sliders on scale to change the threshold to be counted as weak area.

Preparing and Manipulating the Results

Once the analysis is complete, the next step is to prepare and manipulate the results as required. The tools to prepare results are available in the **RESULTS** drop-down in the **Toolbar**; refer to Figure-64. The tools in this drop-down are discussed next.

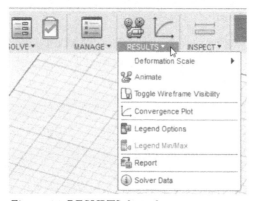

Figure-64. RESULTS drop-down

Deformation Scale

The options in the **Deformation Scale** cascading menu are used to scale up or scale down the deformation caused in the part due to load in the results. By default, the deformation scale is set to **Actual**. To change the scale, select the desired option from the cascading menu; refer to Figure-65.

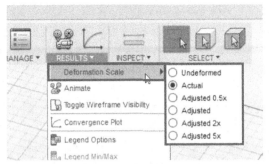

Figure-65. Deformation Scale cascading menu

Animating the Results

Animation is used to represent the analysis results in dynamic motion. The procedure to animate the analysis result is given next.

• Click on the **Animate** tool from the **RESULTS** drop-down in the **Toolbar**. The **ANIMATE** dialog box will be displayed; refer to Figure-66.

Figure-66. ANIMATE dialog box

• Select the **One-way** check box or **Two-way** check box to repeat the animation.
• Set the other parameters as required and click on the **Play** button.
• Click on the **OK** button to exit the dialog box.

Toggle Wireframe Visibility

The **Toggle Wireframe Visibility** tool in the **RESULTS** drop-down is used to toggle the wireframe display of the model in results.

Legend Options

The **Legend Options** tool is used to modify the appearance of legends displayed in the results. The procedure to use this tool is given next.

• Click on the **Legend Options** tool from the **RESULTS** drop-down in the **Toolbar**. The **LEGEND OPTIONS** dialog box will be displayed; refer to Figure-67.

Figure-67. LEGEND OPTIONS
dialog box

- Select the desired legend size and color transition method from the **Legend Size** drop-down and **Color Transition** drop-down respectively.
- Click on the **OK** button to apply changes.

Generating Reports

Once you find the analysis results as expected, it is the time to generate reports. The procedure to generate report is given next.

- Click on the **Report** tool from the **RESULTS** drop-down in the **Toolbar**. The **Report** dialog box will be displayed; refer to Figure-68.

Figure-68. Report dialog box

- Click on the **Save** button to save the report. The **Save Report As** dialog box will be displayed; refer to Figure-69.
- Specify the desired name of the report and click on the **Save** button. The file will be saved and displayed in the web browser.

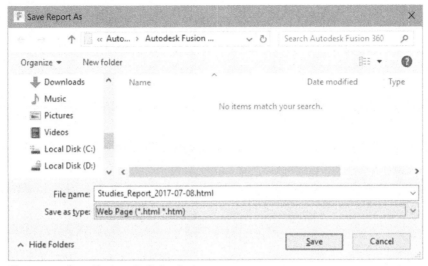

Figure-69. Save Report As dialog box

In this chapter, we have worked on a Static Stress analysis. We have also gone through the basic process of analysis in Fusion 360. In the next chapter, we will work through the other types of analysis available in Autodesk Fusion 360 Ultimate.

Chapter 21

Simulation Studies in Fusion 360

Topics Covered

The major topics covered in this chapter are:

- *Introduction*
- *Nonlinear Static Stress Analysis*
- *Modal Frequencies Analysis*
- *Buckling Analysis*
- *Thermal Analysis*
- *Event Simulation*
- *Shape Optimization*

INTRODUCTION

In the previous chapter, you have learned about the basics of the analysis. In this chapter, you will learn the procedure of applying different analyses on the part.

NONLINEAR STATIC STRESS ANALYSIS

Non-linear static stress analysis is used to check the effect of load on part when three common forms of nonlinearity like material, geometric, and boundary conditions nonlinearity are applicable in analysis. The procedure to apply non-linear static stress analysis is given next.

- Click on the **STUDY** button from the **Toolbar**. The **New Study** dialog box will be displayed.
- Double-click on the **Nonlinear Static Stress (Preview)** button from the dialog box. The analysis environment will be displayed. Note that at the time of writing this book, this option was in preview mode.
- Click on the **Settings** button from the **MANAGE** drop-down in the **Toolbar**. The **Settings** dialog box will be displayed; refer to Figure-1.

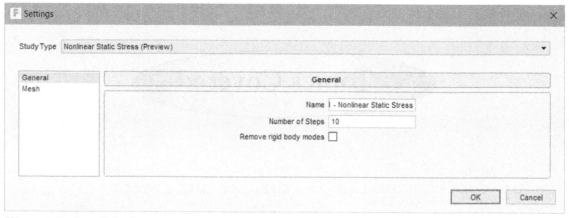

Figure-1. Settings dialog box

- In the **Number of Steps** edit box, specify the desired number of steps in which the total load will be applied.
- Select the **Remove rigid body modes** check box to exclude linear component of force effects. Click on the **OK** button to apply the parameters.
- Apply the elastic material, load, and constraint as required on the model; refer to Figure-2.

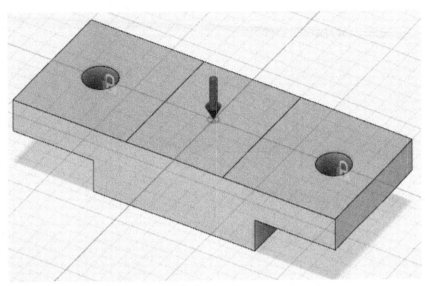

Figure-2. Load and constraint applied

- Click on the **Solve** button from the **SOLVE** drop-down in the **Toolbar**. The **Solve** dialog box will be displayed.
- Click on the **Solve** button from the dialog box. Make sure **Cloud** radio button is selected while solving the analysis. The result will be displayed in the screen; refer to Figure-3.

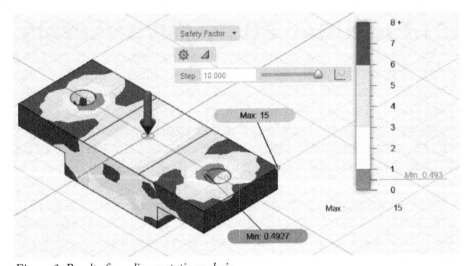

Figure-3. Result of non linear static analysis

- Click on the **2D Chart** button ⊡ in the results to check the transient behavior. The **TRANSIENT RESULTS PLOT** dialog box will be displayed with the results; refer to Figure-4.

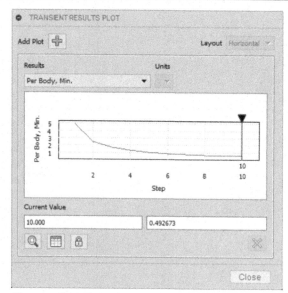

Figure-4. TRANSIENT RESULTS PLOT dialog box

- Move the slider in the graph to check the results of load at different steps. Select the desired result type from the **Results** drop-down in the dialog box.

You can generate the report as discussed earlier.

STRUCTURAL BUCKLING ANALYSIS

Slender models tend to buckle under axial loading. Buckling is defined as the sudden deformation which occurs when the stored membrane (axial) energy is converted into bending energy with no change in the externally applied loads. Mathematically, when buckling occurs, the stiffness becomes singular. The Linearized buckling approach, used here, solves an eigenvalue problem to estimate the critical buckling factors and the associated buckling mode shapes.

A model can buckle in different shapes under different levels of loading. The shape the model takes while buckling is called the buckling mode shape and the loading is called the critical or buckling load. Buckling analysis calculates a number of modes as requested in the Buckling dialog. Designers are usually interested in the lowest mode (mode 1) because it is associated with the lowest critical load. When buckling is the critical design factor, calculating multiple buckling modes helps in locating the weak areas of the model. The mode shapes can help you modify the model or the support system to prevent buckling in a certain mode.

A more vigorous approach to study the behavior of models at and beyond buckling requires the use of nonlinear design analysis codes. In a laymen's language, if you press down on an empty soft drink can with your hand, not much will seem to happen. If you put the can on the floor and gradually increase the force by stepping down on it with your foot, at some point it will suddenly squash. This sudden scrunching is known as "buckling."

Models with thin parts tend to buckle under axial loading. Buckling can be defined as the sudden deformation, which occurs when the stored membrane (axial) energy is converted into bending energy with no change in the externally applied loads. Mathematically, when buckling occurs, the total stiffness matrix becomes singular.

In the normal use of most products, buckling can be catastrophic if it occurs. The failure is not one because of stress but geometric stability. Once the geometry of the part starts to deform, it can no longer support even a fraction of the force initially applied. The worst part about buckling for engineers is that buckling usually occurs at relatively low stress values for what the material can withstand. So they have to make a separate check to see if a product or part thereof is okay with respect to buckling.

Slender structures and structures with slender parts loaded in the axial direction buckle under relatively small axial loads. Such structures may fail in buckling while their stresses are far below critical levels. For such structures, the buckling load becomes a critical design factor. Stocky structures, on the other hand, require large loads to buckle, therefore buckling analysis is usually not required.

Buckling almost always involves compression. In civil engineering, buckling is to be avoided when designing support columns, load bearing walls and sections of bridges which may flex under load. For example an I-beam may be perfectly "safe" when considering only the maximum stress, but fail disastrously if just one local spot of a flange should buckle! In mechanical engineering, designs involving thin parts in flexible structures like airplanes and automobiles are susceptible to buckling. Even though stress can be very low, buckling of local areas can cause the whole structure to collapse by a rapid series of 'propagating buckling'.

Buckling analysis calculates the smallest (critical) loading required for buckling a model. Buckling loads are associated with buckling modes. Designers are usually interested in the lowest mode because it is associated with the lowest critical load. When buckling is the critical design factor, calculating multiple buckling modes helps in locating the weak areas of the model. This may prevent the occurrence of lower buckling modes by simple modifications.

USE OF BUCKLING ANALYSIS

Slender parts and assemblies with slender components that are loaded in the axial direction buckle under relatively small axial loads. Such structures can fail due to buckling while the stresses are far below critical levels. For such structures, the buckling load becomes a critical design factor. Buckling analysis is usually not required for bulky structures as failure occurs earlier due to high stresses. The procedure to use buckling analysis in Fusion 360 is given next.

- Open/create the part on which you want to perform buckling analysis. Click on the **Study** button from the **Toolbar**. The **Study** dialog box will be displayed.
- Double-click on the **Structural Buckling** tool from the **Study** dialog box. The analysis environment will be displayed.
- Apply material, constraint, load, and other parameters as required and solve the study as discussed earlier. The results of buckling will be displayed; refer to Figure-5.

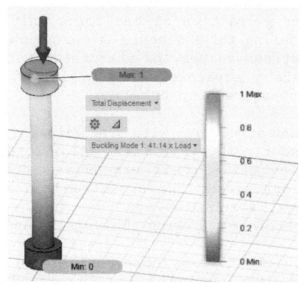

Figure-5. Result of buckling analysis

- Click in the **Buckling Mode** drop-down in the results and select the desired buckling mode to check the effect; refer to Figure-6.

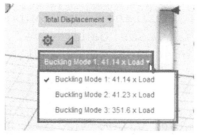

Figure-6. Buckling Mode drop-down

MODAL FREQUENCIES ANALYSIS

Every structure has the tendency to vibrate at certain frequencies, called **natural or resonant frequencies**. Each natural frequency is associated with a certain shape, called **mode shape**, that the model tends to assume when vibrating at that frequency.

When a structure is properly excited by a dynamic load with a frequency that coincides with one of its natural frequencies, the structure undergoes large displacements and stresses. This phenomenon is known as **Resonance**. For undamped systems, resonance theoretically causes infinite motion. **Damping**, however, puts a limit on the response of the structures due to resonant loads.

A real model has an infinite number of natural frequencies. However, a finite element model has a finite number of natural frequencies that are equal to the number of degrees of freedom considered in the model. Only the first few modes are needed for most purposes.

If your design is subjected to dynamic environments, static studies cannot be used to evaluate the response. Frequency studies can help you design vibration isolation systems by avoiding resonance in specific frequency band. They also form the basis for evaluating the response of linear dynamic systems where the response of a system to a dynamic environment is assumed to be equal to the summation of the contributions of the modes considered in the analysis.

Note that resonance is desirable in the design of some devices. For example, resonance is required in guitars and violins.

The natural frequencies and corresponding mode shapes depend on the geometry, material properties, and support conditions. The computation of natural frequencies and mode shapes is known as modal, frequency, and normal mode analysis.

When building the geometry of a model, you usually create it based on the original (undeformed) shape of the model. Some loads, like the structure's own weight, are always present and can cause considerable effects on the shape of the structure and its modal properties. In many cases, this effect can be ignored because the induced deflections are small.

Loads affect the modal characteristics of a body. In general, compressive loads decrease resonant frequencies and tensile loads increase them. This fact is easily demonstrated by changing the tension on a violin string. The higher the tension, the higher the frequency (tone).

You do not need to define any loads for a frequency study but if you do their effect will be considered. By having evaluated natural frequencies of a structure's vibrations at the design stage, you can optimize the structure with the goal of meeting the frequency vibro-stability condition. To increase natural frequencies, you would need to add rigidity to the structure and (or) reduce its weight. For example, in the case of a slender object, the rigidity can be increased by reducing the length and increasing the thickness of the object. To reduce a part's natural frequency, you should, on the contrary, increase the weight or reduce the object's rigidity.

Note that the software also considers thermal and fluid pressure effects for frequency studies.

The procedure to perform the frequency analysis is given next.

- Open/create the part on which you want to perform modal analysis. Click on the **Study** button from the **Toolbar**. The **Study** dialog box will be displayed.
- Double-click on the **Structural Buckling** tool from the **Study** dialog box. The analysis environment will be displayed.
- Click **Settings** tool from the **MANAGE** drop-down in the **Toolbar**. The **Settings** dialog box for **Modal Frequencies** option will be displayed; refer to Figure-7.

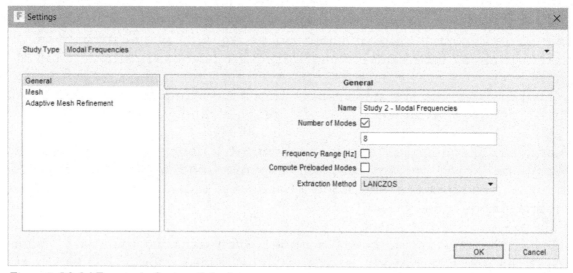

Figure-7. Modal Frequencies Settings dialog box

- Select the **Number of Modes** check box and specify the total number of frequencies that you want to test for resonance. Select the Frequency Range (Hz) and specify the minimum & maximum frequencies within which the natural frequencies are to be found.
- Select the **Compute Preloaded Modes** check box if you include the effect of structural loads in modal frequency analysis.
- Select the desired calculation method from the **Extraction Method** drop-down in the dialog box. **LANCZOS** is used when a large number of modes are being solved (greater than 20). **SUBSPACE** has high computing time when solving for a large number of modes so it is better used for low number of modes. After specifying settings, click on the **OK** button.
- Apply material, constraint, load, and other parameters as required and solve the study as discussed earlier. Note that you can perform modal analysis without applying any load but if apply a load then its effects will also be counted in the results if **Compute Preloaded Modes** check box is selected in the **Settings** dialog box. The results of buckling will be displayed; refer to Figure-8.

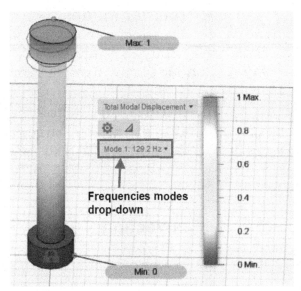

Figure-8. Frequencies drop-down

- Click in the **Frequencies modes** drop-down in **Results** area to check different natural frequencies of the model.

THERMAL ANALYSIS

Thermal analysis is a method to check the distribution of heat over a body due to applied thermal loads. Note that thermal energy is dynamic in nature and is always flowing through various mediums. There are three mechanisms by which the thermal energy flows:

- Conduction
- Convection
- Radiation

In all three mechanisms, heat energy flows from the medium with higher temperature to the medium with lower temperature. Heat transfer by conduction and convection requires the presence of an intervening medium while heat transfer by radiation does not.

The output from a thermal analysis can be given by:
1. Temperature distribution.
2. Amount of heat loss or gain.
3. Thermal gradients.
4. Thermal fluxes.

This analysis is used in many engineering industries such as automobile, piping, electronic, power generation, and so on.

Important terms related to Thermal Analysis

Before conducting thermal analysis, you should be familiar with the basic concepts and terminologies of thermal analysis. Following are some of the important terms used in thermal analysis:

Heat Transfer Modes

Whenever there is a difference in temperature between two bodies, the heat is transferred from one body to another. Basically, heat is transferred in three ways: Conduction, Convection, and Radiation.

Conduction

In conduction, the heat is transferred by interactions of atoms or molecules of the material. For example, if you heat up a metal rod at one end, the heat will be transferred to the other end by the atoms or molecules of the metal rod.

Convection

In convection, the heat is transferred by the flowing fluid. The fluid can be gas or liquid. Heating up water using an electric water heater is a good example of heat convection. In this case, water takes heat from the heater.

Radiation

In radiation, the heat is transferred in space without any matter. Radiation is the only heat transfer method that takes place in space. Heat coming from the Sun is a good example of radiation. The heat from the Sun is transferred to the earth through radiation.

Thermal Gradient

The thermal gradient is the rate of increase in temperature per unit depth in a material.

Thermal Flux

The Thermal flux is defined as the rate of heat transfer per unit cross-sectional area. It is denoted by q.

Bulk Temperature

It is the temperature of a fluid flowing outside the material. It is denoted by T_b. The Bulk temperature is used in convective heat transfer.

Film Coefficient

It is a measure of the heat transfer through an air film.

Emissivity

The Emissivity of a material is the ratio of energy radiated by the material to the energy radiated by a black body at the same temperature. Emissivity is the measure of a material's ability to absorb and radiate heat. It is denoted by e. Emissivity is a numerical value without any unit. For a perfect black body, e = 1. For any other material, e < 1.

Stefan–Boltzmann Constant

The energy radiated by a black body per unit area per unit time divided by the fourth power of the body's temperature is known as the Stefan-Boltzmann constant. It is denoted by s.

Thermal Conductivity

The thermal conductivity is the property of a material that indicates its ability to conduct heat. It is denoted by K.

Specific Heat

The specific heat is the amount of heat required per unit mass to raise the temperature of the body by one degree Celsius. It is denoted by C.

PERFORMING THERMAL ANALYSIS

- Click on the **STUDY** button from the **Toolbar**. The **New Study** dialog box will be displayed. Double-click on the **Thermal** button from the dialog box. The thermal analysis environment will be displayed.
- Apply the desired material to the mode as discussed earlier.
- Click on the **Thermal Loads** tool from the **LOAD** panel in the **Toolbar**. The **THERMAL LOADS** dialog box will be displayed; refer to Figure-9.

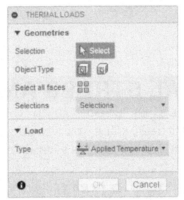

Figure-9. THERMAL LOADS
dialog box

- Select the desired thermal load type from the **Type** drop-down in the **Load** area of the dialog box. Select the **Applied Temperature** option from the drop-down and enter the desired temperature if you want to apply a fix temperature to the selected geometry. Select the **Heat Source** option if you want to specify the amount of heat energy to be applied on selected geometry. Select the **Radiation** option if heat is transferred through radiation to the selected face. Select the **Convection** option from the drop-down if heat is transferred through convection. Select the **Internal Heat** option from the drop-down if heat is generated inside the model.
- Select the face/edge/vertex on which you want to apply thermal load. If you want to select a body then click on the **Bodies** button from the **Object Type** section of the dialog box and then select the body.
- Specify the desired values of thermal load in edit boxes as per the option selected in the **Type** drop-down.

- Specify the contacts as discussed earlier for heat flow between different components of the assembly.
- Click on the **Settings** button from the **MANAGE** drop-down in the **Toolbar**. The **Settings** dialog box will be displayed. Specify the desired value of atmospheric temperature in the **Global Initial Temperature** edit box. Set the other parameters as discussed earlier. Click on the **OK** button from the dialog box to apply settings.
- After specifying all the parameters, click on the **Solve** button. The **Solve** dialog box will be displayed. Click on the **Solve Study** button in the dialog box. The results of analysis will be displayed; refer to Figure-10.

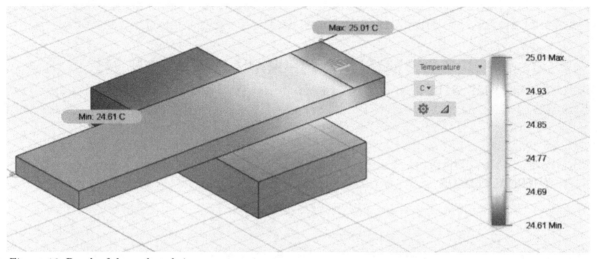

Figure-10. Result of thermal analysis

THERMAL STRESS ANALYSIS

The thermal stress analysis is used to check the effect of thermal and structural loads on the model. The procedure to perform the analysis same as discussed earlier for thermal and stress analyses.

EVENT SIMULATION

Event Simulation is used to study the effect of motion/load on different parts/bodies in model. This analysis is similar to dynamic non-linear analysis you may have studied in engineering. Note that event simulation is in the preview mode while we are writing this book. It may get new features in the later versions. In this example, we will simulate the collision of a ball on the plate. The procedure to perform event simulation is given next.

- Click on the **Study** button from the **Toolbar**. The **Study** dialog box will be displayed. Double-click on the **Event Simulation** button from the dialog box. The environment to solve event simulation will be displayed.
- Specify the material, constraints, and load as applied earlier.
- Click on the **Prescribed Translation** tool and specify the displacement, velocity, or acceleration of the selected body as required; refer to Figure-11.

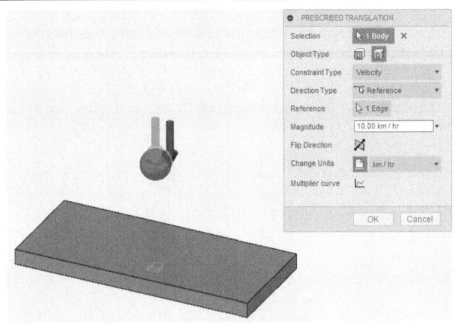

Figure-11. PRESCRIBED TRANSLATION dialog box

- Click on the **Solve** button from the **Toolbar** and then click on the **SOLVE** button from the dialog box displayed. The results of analysis will be displayed. Set the slider to desired step to check the analysis result at intermediate steps.

SHAPE OPTIMIZATION

The Shape optimization study is used to reduce the mass of part while satisfying all the design requirements of the part. The procedure to use shape optimization is given next.

- Create/open the model on which you want to perform shape optimization study. Click on the **Study** tool from the **Toolbar**. The **New Study** dialog box will be displayed.
- Double-click on the **Shape Optimization** button from the dialog box. The environment to perform shape optimization will be displayed.
- Set the material, constraint, load, and contacts as required.

Preserve Region

- Click on the **Preserve Region** tool from the **SHAPE OPTIMIZATION** dropdown in the **Toolbar**. The **PRESERVE REGION** dialog box will be displayed; refer to Figure-12.

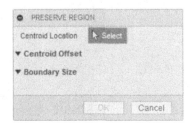

Figure-12. PRESERVE RE-GION dialog box

- Select the face that you want to be preserved after shape optimization. The options for preserving region based on selected faces will be displayed in the dialog box.
- Set the desired options like if you have select a cylindrical face then specify the radius up to which you want to preserve the region; refer to Figure-13. Click on the **OK** button from the dialog box to apply the parameters.

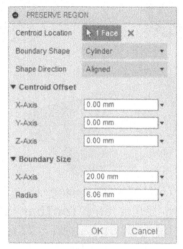

Figure-13. PRESERVE RE-GION dialog box with cylindrical face selected

Setting Shape Optimization criteria

- Click on the **Shape Optimization Criteria** tool from the **SHAPE OPTIMIZATION** drop-down in the **Toolbar**. The **SHAPE OPTIMIZATION CRITERIA** dialog box will be displayed; refer to Figure-14.

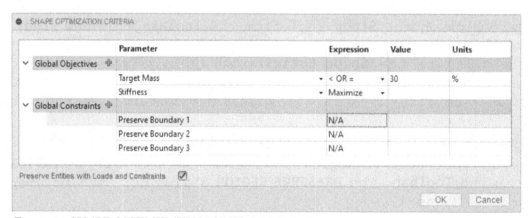

Figure-14. SHAPE OPTIMIZATION CRITERIA dialog box

- Set the desired conditions in the dialog box like, Target Mass less than or equal to 40%.
- Click on the **OK** button.

- Click on the **Solve** tool from the **SOLVE** drop-down in the **Toolbar**. The Solve dialog box will be displayed.
- Click on the **Solve** button. The result will be displayed as shown in Figure-15.

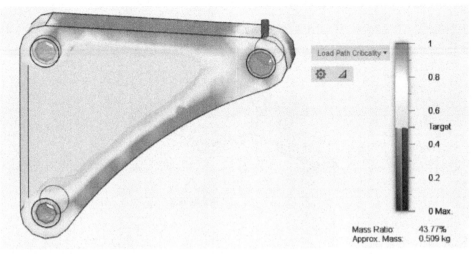

Figure-15. Result of shape optimization

PRACTICE 1

Consider a rectangular plate with cutout. The dimensions and the boundary conditions of the plate are shown in Figure-16. It is fixed on one end and loaded on the other end. Under the given loading and constraints, plot the deformed shape. Also, determine the principal stresses and the von Mises stresses in the bracket. Thickness of the plate is 0.125 inch and material is **AISI 1020**.

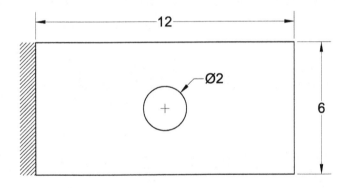

Figure-16. Drawing for Practice 1

PRACTICE 2

Open the model for this exercise from the resource kit and perform the static analysis using the conditions given in Figure-17. Find out the **Factor of Safety** for the model.

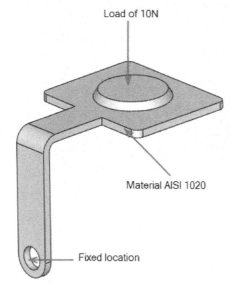

Figure-17. Practice 2

PRACTICE 3

Check out what happens to the fork under the conditions specified in Figure-18. Note that the material of fork is **Alloy Steel** and it is in the hands of a nasty kid. (You know kids!! they don't exert linear forces.)

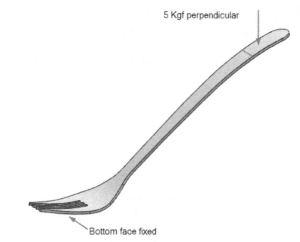

Figure-18. Practice 3 Non Linear Static Analysis

PROBLEM 1

A metal sphere of diameter d = 35mm is initially at temperature Ti = 700 K. At t=0, the sphere is placed in a fluid environment that has properties of $T\infty$ = 300 K and h = 50 W/m2-K. The properties of the steel are k = 35 W/m-K, ρ = 7500 kg/m3, and c = 550 J/kg-K. Find the surface temperature of the sphere after 500 seconds.

PROBLEM 2

A flanged pipe assembly; refer to Figure-19, made of plain carbon steel is subjected to both convective and conductive boundary conditions.

Fluid inside the pipe is at a temperature of 130°C and has a convection coefficient of hi = 160 W/m²-K. Air on the outside of the pipe is at 20°C and has a convection coefficient of ho = 70 W/m²-K. The right and left ends of the pipe are at temperatures of 450°C and 80°C, respectively. There is a thermal resistance between the two flanges of 0.002 K-m²/W. Use thermal analysis to analyze the pipe under both steady state and transient conditions.

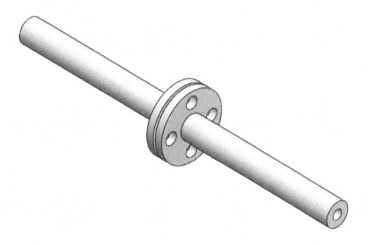

Figure-19. Flanged pipe assembly

FOR STUDENT NOTES

Index

Ethics of an Engineer

- Engineers shall hold paramount the safety, health and welfare of the public and shall strive to comply with the principles of sustainable development in the performance of their professional duties.

- Engineers shall perform services only in areas of their competence.

- Engineers shall issue public statements only in an objective and truthful manner.

- Engineers shall act in professional manners for each employer or client as faithful agents or trustees, and shall avoid conflicts of interest.

- Engineers shall build their professional reputation on the merit of their services and shall not compete unfairly with others.

- Engineers shall act in such a manner as to uphold and enhance the honor, integrity, and dignity of the engineering profession and shall act with zero-tolerance for bribery, fraud, and corruption.

- Engineers shall continue their professional development throughout their careers, and shall provide opportunities for the professional development of those engineers under their supervision.

OTHER BOOKS BY CADCAMCAE WORKS

Autodesk Inventor 2018 Black Book

Autodesk Fusion 360 Black Book

AutoCAD Electrical 2015 Black Book
AutoCAD Electrical 2016 Black Book
AutoCAD Electrical 2017 Black Book
AutoCAD Electrical 2018 Black Book

SolidWorks 2014 Black Book
SolidWorks 2015 Black Book
SolidWorks 2016 Black Book
SolidWorks 2017 Black Book

SolidWorks Simulation 2015 Black Book
SolidWorks Simulation 2016 Black Book
SolidWorks Simulation 2017 Black Book

SolidWorks Electrical 2015 Black Book
SolidWorks Electrical 2016 Black Book
SolidWorks Electrical 2017 Black Book

Mastercam X7 for SolidWorks 2014 Black Book
Mastercam 2017 for SolidWorks Black Book

Creo Parametric 3.0 Black Book
Creo Parametric 4.0 Black Book

Creo Manufacturing 4.0 Black Book